T0298777

Asian Migrants and European Labour Markets

In an era of globalization and demographic transition international migration has become an important issue for European governments. The past decades have seen an increasing and diversifying flow of migrants from different parts of the world, including many from South, South East and East Asia. It has become apparent that in several countries of Europe the demand for workers in certain sectors of the labour market is increasing and that Asia has become an important source for these workers.

This collection explores the phenomenon of Asian immigration in Europe and particularly the ways in which Asian immigrants gain access to local labour markets, including analyses of the factors fostering and sustaining integration. The book includes studies of nine European countries, including Germany, France, Spain, Italy and the United Kingdom, shedding light on the labour market positions of different ethnic groups within Europe with a focus on immigrant entrepreneurship and economic strategies. In the analysis, socio-cultural, economic, institutional and policy aspects are taken into account.

Asian Migrants and European Labour Markets will interest scholars in the field of labour economics, population and migration studies, geography and international business.

Ernst Spaan is Senior Researcher at the Netherlands Interdisciplinary Demographic Institute with an interest in population and development issues, international migration and inter-ethnic relations. He has collaborated on a range of research projects concerning international migration. **Felicitas Hillmann** is Senior Lecturer (*Privatdozent*) at the Free University, Berlin. Her academic interests are in the fields of social geography and development research. For some years her research has focused on migration and gender, ethnic economies and the labour market insertion of immigrants. **Ton van Naerssen** is Associate Professor in the Department of Geography at Radboud University, Nijmegen. His research interests include development theory, international migration and the governance of urban poverty in developing countries.

Routledge research in population and migration
Series Editors: Paul Boyle and Mike Parnwell

Asian Migrants and European Labour Markets

Patterns and processes of immigrant labour market insertion in Europe

Edited by Ernst Spaan, Felicitas Hillmann and Ton van Naerssen

 Routledge
Taylor & Francis Group

LONDON AND NEW YORK

First published 2005
by Routledge
2 Park Square, Milton Park, Abingdon, Oxon OX14 4RN

Simultaneously published in the USA and Canada
by Routledge
711 Third Ave, New York, NY 10017

Routledge is an imprint of the Taylor & Francis Group

© 2005 Ernst Spaan, Felicitas Hillmann and Ton van Naerssen
editorial matter and selection; the contributors their contributions

Typeset in Garamond by Wearset Ltd, Boldon, Tyne and Wear

British Library Cataloguing in Publication Data
A catalogue record for this book is available from the British Library

Library of Congress Cataloging in Publication Data
A catalog record for this book has been requested

ISBN 0-415-36502-3

Contents

Figures

Tables

Contributors

Fauzia Ahmad is a senior researcher in the Department of Sociology of the University of Bristol. With Tariq Modood, she has finalized a study of the diversity in the employment profiles of South Asian women. *Inter alia* she has published 'Modern traditions? British Muslim women and academic achievement', *Gender and Education* (2001).

Joaquín Beltrán Antolín (Ph.D. in sociology and political sciences, Universidad Complutense de Madrid) is a social anthropologist and specialist in Chinese studies. He is professor and researcher at the Centre for International and Intercultural Studies, Universidad Autónoma de Barcelona. He has spent a number of years in the People's Republic of China, where he conducted research on Chinese international migration. His publications include articles and chapters of books on topics related to China, Chinese international migration and intercultural exchange. His most recent books include (with Amelia Sáiz López) *Els xinesos a Catalunya. Família, educació i integració* (2001) and the forthcoming *The Eight Immortals cross the Ocean. Chinese in the Far West*. He is the director of the series Biblioteca de China contemporánea. He is conducting research on Chinese immigrants' labour and settlement strategies in Barcelona and its metropolitan area.

Giles A. Barrett is Senior Lecturer in Human Geography at Liverpool John Moores University. He was the British partner in the EU Fourth Framework Targeted Socio-economic Research network 'Working on the Fringes'. He is founder of the Baseline consultancy at Liverpool John Moores University. His publications include papers in *Urban Studies*, *Journal of Ethnic and Migration Studies*, *International Journal of Entrepreneurial Behaviour and Research* and several edited volumes.

Daniele Cologna is senior researcher at the Italian social research, training and consulting firm Synergia, based in Milan. His research is concerned, among other things, with immigrant integration and the labour market insertion of Asians and Africans. He is co-author of *China in Milan. Family, Place and Work in Milan's Chinese Population* (in Italian) and *Africa in Milan. Family, Place and Work in Milan's African Populations* (in Italian).

Maria Josè Compiani specializes in Hindi language and culture. She obtained a certificate at Isiao and graduated in political science with a degree thesis about the Sikh community in Lombardy. She has done research on immigration for Italian private research institutes and the University of Milan, Bicocca. She collaborates with the provincial administration of Cremona on a research project concerning the insertion of the Sikh community in the local labour market.

Felicitas Hillmann is assistant professor in social geography at the Free University, Berlin, specializing in migration, gender and development studies. She did her Ph.D. at the University of Freiburg on women from developing countries migrating to Italy, then taught at Humboldt University in Berlin. She worked with the Social Science Research Centre (WZB) in Berlin on the mobility of highly qualified workers from Western countries to the transition States and contributed to the green card debate in Germany. Another research focus is on ethnic economies on urban labour markets with special emphasis on the gender dimension. Her publications include 'Are ethnic economies the revolving doors on labour markets in transition?' *International Journal of Urban and Regional Research* (forthcoming); 'Vita e lavoro degli immigrati nella Milano degli anni '90', *Studi Emigrazione/Migration Studies* (1999); (with H. Rudolph) 'The invisible hand needs visible heads: managers, experts, and professionals from Western countries in Poland', in *The new Migration in Europe: Social Constructions and Social Realities* (1998).

Christiane Hintermann obtained her Ph.D. in geography from the University of Vienna. She is responsible for basic research at the Vienna Fund for Integration. Formerly, she worked as researcher at the Institute of Demography of the Austrian Academy of Sciences and the Institute of Urban and Regional Research in Vienna. Her fields of interest are international migration, migrant integration and migration theory and policy. Her publications include 'Potential East–West migration: demographic structure, motives and intentions', *Czech Sociological Review* (with Heinz Fassmann); 'Inderinnen in Wien. Zur Rekonstruktion der Zuwanderung einer "exotischen" MigrantInnengruppe' in H. Haussermann and I. Oswald (eds) *Zuwanderung und Stadtentwicklung* (1997) and 'Die "neue" Zuwanderung nach Österreich. Eine Analyse der Entwicklungen seit Mitte der 80er Jahre', *SWS-Rundschau* (2000).

Krystyna Iglicka is an economist and social demographer, associate professor at the LK Academy of Management. Polish government expert on issues concerning migration policy. She has been a visiting professor at several American and European universities. Senior Fulbright Fellow at the University of Pennsylvania 1999–2000. Between 1996 and 1999 she served as a co-ordinator of the Polish Migration Project at the School of Slavonic and East European Studies, University College London.

Co-editor of such books as *The Challenge of East–West Migration for Poland* (with Keith Sword, 1999) and *From Homogeneity to Multiculturalism: Minorities Old and New in Poland* (with F.E.I. Hamilton, 2000). Author of *Poland's Post-war Dynamic of Migration* (2001) and more than twenty papers on different aspects of migration published in Poland, the United States and Britain.

Trevor P. Jones is visiting professor in the Department of Corporate Strategy at De Montfort University, Leicester, and was formerly Reader in Social Geography at Liverpool John Moores University. His papers on ethnic minority business have appeared in *Journal of Ethnic and Migration Studies*, *New Community*, *New Economy*, *Revue Européenne des Migrations Internationales*, *Social Forces*, *Sociological Review*, *Urban Studies*, *Work Employment and Society* and several edited books. He was senior author of *Geographical Issues in Western Europe* (1988) and *Social Geography* (1989).

Maggi W.H. Leung is assistant professor at the Chinese University of Hong Kong. Her dissertation, published as *Chinese Migration in Germany: Making Home in Transnational Space* (2004), concerns the life of ethnic Chinese in Germany. She has also written on migrant entrepreneurship, migrants' concept of home and diasporic identities.

Stephen Lissenburgh is Senior Researcher at the Policy Studies Institute, London. His expertise pertains to questions of unemployment, wage determination and discrimination and labour market participation. His research includes analysis of South Asian women and employment, focusing on the factors that influence South Asian women's labour market participation decision. This study uses the Institute's Fourth Survey of Ethnic Minorities and a study of employment transitions after fifty, using the British Labour Force Survey.

Emmanuel Ma Mung is a geographer and Director of Research at the CNRS, the French National Centre of Scientific Research. He has carried out studies of ethnic economy and entrepreneurship, the mobility of international migrants and the Chinese diaspora. Since 1995 he has been in charge of Migrinter (International Migrations, Spaces and Societies) research unit of the CNRS at the Universities of Poitiers and Bordeaux. He is author of *Commerçants maghrébins et asiatiques en France* (1990), *Mobilités et Investissements des Emigrés, Maroc, Tunisie, Turquie, Sénégal* (Paris, L'Harmattan, 1996), *La Diaspora chinoise: géographie d'une migration* (2000). He is editor of special issues of *the Revue européenne des migrations internationales: entrepreneurs entre deux mondes* (1992) and *La Diaspora chinoise en Occident* (1992).

David McEvoy is Emeritus Professor of Urban Geography at Liverpool John Moores University. He was Director of the School of Social Science in that university from 1991–98. His research on ethnic minority residence and on ethnic minority business in Britain and Canada stretches over twenty years. It has been funded by the Social Science Research Council (UK), the Economic and Social Research Council (UK), the Commission for Racial Equality (UK) and the Canadian High Commission in London.

Tariq Modood is the Director of the Centre for the Study of Ethnicity and Citizenship and a founding editor of the new international journal *Ethnicities*. He is at the Department of Sociology, University of Bristol. He is a leading authority in the field of ethnicity and was the principal researcher of the Fourth National Survey of Ethnic Minorities in Britain published as *Ethnic Minorities in Britain: Diversity and Disadvantage* (1997). His other publications include *Changing Ethnic Identities* (with S. Beishon and S. Virdee, 1994) and *Asian Self-employment* (with H. Metcalf and S. Virdee). He also edited *Church, State and Religious Minorities* (1997) and *Debating Cultural Hybridity and the Politics of Multiculturalism in the New Europe* (1997), both with P. Werbner. He has completed a Nuffield Foundation-sponsored project, South Asian Women and Employment, with Fauzia Ahmad. He is working on several comparative cross-national projects on ethnicity and public policy with colleagues in the United States and Canada.

Liane Mozère is Professor of Sociology at the University of Metz. She has been working for many years on migrant populations and on informality (illegal work) in France as well as on Philippine women working illegally as domestics in Paris. She has published a reader on the subject entitled *Travail au noir: informalité, liberté ou sujétio?* (1999). Another book is forthcoming on the theme of family child care in a suburb of Paris, with particular attention to the handling of foreign and migrant women caring for children and how they 'manage' their 'business'.

Ton van Naerssen is senior lecturer and co-ordinator of the Master's programme 'Globalisation and urban development' of the Nijmegen School of Management, Radboud University, Nijmegen. He is a development geographer with a special interest in development theory, urban issues and South East Asia. He was co-editor of *Urban Social Movements in the Third World* (1989), *The Diversity of Development* (1997) and *Healthy Cities in Developing Countries* (2002).

Knut Onsager (dr. polit. in social geography) is a geographer and Research Officer at the Norwegian Institute of Urban and Regional Research. He specializes in industrial systems, innovation systems, entrepreneurship and labour markets. His research includes work on immigration and the labour market insertion of immigrants in Norway. His publications include *The Division of Labour, Innovation and the Territorial Dimension* (2000) and (with B. Sæther), *Immigrant Enterprises and Ethnic Entrepreneurs* (2001).

Geir Inge Orderud is a geographer and Research Officer at the Norwegian Institute of Urban and Regional Research. His interests can be subsumed under the following headings: migration and territorial belonging; social structuring and poverty; industrial analysis; environmental studies. Among his publications are *Explaining Urban Socio-spatial Structuring*, *Immigration to Norway in a European Perspective, Immigrant Self-employment in Norway, Property Development in Nydalen, Oslo* (with Per Gunnar Røe) and *The Housing Industry. Actors and Networks*. He is undertaking research on migration, unemployment and poverty in Riyadh, Saudi Arabia.

Laura Oso has been awarded doctor's degrees from the Sorbonne and the University of La Coruña. She lectures and researches in the Faculty of Sociology at the University of La Coruña. She has acted as adviser to a number of international organizations (the OECD, INSTRAW). Her research has focused mainly on the issue of female immigration in Spain. She has presented papers at national and international congresses dealing with this issue, and has published two monographic studies, as well as chapters in books and articles for specialist journals. Two of her most important publications were the monographic studies *La migración hacia España de mujeres jefas de hogar* (1998) and *Españolas en Paris* (2004).

Fabio Quassoli is researcher in the Department of Sociology and Social Research of the University of Milan, Biccocca. He holds a Ph.D. in sociology from the University of Milan and has participated in a number of EU-sponsored research projects on migration-related issues. He is co-author of *La formazione linguistica per lavoratori stranieri: dare voce ai diritti e alle risorse* (with Cristina Venzo) and *La comunicazione degli immigrati a Milano: reti sociali, rappresentazioni e modalità di accesso ai servizi nell' metropolitana milanese* (with Paolo Barbesino); and of various articles in scientific journals on migrants in the Italian labour market, migrant criminalization and multiculturalism.

Monder Ram is Professor of Small Business Research at De Montfort University, Leicester. He is an experienced researcher in the field of ethnic minority businesses, employment relations, management and small business policy. He is co-author (with Trevor Jones) of *Managing to Survive. Working Lives in Small Firms and Ethnic Minorities in Business* (1994) and *Ethnic Minorities in Business* (1998) besides many articles in scientific journals.

Ursula Reeger has been a researcher at the Institute of Urban and Regional Research since 1989. She studied geography (spatial research and regional planning) at the University of Vienna; her Ph.D. thesis was entitled 'Xenophobia in Vienna. An Empirical Analysis'. She is working on housing and the segregation of immigrants in Vienna, on the transnational mobility of Poles and on the situation of foreign students at Vienna's universities. Her general research interests include integration and segregation, social mobility, xenophobia and east–west migration. She has published on these subjects in a range of academic journals and books. From 1997 to 2000 she was a member of the Group of Specialists on Innovatory Social Policies in the City at the Council of Europe in Strasbourg.

Natalia Ribas-Mateos is a Marie Curie Research Fellow at the Mediterranean Laboratory of Sociology, Université de Provence–CNRS, France). Among her recent books in Spanish are *Una invitación a las sociología de las migraciones* (2004) and *El debate sobre la globalización* (2002). She is publishing in 2005 *The Mediterranean in the Age of Globalization. Migration, Welfare, and Boundaries*.

Ernst Spaan is social anthropologist specializing in migration studies. He holds a Ph.D. in spatial sciences from the University of Groningen. He has been involved in research projects on international migration and is senior researcher at the Netherlands Interdisciplinary Demographic Institute in The Hague. His interests cover population and development issues, international migration and inter-ethnic relations. His publications include 'Immigrant workers in Arab States', in *The Encyclopedia of the World's Minorities*, edited by Carl Skutsch (2004), *Labour Circulation and Socioeconomic Transformation. The Case of East Java, Indonesia* (1999) and *Labour Migration from Sri Lanka to the Middle East* (1989).

Harry van den Tillaart is a sociologist working at the Institute of Applied Social Sciences, Radboud University of Nijmegen. His work focuses on small and medium-size businesses and entrepreneurship. He is researching immigrant entrepreneurs and issues of urban development. He has documented the increase in immigrant entrepreneurs and the changes in choice of industry these entrepreneurs – in particular those of the second generation – made in the Netherlands 1985–2004. He has published several books and articles on these topics.

Preface

In the current era of globalization, a conspicuous feature has been the increasing migration flows from Asia to Europe, both highly dynamic migration regions and increasingly interconnected. In contrast to the former 'guest workers' from Mediterranean countries to selected countries in northern Europe, who found work in industry and manufacturing mainly through bilateral agreements between governments and employers, recently arrived Asians have carved out niches on the European labour market as, for instance, Indian and Pakistani retailers, Vietnamese caterers, Filipino domestic workers or Sikh milkers. Likewise conspicuous is that, up to now, relatively little research has been devoted to the modes of incorporation of Asians into the European labour market. To help fill this void, we decided to initiate a research network on Asian migrants and entrepreneurs in Europe.

This book owes its origin to a conference on 'Asian Immigrants and Entrepreneurs in the European Community' which took place in the city of Nijmegen (the Netherlands) on 9–11 May 2001. The conference was funded by the European Science Foundation (ESF) under the ESF SCSS Exploratory Workshop Scheme. It was the first workshop highlighting especially the role of Asian immigrants in European labour markets, including the development of Asian immigrant entrepreneurship. Initiators were the Netherlands Interdisciplinary Demographic Institute (NIDI), the Nijmegen Centre of Border Research of the University of Nijmegen (now Radboud University Nijmegen) and the Economic and Social Science Institute of the Hans Boeckler Foundation in Dusseldorf (WSI), Germany. They acted on behalf of the European Network on Asian Immigrants. During the conference nineteen papers were presented by conference participants from ten European countries. Moreover a symposium was held during which Dan Hiebert, David Ley (both from the University of British Columbia), Ivan Light (University of California) and Jan Rath (University of Amsterdam) gave their valuable inputs with regard to the conference theme. We thank these persons and institutions for their support for the conference.

We are grateful to all participants whose enthusiasm convinced us to continue with a book project, and to the anonymous referees of the initial book proposal who suggested expanding the subject to encompass the broader theme of Asian immigrants on the European labour market. We want to extend our gratitude to all the contributing authors whose quick acceptance of our invitation to contribute strengthened our idea of the usefulness of the project. Most of the authors updated or rewrote their contributions while some wrote completely new ones. Many thanks also to Tonny Nieuwstraten-Prins and Jacqueline van der Helm of NIDI for the technical editing and preparing the layout of the book, and to Jing Maglunob for editing the English of some chapters. Others whom we would like to thank for their contribution to the preparation of the ESF workshop and the ensuing book project are David McEvoy (Liverpool John Moores University), Natalia Ribas-Mateos (then at Universitat Autònoma de Barcelona), Daniele Cologna (Synergia s.r.l., Milan), Marisha Maas and Marcel Nieling (both from the Department of Geography, Radboud University Nijmegen), Helga de Valk (NIDI) and Leon Vermeulen (NIDI). A special word of thanks goes to Vera Holman (NIDI) for her excellent logistic support during the whole project and for pointing out new information on Asian migrants. We appreciate the institutional support for the project from the Centre for Development Studies (ZELF) of the Geography Department of the Free University, Berlin, next to the Radboud University Nijmegen and NIDI.

Ernst Spaan
Felicitas Hillmann
Ton van Naerssen

1 Landscapes of Asian migrant labour in Europe

Felicitas Hillmann, Ernst Spaan and Ton van Naerssen

Up to the present Asian migration to Europe covers a minor share on the total of immigration to the European countries. However, while immigration from the Asian regions played a very limited role in most European countries during the first decades after World War II, this pattern changed in the 1990s. In this decade new patterns of immigration arose and stricter policies to challenge upcoming new mobility were implemented – the European migratory space transformed rapidly. With such changes the presence of Asian migration also became more prominent. In this book we'll focus on South, South East and East Asian migrants. They stem from a range of countries such as Sri Lanka, India, China, Pakistan, Taiwan, the Philippines, Indonesia and Vietnam. Despite the implementation of more restrictive immigration policies in the European Union, Asian immigration can be expected to continue in the future. As is shown in Figure 1.1, in many European countries Asians comprise quite a large proportion of the non-national population.

However, in contrast to earlier periods, whereby immigration to Europe was in many cases linked up with former colonial bonds and with the demand for temporary 'guest workers' from the Mediterranean basin countries, we are confronted today with new forms of migration, new social realities (Koser and Lutz, 1998). Most evident is that the migration flow has become more diverse in terms of nationalities, destination, gender, age and the legal status of migrants. The recent flows include labour migrants, refugees and increasing numbers of asylum seekers – albeit not in all European countries simultaneously – and undocumented migrants in particular – a mixed group of legal status affiliation. Today there is much greater heterogeneity within the composition of the immigrants than was the case in the 1970s and the 1980s. This tendency to heterogeneity is on the one hand due to economic, social and political developments within the region of origin and on the other hand also connected with tendencies in the labour market and migrations policies of the North American and European migration system.

Asian migration – this becomes extremely evident in the case of Chinese migrants – is a valid example to observe the modification of migration

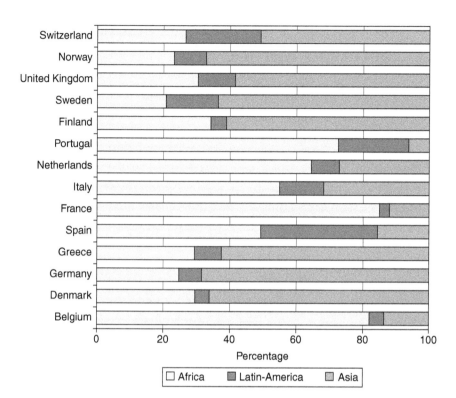

Figure 1.1 Non-national populations by main region of citizenship and country, 1 January 2000 (source: Based on European Social Statistics, *Migration* (2002 edition), Brussels: Office of Official Publications of the European Communities).

patterns adapted to a changing world. Asian migration seems to react in a flexible way to the regional and social changes under the auspices of globalization, economic restructuring and new political configurations in most parts of the world. The case of Asian migrants – and this is the underlying thesis in this book – shows in a very cogent way (and presumably in a much sharper way than it would be the case for the 'traditional' immigrant groups in Europe) the ongoing changes in the European labour markets.

In the decade of the 1990s globalization had an important impact on sending countries as well as on the target countries of migration: deregulation, privatization and less state intervention were the principal instruments of economic restructuring on the global scale. Migration processes take place in this changing economic order, migration that, as is argued here, constitutes a crucial link between globalization and local labour markets. As the European Commission (Commission of the European Communities, 2003: 10) has stated:

While immigration should be recognized as a source of cultural and social enrichment, in particular by contributing to entrepreneurship, diversity and innovation, its economic impact on employment and growth is also significant as it increases labour supply and helps cope with bottlenecks. In addition, immigration tends to have an overall positive effect on product demand and therefore on labour demand.

Some of the rearrangements in the European local labour markets would not be possible if certain forms of immigration did not 'help out'. A clear example is the case of Asian domestic workers in many European countries. New modes of production require certain forms of flexible work, which in many cases is offered by migrants. The way migrants offer their work force on European labour markets says a lot about the more general profound changes on the various labour markets. New modes of immigrant incorporation seem to signalize that larger sections of the working population underlie a transformation process. Assuming that migrant workers are represented usually in the weakest positions on the various national labour markets, their shift towards even more precarious positions during the 1990s indicates ongoing changes on the European labour markets in general.

New forms of labour organization within the labour markets of the countries of arrival, as well as within the immigrant groups itself, emerged in the 1990s. In some cases we could speak of an acceleration and radicalization of informal and illegal practices. In the long run this situation will have an impact also on the overall labour market in the European countries of arrival. It is a new phenomenon that this holds true on different levels of qualification:

1   It is apparent that, in several European countries, the demand for certain categories of skilled labour from outside the European Union is expanding. The EU labour markets themselves open up different spaces and niches for the migrants as compared with the situation a decade ago. In this demand, the immigration of Asians is becoming more conspicuous and shows certain specific characteristics. Examples are for instance Indian IT workers in Germany or Indonesian nurses in the Netherlands.
2   Asian female domestic workers in several EU countries are part of an ongoing feminization of migration on the one hand and of the engendering of labour markets on the other hand. Many highly skilled Asian immigrants, i.e. male and female migrants at various levels of occupation, suffer from dequalification on the labour market of arrival.
3   Among certain Asian groups the propensity for being self-employed is relatively high. Labour force participation and the fostering of entrepreneurship are considered of importance for the successful integration of new immigrant groups in the host societies.

Why this book?

Up to today we know only little about Asian migration to Europe. In comparison with other immigrant groups, Asian immigration is underresearched. Considering the fact that the South, South East and East Asian regions are three of the most dynamic migration regions worldwide and that the impact on the European migration system will increase in the coming few years, more insight into these processes is highly desirable. This book aims at filling a void in the existing literature. Although there are studies of immigration and the labour market in Europe, there are no collective volumes dealing specifically with Asian immigration and the position of Asians on the labour market. Owing to its wide coverage in terms of geography and ethnic groups, and its multidisciplinary approach, the book is addressed to an international readership, including anthropologists, geographers, demographers, economists, business scientists, political scientists and sociologists. It is intended for academics and interested readers as well as for policy makers. This book takes stock of the existing (albeit limited) knowledge pertaining to more recent South, South East and East Asian migrant communities in the European Union and pronounces the rather new aspects for the whole European setting. In particular, the extent to which Asian immigrants gained access to local labour markets and attain a certain measure of socio-economic incorporation within the countries of arrival are reviewed (Part I, ' "Within the system": Asian immigration and the European labour market'), as well as the factors fostering and sustaining such a process (Part II, ' "Coming in": specific incorporation trajectories of Asian migrants'). A special focus is on immigrant entrepreneurship and economic strategies, including the mobilization of social and business networks (Part III, 'Self-employment and entrepreneurship among Asian immigrants in Europe'). There is further an explicit focus on the gender dimension of Asian migration to Europe, considering the crucial importance of gender-specific mobility management in transnational migration processes.

The book does not aim at covering all aspects of Asian migration to Europe. The most important criterion for the selection of the contributions was their relation to labour market problems and their originality. From the outset we did not aim at limiting the theoretical focus of the book to one line of thinking only and we accepted different theoretical traditions as well as different methodological approaches. As migration is a complex and multidimensional event there is not *the one* grand theory in grade to explain all the different outcomes of the labour market integration of Asian immigrants.

This book may be interesting to those who are in a position to formulate and implement adequate policies in the field of e.g. employment, education or integration. It is in these fields that it is imperative to acquire more knowledge of Asian immigrants. In particular, information is needed concerning, among other things, modes of labour market insertion of Asians, their presence in immigrant business (numbers, type of businesses, countries) and the factors contributing to the success, sustainability or failure of immigrant business. Old political recipes may not work, since the ingredients of immigrant integration have changed and also the climate and temperature of the context changed considerably. We feel that migration research is undergoing (again) a new paradigmatic change. Over the past years the perspective of migration research has opened up: in the 1990s the idea of new transnational spaces and migration, as put forward by authors like Vertovec (2001), Pries (1999), Schiller *et al.* (1997) gained ground. The concept of transnationalism sometimes seems to be contrasted by 'traditional' concepts of immigrant integration in the countries of arrival by focusing most scientific attention on the social networks of migrants, on social capital and on transnational practices (see critique by Waldinger and Fitzgerald, 2004). In this perspective the role of diasporas within transnational settings and the consequences of international political practices becomes of great interest (Østergaard-Nielsen, 2003).

More recently, a new dimension has been added to the migration research on transnationalism and questions of migration and development are put on the agenda. This is certainly due to the worldwide background of increasing violence, conflict and crisis in many parts of the world, the ongoing debate on the impact of globalization in many countries of the developing world and the fear of many in the European Union that mass migrations are no longer controllable. It certainly holds true that there is now more awareness of the scientific community on the importance of diasporas and ethnic minorities as agents of change *here* and *there*. Both debates, the debate on globalization and migration and the one on transnationalism, have in common that they have shifted the focus away from the national scene. The nexus between migration and development becomes now a top-of-the-agenda item in migration research (Van Hear and Nyberg Sørensen, 2001). Local developments fell easily out of sight in the debate on globalization – in this book we cope with that lack by presenting a range of local realities without expecting them to fit neatly in a determined theoretical perspective.

We invited the contributors to introduce a gender-sensitive perspective in their contribution; acknowledging that gender plays a decisive role in migration processes. Some chapters exclusively take a gender perspective. The emerging debate on 'gendered geographies of power' within migration research (Pessar and Mahler, 2003; Zlotnik, 2003) and on the existence of gendered 'counter-geographies of globalization' (Sassen, 2003) as well as the existing literature on gender and migration (Morokvasic, 2003) was at its inception and few authors seemed to be aware of the strong impetus that gender has on migration when first working on their articles. The ongoing debate on gender, global care chains and migrant nurseries does not refer to new phenomena (Hochschild, 2000; Nyberg-Sørensen, 2004), but finally finds appropriate, somewhat eye-catching, conceptualizations and acceptance.

Theoretical outline: globalization, migration and labour markets

The contributions in this book take very different perspectives on the Asian migration processes. By reading through the chapters it becomes clear, though, that the process of globalization and labour market fragmentation *helps* in understanding the recent developments. Migration constitutes on one hand an outcome of globalization processes that provokes changes in the socio-economic setting in most countries. Migration is, on the other hand, a driving force behind many changes in the economic order in the decade of the 1990s. New technologies eased migrant networks all over the planet and certain sectors in the economy can hardly do without migrant work. On the national scale some publications take account of the 'mixed embeddedness approach', as proposed by Jan Rath (2000), which is ought to be a synthesis of various earlier theories and emphasizes the need to relate socio-cultural factors to structural circumstances and transformations in the (urban) economy and the linkages with the politico-institutional environment in the study of ethnic minority businesses. Ethnic minorities are conceptualized as situated on the fringes of formal and informal labour markets, a position providing new immigrants with employment opportunities (Kloosterman, 2000). Possibilities and regulations with respect to employment and setting up in business differ between the countries of the European Union. The welfare state regimes in the different countries have a clear influence on entrepreneurship among migrants too.

On the methodological side it is important to mention that many contributions to the book are based on smaller survey samples and/or on qualitative data. This is due to the nature of the topics that the authors address. Often the structure of the subject does not allow, neither recommend, a quantitative approach. The book also does not present, except for the introduction, any cross-national data. The availability and the reliability of data on Asian migrants are limited. Data pertaining to immigrants and immigrant labour markets are structured by the nation state, suited to its image of society, its regulatory regimes and priorities. Thus the cross-national use of national statistics on migration is difficult and sometimes impossible or misleading – so we did not include comparative studies by choice. Furthermore, the ethnic groups analysed in the various chapters are not necessarily conceptualized as homogeneous, stable groups. The term 'ethnic minority' is based upon a political point of view and ethnicity is often mixed up with foreign nationality, while sub-national groups are often glossed over (e.g. Sikhs, Tamils, ethnic Chinese from Vietnam). The heterogeneity and dynamism within these groups are reflected in the contributions to this book.

Generally, there is strong heterogeneity among the ethnic groups and within the various European countries. In Spain, for instance, two distinct flows are apparent: first, Chinese immigration concerns both men and

women, often married couples, indicating the existence of family reunification. Second, Filipino immigration is highly feminized, mainly concerns
single women, working as domestic workers or, as is the case in Italy, as
nurses and care persons for the children and the elderly. The European
migration space at present is marked by different sub-systems that are characterized by certain common features and different patterns of labour market
integration. The distribution of Asian immigration to Europe and the mode
of labour market integration of the immigrants, as well as the internal
organization of labour markets, reflect underlying hierarchies in the European migration system (Hillmann, 2000). On this view southern European
countries show clearly different patterns compared with the countries of
north/eastern Europe or the former colonial states (the United Kingdom,
France and the Netherlands). While all such countries show very different
migrant characteristics for the various Asian immigrant groups, certain
groups, like the Chinese, seem to follow a more homogeneous migration
pattern. They appear to be the most globalized groups by relying on transnational family networks. What seems to emerge from the contributions
collected in this volume is that the underlying European migration hierarchies pre-structure the settlement and the modes of incorporation of the
globalized Asian migrants in the European labour markets. It is not by
chance that certain types of migration are significant in one country but
insignificant in another. Before we limit our introductory outline on the
European scale and on the chapters in the book we sketch, for a better
understanding of the situation, the regional scale on both sides: the Asian
migratory space and the globalization and flexibilization of the European
labour markets.

Asian migratory spaces

The international emigration pattern of Asian countries and regions show
remarkable differentiation in intensity and characteristics concerning ethnicity, gender and professions. After World War II migration flows were
mainly directed towards the Gulf states or towards the United States,
Canada and Australia. During the 1980s and 1990s increased intra-regional
migration took place towards the Newly Industrialized Countries in the
region like Malaysia and Thailand as well as towards the 'Tiger' economies
like South Korea, Taiwan, Hong Kong and Singapore (Massey *et al.*, 1998;
Spaan, 2000; Stahl, 2001). The decade of the 1990s brings important
changes to the Asian migration system. Asia has seen the highest rate of economic growth of any regions. This growth has been accompanied by an
increase in capital flows, trade and by mass migrations. Strong economic
growth – intensive in the use of first unskilled and, later, skilled labour, led
to a reinforcement of the global scale instead of an orientation towards
limited national markets (Fong and Lim, 1996; Kim, 1996). Asia now
turns out to be one of the largest sending regions in the world. A specific

'migration industry' developed, in which all sorts of intermediary agents within the region play an important role (Spaan, 1994; Skeldon, 2000; Husa and Wohlschlägl, 2000). Asian immigration to OECD countries continued to increase and to diversify in that decade – while there was persistence of flows to the United Kingdom and France (OECD, 1997).

Moreover, between Asia and Europe a great diversity in migratory spaces exists. Traditionally this diversity is associated with the colonial period. Asian countries were linked with specific European countries, e.g. the United Kingdom to the countries of South Asia, France to Indo-China and the Netherlands to Indonesia, which naturally had its impact on direction of the Asian emigration. The colonial ties explain why the majority of immigrants from South Asia in Europe are in the United Kingdom, why Vietnamese emigrants prefer France and why Indonesians migrants particularly live in the Netherlands.

Asian–European migratory spaces however, concern spatially even more specific regions than countries. For example, the majority of the Indian emigrants in the United Kingdom originate from the states of Gujarat, Rajastan and the Punjab. With regard to the latter, migrants come from areas that had long provided recruits for the British army and merchant navy. Migrants from Gujarat do not originate from all the districts but from the districts of Baroda, Surat and Baruch especially, which have a long tradition of migration and trading. Chinese migrants in Europe traditionally originate mainly from the provinces of Zhejiang (the port city of Wenzhou and the nearby rural area around the town of Qingtian) and Guangdong (Hong Kong and the Pearl River Delta). After the reforms of 1978, emigration from these areas resumed while new emigration started from the province of Fujian, particularly from the Fuzhou/Fuqing area that had a long tradition of outmigration to South East Asia. So as for Europe, Chinese migratory space is concentrated in the coastal provinces of south China. Another source area of Chinese emigrants have been the former colonies of South East Asia, where most of the ancestors of these migrants also originated from the coastal provinces in south China (Pieke, 2002). A more recent trend, and apparently a regionally more diverse one, concerns the brain drain from Chinese students and skilled professionals.

As for the region of South East Asia, in 1975, after the end of the war in Vietnam, many Vietnamese from the South departed the country. The number of Vietnamese residing in Europe is substantially more than the number of migrants from any other country in South East Asia. The wave from 1975 to 1985 concerned refugee and permanent migration. The picture changes when we consider temporary and labour migration. In this sector it is clearly the Philippines that dominate. With around seven million overseas contract workers (OCWs), the country is the second largest exporter of labour in the world, a forerunner of what some call a migration industry. Although the majority of these overseas workers stay in the nearby countries of Pacific Asia and the Middle East, it is estimated that at present around

0.8 million Filipino migrants live in European countries. In the specific case of mail brides, both the Philippines and Thailand are the major source countries in South East Asia. As for Indonesians in the Netherlands, owing to the political situation shortly after independence, a relatively large group concerns the people from the Moluccas.

Globalization and flexibilization of the European labour market

Globalization has a profound impact on the structuring of European labour markets and the mode of labour market insertion of migrant labour. Among other things the emergence of a global economy with a shift from manufacturing to services and the emergence of cities functioning as nodes in the world economy have led to a change in the demand for low-skilled labour. As in many other countries all around the world European labour markets were restructured in the neo-liberal spirit of privatization, deregulation and less state intervention. In Europe, the male guest worker from the 1970s who worked in factories has been replaced by a low-skilled labour force of both sexes in services such as catering and domestic services. In the 1990s the decline of the big old industries led to severe unemployment in many countries, often concerning exactly the guest-worker generation of immigrants. Some countries transformed in this period from emigration to immigration countries; in most countries a second and third generation of immigrants already resides. Such immigrant and minority population faces certain common features on the labour market.

Currently European labour markets are characterized by the following processes. First, a convergence of markets is taking place which is leading to more mobile work forces in the European Union, less restricted by immigration policies at least for those immigrant groups that form part of the system, while restrictions on the entrance of Non-EU foreigners is strengthened. Next, a trend towards increasing flexibility can be observed: since the 1980s there is a shift away from fixed long-term employment contracts toward part-time, on-call and self-employment, depending on sector and production cycles. This is linked with the emergence of complex business systems characterized by strategic alliances of large-scale enterprises delimited by space, time or specific product. Women are increasingly participating in the labour market, particularly in part-time employment. Furthermore, European work forces are ageing, with increasing skill gaps, leading to a larger demand for specific categories of immigrant labour, e.g. in the ICT and health sectors. Demographic changes are expected to have an important impact in the future: certain regions, e.g. large parts of Eastern Germany, suffer outmigration of the younger generation.

European labour markets are segmented to a certain degree, with migrant labour predominant in low-skilled work and self-employment, but becoming increasingly marginalized in the face of deindustrialization and the development of a knowledge-based and skill-intensive service economy. This has translated into more difficult labour market access for minority groups and higher unemployment rates of immigrant workers as compared with the native work forces. At present, there are large informal economies coupled with alternative modes of labour recruitment (migrant/ethnic networks, trafficking, illegal immigration, within-company transfers) and the development of 'enclave' or 'ethnic economies'. The informal sector of the European economies grew fast in the 1990s and is estimated to make up on average about 16.7 per cent of gross domestic product in the OECD countries (Deutscher Bundestag, 2002: 247).

The pattern of migration is influenced by specific policy developments within the European Union, i.e. the free mobility of people in the EU zone on the one hand and restricting entry for non-EU citizens on the other. Asians who already have gained access within the European Union face fewer restrictions on labour mobility or setting up businesses in different EU countries. Asians from outside the European Union are confronted with increasing barriers to entry, which are circumvented through informal forms of labour recruitment and migration. The policies adopted by EU countries are diverse. For instance, family reunification rules have become stricter in some countries (e.g. the Netherlands, Austria). Simultaneously, some EU countries' policies (the United Kingdom, Germany, Norway, France, Netherlands) aim at greater selectivity of immigrant workers, benefiting those migrants with higher qualifications and specific expertise. In particular for ICT specialists and (female) workers in the health, care and domestic services industries, for which there is large demand, labour immigration is stimulated. Furthermore, most countries have raised concern over the increasing flow of immigrants, but the Mediterranean countries in particular have chosen to regularize illegal workers (OECD, 2003).

Earlier studies among minority groups in European countries have indicated that the labour market incorporation of immigrants shows variation across countries, specific locations and ethnic groups. These differences not only show in diverging participation rates and unemployment, but also in terms of their distribution over various economic sectors and in the degree of self-employment and entrepreneurship. For instance., in Greater London, as compared with other minority groups from Africa and the Caribbean, South Asians have lower participation rates as employees owing to their higher self-employment rate. South Asians are overrepresented in semi-skilled work and professional jobs (Cross and Waldinger, 1997).

In the Netherlands, the main minority groups (Turks, Moroccans, Surinamese and Antilleans) show higher unemployment rates as compared with native workers or immigrant workers from other EU countries or from the former colony of Indonesia. Turks show much higher self-employment rates than the Surinamese or Antilleans. Immigrants have a significantly lower educational level and are more often un(der)employed. Relatively little is known about the self-employment of Asian groups in the Netherlands, but the limited evidence available seems to indicate that there is quite some variation in terms of their geographical spread, economic activities and the labour market niches in which they are involved.

In Italy, the labour market for immigrants is highly segmented and gendered, with a large informal economy attracting large inflows of illegal migrants. Generally the entry point for immigrants is the lowest segments of the Italian labour market (Ambrosini, 2001). First immigration to Italy was limited to some female domestic workers from the Philippines and decolonized Somalia. In many cases immigration by women was channelled by institutions of the Catholic Church. Only in the 1980s did many male and single immigrant workers come from the Maghreb and West African countries, outnumbering female migration, after which their dependants eventually came over. These immigrants found employment in the tertiary sector, small and medium-size firms and in agriculture (Hillmann, 1996; Chell, 1997). During the 1990s, immigration from the Philippines and the Indian subcontinent increased, in addition to (clandestine) immigration from Eastern Europe, the Balkans and Latin America. The Asian female immigrants, in particular from the Philippines and Sri Lanka, were drawn into the domestic service industry, for which there has not been for many years indigenous local labour supply. Similarly, in Spain, two distinct migration flows are apparent: first, Chinese immigration concerns both men and women, often married couples, indicating the existence of family reunification (Beltrán and Sáiz López, 2002). Second, Filipino immigration is highly feminized and mainly concerns single women, working as domestic servants or, as is the case in Italy, as nurses and carers of children and the elderly (Ribas-Mateos, 2000).

Landscapes of Asian migrant labour

As stated above, most contributions in this book address more than one Asian immigrant group and many Asian immigrant groups can be found in more than one chapter. We decided in favour of a structure which articulates the specific position of the Asian immigrants on the labour markets in Europe, rather than structuring the book around the various ethnic groups – which sometimes might lead to a rather 'ethnical reading' of labour-market processes. This does not mean that there is no relation between the cultural background of migrants and their performance on the labour market of the country of arrival, but it might be misleading to put the focus too heavily on that one component of the economic behaviour of immigrant groups. Migration is thought to be one important link between processes on the global scale and the local economies. Migration is a double-faced phenomenon: it is an outcome of economic change as well as a driving force for economic change at the same time. Even if none of the contributions in this book refers specifically to the importance of globalization for the national labour markets and the respective immigrant incorporation, the synopsis indicates similarities of migrant incorporation in different European countries.

The book is divided into three parts. The first, '"Within the system": Asian migration and the European labour market', is all about changes in the overall political and economic context and the implications for the incorporation of Asian immigrants. Each chapter touches upon specific developments. The first part of the book analyses such cases of Asian immigration that are characterized in terms of 'old recipes'. These contributions focus on the Asian migrants who are already 'within the system', i.e. in highly regulated countries without minority traditions (such as Germany and Austria) and on countries that followed in the past years more deregulated labour market policies, but have long-standing minorities (the Netherlands, France and the United Kingdom).

The first chapter in this part, by Giles Barrett and David McEvoy, points out that the changes in the policy setting, i.e. the flexibilization of the labour market, have not translated into better wage employment opportunities for Asian immigrants. Asians have worse jobs than one would expect, even allowing for their education, age, place of residence and other variables. They perform better in self-employment but much self-employment can be considered as an attempt to cope with the increased flexibility of the labour market. In comparing Asian groups with British-born whites, increasing similarity across generations with the distribution of jobs held by UK-born whites is demonstrated, but there are exceptions. Focusing on the Chinese immigrants in France, as Emmanuel Ma Mung does in the second chapter in this part, self-employment appears as an overly important aspect of the labour market integration of this group. Chinese immigrants in France are overrepresented in the sectors that match those where Chinese companies are numerous, such as consumer goods, manufacturing, trade, services to individuals (including hotels, restaurants and personal and house work). Hence the economic configuration, as Ma Mung puts it, creates an ethnic economy. Contrary to the tendencies to move into self-employment that developed in the United Kingdom and in France, the chapter by Christiane Hintermann and Ursula Reeger illustrates how a specific labour market regime fosters the development of a highly segmented labour market for immigrants. Self-employment here turns out to generate completely new forms: the Indian newsvendors in Vienna might be interpreted as signalling a trend toward a highly restrictive and segmented labour market that makes the worker strictly dependent on the employer.[1] Flexibilization in this case leaves no room for eventual social and horizontal mobility. In this specific case the term 'self-employment' is somewhat ambivalent since being a newsvendor on the one hand means to rely on one's own initiative to find customers but it does not mean to *be the boss of your business*, but to *be highly dependent* on one employer.

Concentrating on another country with a strong guest-worker history, Felicitas Hillmann presents us with the situation of the Vietnamese in

Berlin. Here the changes that took place after the reunification of the two Germanys resulted in the predominance of the 'Westernized' perception of the immigrant situation in the country. The clash of the two migration systems (contract worker in the East versus guest worker in the West) in one city led to the marginalization of the Vietnamese, who today are to be found strongly in self-employment, and, so some experts suggest, in parallel economies and realities. Krystyna Iglicka shows that Poland is no longer a country of transit for the Vietnamese. In Poland the Vietnamese are caught up into a certain segment of the labour market that is characterized by vulnerability. Contrary to other groups of foreigners from the 'East' in Poland, Vietnamese find employment not only in the secondary sector but on the primary markets as well. A parallel labour market structures also exists here – even if the author expects this to disappear gradually once the private market economy has come to full fruition.

Part II, '"Coming in": specific incorporation trajectories of Asian migrants', consists of contributions that highlight the strategies of immigrants in order to enter (certain segments of) the labour market or to cope with their difficult labour-market position. Specific incorporation trajectories of Asian immigrants, the gender aspect and the flexible strategies of Asian immigrants, searching for alternatives to old path dependences in response to changing labour-market conditions, are highlighted. Three out of four case studies are, certainly not coincidentally, located in the new European migratory space, the southern European countries where there is quite some elasticity in terms of new forms of migrant labour and for migrant networks.

Part II opens with a chapter on South Asian women in the United Kingdom by Fauzia Ahmad, Stephen Lissenburgh and Tariq Modood. The chapter aims at explaining the diversity within South Asian ethnic and religious groups in terms of educational and employment participation of Asian women, with Indian and East African Asians better represented compared with Pakistani and Bangladeshi women. Religious factors that are often used as an argumentative basis for lower participation rates of Muslim women in the labour market turn out to be rather an 'etiquette', a monocausal explanation for a more complex picture that needs more sensitive interpretation. Another gendered trajectory into the labour market is analysed by Josè Compiani and Fabio Quassoli. After having identified different models of immigrant labour market insertion in Italy, the authors present the exceptional case of the male Indian milkers in the Padania plain of Cremona. Punjabis, who migrate since many years to Italy, first working in the greenhouses of Tuscany, are thought by the local population to be qualified for the hard work of milking through their veneration for cows. Beside this misconception of cultural heritage and work ethics, the case study shows that the capacity to combine formal and informal working arrangements and their mutual advantages are typical for both sides; for the employee and the employer.

Natalia Ribas-Mateos and Laura Oso highlight the almost classical female side of the immigrant labour market. By asking why Philippine women are captured in the domestic labour market, they point to the importance of migration as a family project. The value and social norms emphasizing family and community cohesion, strengthened by their social isolation in Spain, causes them to stay in this specific niche. The stagnation of these often highly qualified women on that level seems to be due also to the absence of business networks and the development of an ethnic economy that could provide some employment opportunities. While they stress the case of Filipina women as an 'anti-entrepreneurial example', Liane Mozère in her contribution on Filipinas in Paris, maintains that they embody the type of 'entrepreneur'. By coming to Paris they make use of a variety of resources and skills and during the years, whether their families join them or not, they gain in autonomy. Migration becomes a life project, whereby they make instrumental use of their competences as well as their skills to attain their goal. In this sense the Philippine domestic workers are 'entrepreneurs of themselves' in a transnational world market.

Part III, 'Self-employment and entrepreneurship among Asian immigrants in Europe', provides a variety of case studies on immigrant entrepreneurship of Asian immigrants in Norway, the United Kingdom, the Netherlands, Italy, Spain and Germany. Here immigrant entrepreneurship is analysed, with the focus on opportunity structures, business strategies and transnational linkages and immigrant networks. Moreover, a number of authors stress the usefulness of the concept of 'mixed embeddedness'.

Part III starts with a paper on the Norwegian setting. Geir Inge Orderud and Knut Onsager attempt to identify the scope of and the distinctions in recruitment and career parts for entrepreneurs from Asia in Norway. As in the preceding parts of the book they give us an overview of the whole range of labour-market integration of Asians in Norway, contrasting South Asians and Chinese and, to a certain extent, comparing their self-employment activities with Danish immigrants, who are close to Norwegian culture. Although there are differences among the several groups regarding specific sub-sectors they are mostly employed in food retailing and catering. Interestingly the authors state that quite a substantial number of Asian migrants switch from self-employment to waged labour and vice versa, mainly within the same sub-sector.

The chapter by Monder Ram and Trevor Jones leads us into the logic and management strategies within ethnic business. The authors advocate looking beyond culturalist explanations of business dynamics. They illustrate, by focusing on the internal management of businesses, i.e. the recruitment, patterns of control, remuneration, that making use of the broader mixed embeddedness approach is more fruitful. They remind us that ethnicity may not be as determinant as other more general factors within labour relations such as informal processes, conflict and market environments. The contribution by Ernst Spaan, Ton Van Naerssen and Harry van den Tillaart shows that although Asian migrants have the reputation to be fairly active on the labour market and are supposed to be well represented in small business, this is certainly not true for all Asian groups. By focusing on two immigrant groups in the Netherlands, the Moluccans and the Filipinos, that are underrepresented in immigrant business, it is argued that the specific immigration histories, cultural preferences and the policy environment are factors that have a strong bearing on their labour market insertion and inclination to entrepreneurship.

Daniele Cologna in his article on Chinese immigrant entrepreneurs in Milan is in favour of an enclave approach to Asian immigration. Contrary to conventional wisdom the Chinese community is no longer characterized by strong bonds of ethnic solidarity. Mutual help and support go first to the family but not to co-ethnics. During the 1990s the recruitment of immigrants from China into sweatshops run by Chinese and producing for Italian garment factories and in restaurants went on, but has become less prominent nowadays. Owing to the saturation of the Chinese restaurant and garment sector Chinese entrepreneurs are moving into new sector like an ethnic service industry. While Milan is a good example of a declining enclave Joaquin Beltrán Antolín argues that while Madrid and Barcelona do not yet have 'Chinatowns' they are in the early stages of formation. The author also argues that the Chinese are moving out of the traditional catering business and into the garment industry, following a particular geographical pattern. Initially the first garment workshops, set up with capital from Chinese restaurant owners, were geographically dispersed all over Spain and of an informal nature. The latest trend is that they have started to concentrate in the two cities when opening up the new line of business of wholesale trade and garment production, which he calls 'the seeds of Chinatown'. Maggi Leung focuses on ethnic networks among Chinese migrant businesses but she does not fall into the trap of cultural determinism by presenting examples from three different sectors of activities within German cities. The traditional sector, food, and the new ones, travel and computer wholesale and retail, have each specific networks at different geographical scales and this dynamism cannot be explained by culture. As in the case of Italy and Spain new, somewhat globalized forms of ethnic business became more significant in the 1990s.

All in all, this book presents the many various forms of Asian migrant experiences on European labour markets. Hence the title of this introduction. International migration and labour market structures are in constant flux. The incorporation of ten new member states into the European Union and its consequences for Asian immigration and national and regional labour markets in the new Europe call for further research. Another topic concerns the labour market insertion and socio-economic mobility of second-generation Asians compared with the first generation. Furthermore, as mentioned before, there is increasing interest in global migrant networks' support of fragile economic and social arrangements in the countries of origin. These are three themes that apply to the various Asian immigrant groups analysed in this book and that need to be explored and researched in future.

Note

1 Cf. the concept of dependent self-employment discussed by Bromley (1985).

References

Ambrosini, M. (2001) 'The role of immigrants in the Italian labour market', *International Migration*, 39 (3): 61–84.

Basch, L., Glick Schiller, N. and Szantos Blanc, C. (1997) 'From immigrant to transnational migration', in Ludger Pries (ed.) *Transnational Migration*. Baden-Baden: Nomos, 121–40.

Beltrán, A.J. and Sáiz López, A. (2002) *Comunidades asiáticas en España*. Barcelona: CIDOB.

Bromley, R. (ed.) (1985) *Planning for Small Enterprises in Third World Cities*. Oxford: Pergamon Press.

Chell, V. (1997) 'Gender-selective migration: Somalian and Filipina women in Rome', in *Southern Europe and the New Immigration*. Brighton: Academic Press, 75–92.

Commission of the European Communities (2003) Communication from the Commission to the Council, the European Parliament, the European Economic and Social Committee and the Committee of the Regions on Immigration, Integration and Employment. Brussels, COM (2003) 336 final, 3 June 2003.

Cross, M. and Waldinger, R. (1997) 'Economic integration and labour market change: a review and re-appraisal'. Discussion paper prepared for the Second International Metropolis Conference, Copenhagen, 25–27 September.

Deutscher Bundestag (2002) *Globalisierung der Weltwirtschaft*. Opladen: Leske & Budrich.

Fong, Pa.E. and Lim, L.Y.C. (1996) 'Structural change in the labour market, regional integration and international migration', in *Migration and the Labour Market in Asia: Prospects to the Year 2000*. Paris: OECD, 61–74.

Hillmann, F. (1996) *Jenseits der Kontinente. Migrationsstrategien von Frauen nach Europa*. Opladen: Leske & Budrich.

Hillmann, F. (2000) 'Von internationalen Wanderungen zu transnationalen Migrationsnetzwerken? Der neue Europäische Wanderungsraum', in Maurizo Bach (ed.) *Die Europäisierung Nationaler Gesellschaften*. Cologne: Kölner Zeitschrift für Soziologie und Sozialpsychologie, 363–85.

Hochschild, A. (2000) 'Global care chains and emotional surplus value', in W. Hutton and A. Giddens (eds) *On the Edge: Living with Global Capitalism*, London: Cape, 1–51.

Husa, K. and Wohlschlägl, H. (2000) 'Aktuelle Entwicklungstendenzen der internationalen Arbeitsmigration in Südost- und Ostasien vor dem Hintergrund von Wirtschaftsboom und Asienkrise', in Karl Husa *et al.* (eds) *Internationale Migration. Die globale Herausforderung des 21. Jahrhunderts*. Frankfurt am Main: Brandes & Apsel, Vienna: Südwind, 247–80.

ILO (International Labour Office) (2004) *Towards a Fair Deal for Migrant Workers in the Global Economy*. Report VI. Geneva: ILO.

IOM (International Organization for Migration) (2003) *World Migration: Managing Migration*, Geneva: IOM.

Kim, W.B. (1996) 'Economic interdependence and migration dynamics in Asia', *Asian and Pacific Migration Journal*, 5 (2–3): 303–17.

Kloosterman, R. (2000) 'Immigrant entrepreneurship and the institutional context: a theoretical exploration', in Jan Rath (ed.) *Immigrant Businesses*. London: Macmillan, 90–106.

Koser, K. and Lutz, H. (1998) *The New Migration in Europe*. Basingstoke: Macmillan.

Lewis, P. (2002) *Islamic Britain*. London: Tauris, 16–17.

Massey, D., Arango, J., Hugo, G,, Kouaouci, A., Pelegrino, A. and Taylor, E.J. (1998) *Worlds in Motion: Understanding International Migration at the End of the Millennium*. Oxford: Clarendon Press.

Morokvasic, M. (2003) 'Transnational mobility and gender: a view from post-war Europe', in M. Morokvasic, U. Erel and K. Shinozaki (eds) *Crossing Borders and Shifting Boundaries*, I, *Gender on the Move*. Opladen: Leske & Budrich, 101–33.

Nyberg-Sørensen, N. (2004) 'Transnational family life: the gendered "nature" of transnational parenthood, family values and family functions'. Paper presented at the World Congress on Human Movement and Immigration, Barcelona, 2–5 September.

OECD (1997) *Trends in International Migration*. Paris: OECD, SOPEMI.

OECD (2003) *Trends in International Migration: SOPEMI 2002 Edition*. Paris: OECD.

Østergaard-Nielsen, E. (2003) *Transnational Politics: Turks and Kurds in Germany*. London: Routledge.

Pessar, P.R. and Mahler, S. (2003) 'Transnational migration: bringing gender in', *International Migration Review*, 37 (3): 812–46.

Pieke, F.N. (2002) *Recent Trends in Chinese Migration to Europe: Fujianese Migration in Perspective*. International Organization for Migration (IOM) Research Series 6. Geneva: IOM.

Pries, L. (1999) 'Transnationale Räume zwischen Nord und Süd. Ein neuer Forschungsansatz für die Entwikclungssoziologie', in Karin Gabbert *et al.* (eds) *Migrationen, Lateinamerika. Analysen und Berichte*. Bad Honnef: Horlemann, 39–54.

Rath, J. (ed.) (2000) *Immigrant Businesses: the Economic, Political and Social Environment*. Basingstoke: Macmillan.

Ribas-Mateos, N. (2002) 'Female birds of passage: leaving and settling in Spain', in Anthias Flora and Gabriella Lazardis (eds) *Women in the Diaspora: Gender and Migration in Southern Europe*. Oxford: Berg.

Sassen, S. (2003) 'The feminisation of survival: alternative global circuits', in M. Morokvasic, U. Erel and K. Shinozaki (eds) *Crossing Borders and Shifting Boundaries*, I, *Gender on the Move*. Opladen: Leske & Budrich, 59–97.

Skeldon, R. (2000) 'Trafficking: a perspective from Asia', *International Migration*, 38 (3): 7–30.

Spaan, E.J. (1994) 'Taikongs and Calo's: the role of middlemen and brokers in Javanese international migration', *International Migration Review*, 23 (1): 93–113.

Spaan, E.J. (2000) 'Les migrations internationales en Asie', *Revue européenne des migrations internationales*, 16 (1): 11–35.

Stahl, C.B. (2001) 'The impacts of structural change on APEC labor markets and their implications for international labor migration', *Asian and Pacific Migration Review*, 10 (2–3): 349–77.

Van Hear, Nicholas and Nyberg-Sørensen, Ninna (eds) (2003) *The Migration–Development Nexus*. Geneva: IOM.

Vertovec, S. (2001) 'Transnational social formations: towards conceptual cross-fertilization', in *Transnational Communities Programme*, Working Papers series ed. Ali Rogers WPTC-01-16, Oxford: Transcom www.transcomm.ox.ac.uk/working_papers.htm (12 July 2004).

Waldinger, R. and Fitzgerald, D. (2004) 'Transnationalism in question', *American Journal of Sociology*, 109 (5): 1177–95.

Zlotnik, H. (2003) 'The global dimensions of female migration', *Migration Information Source*. Washington DC: Migration Policy Institute.

Part I

'Within the system'

Asian immigration and the European labour market

2   Not all can win prizes

Asians in the British labour market

Giles A. Barrett and David McEvoy

1979 was an important year for British politics and for the British labour market. Margaret Thatcher's first government came to power and began a regime change in the regulation of the economy. This was maintained under Thatcher's Conservative successor, John Major, and, with some exceptions, under Tony Blair's 1997 Labour government. Public utilities and other state-owned industries were transferred to the private sector. The right to strike was restricted, and Wages Councils, which regulated pay in low-paying industries, were abolished. State benefits for the unemployed were reduced, particularly for the young. Limits on shop opening hours in England and Wales, introduced by a Labour government strongly influenced by the shop workers' trade union in 1950, were eliminated. These and other measures exposed employers and workers to the rigours of the market in a way which would have been politically unthinkable, whatever the party in power, between 1945 and 1979. Unemployment was allowed to more than double to well over three million. Moreover official definitions of unemployment were revised over twenty times so that figures in newspaper headlines became less dramatic.

During the 1990s, however, Britain's unemployment fell again to below the level applying in most other EU members. This is widely interpreted as a measure of the flexibility introduced into the economy by the post-1979 reforms. In contrast to many of its major European partners the United Kingdom is no longer seen as a corporatist state in which the rights of workers and their trade unions are strong determinants of national policy. The restructuring of industries has proceeded rapidly. Coal mining for example has disappeared as a significant employer. Manufacturing too has suffered many job losses. Meanwhile many service industries have been growing, including health, education, hotels, catering, finance, and professional services.

The growing ethnic minority populations have not therefore entered a stable opportunity structure. The number and nature of employment openings have been constantly changing. More jobs have become temporary. Success in this environment may be dependent on a dynamic mix of endeavour, experience, qualifications and adaptability. Britain's Asian minorities are popularly thought to have performed well in this context, particularly in comparison with the African-Caribbean community. This is a viewpoint based partly on the multiplication of Asian-owned retail and restaurant businesses, activities which are highly visible in urban areas.

In order to assess the impression of Asian success this chapter looks first at the migration history and population characteristics of the main Asian groups. Consideration is then given to their educational attainments as preparations for employment. Attention then turns to engagement with the labour market, including levels of participation and unemployment. After this the differential presence of Asian communities in broad economic sectors is considered. The distinction between male and female experience is kept in view throughout. Particularly in the examination of sectors, the experiences of the immigrant generation are contrasted with those of their British-born descendants.

Demographic and educational background

Britain's Asians have their origins in the imperial past. In the twenty years following the Second World War streams of migrants from the Caribbean, from South Asia, from Hong Kong and from other quarters of the disappearing empire were established. Most of the newcomers, as citizens of former or current British colonies, held British passports. In spite of recurring moral panics, most notoriously characterized by the 'Rivers of Blood' speech of Enoch Powell, these migrations were a response to the economic needs of British society. Immigrants came to fill the vacant job niches in the economy. In a period of economic growth and full employment, positions which were poorly paid, or involved long hours or unpleasant conditions, were no longer attractive to the indigenous population. Industries such as the foundries of the Midlands and the textile factories of Lancashire and Yorkshire relied on immigrants to maintain their competitiveness. Male newcomers, sometimes actively recruited from overseas, often staffed night shifts which were unattractive to white males and illegal for female workers (Kalra, 2000: 96). Similarly the National Health Service and train and bus operators plugged staffing gaps with immigrants of both sexes.

This replacement labour phase did not, however, continue. Primary immigration became much more difficult under laws of 1962, 1968 and 1971 which changed the passport entitlement of colonial and former colonial citizens. Nevertheless family reunification continued to be allowed, so that wives and children were able to join men who, for reasons of economy, had originally migrated alone. Now the passage of decades, with births and education in Britain, had the unintended consequence of turning possibly temporary migrant populations into settled ethnic minority communities. The sojourner mentality of the original migrants, and the associated 'myth of return' (Anwar, 1979), were supplanted by the recognition that minorities were 'here to stay' (Bradford Heritage Recording Unit, 1994). Meanwhile immigration on the basis of work permits, for those with professional qualifications or specialist skills, continued on a restricted basis, and some foreign students and asylum seekers were also able to gain admission.

Government data have struggled to define these evolving populations. Until the 1971 census only birthplace and previous residence were recorded. Continuing this practice would have rendered the growing number of children, born in Britain of immigrant parentage, statistically invisible. In 1971 a census question was asked about parental birthplace. The 1981 census dealt with the matter by inference: ethnic minorities were identified on the basis of the birthplace of the head of the household in which a person resided (Coleman and Salt, 1992: 483–6). Only in 1991 were census respondents asked directly about ethnicity. This was repeated in 2001, although the results have been classified slightly differently from ten years earlier. Additional complications arise from differences between the categories enumerated and reported in Scotland and those used in England and Wales. Northern Ireland even has a separate census, resulting in further variations in ethnic classification (NISRA, 2003).

It is nevertheless possible to compile comparative figures for ethnic groups for the United Kingdom's constituent countries, and for the government regions of England. These geographical differences are important to Asian labour markets because variations in economic structure and prosperity provide different opportunity structures in different places. Table 2.1 presents population data for the Asian groups identified by the census. For purposes of comparison it also includes figures for the white majority. A residual category, consisting partly of persons of black descent, partly of those of mixed ethnicity, and partly of persons of other non-Asian descent, is omitted. The white group is predominantly British in origin, but also includes the Irish, other Europeans, and some from former colonies of white settlement.

The figures show that Britain is a predominantly white country, with ethnic minorities comprising about 8 per cent of the population. Rather more than half of these, 4.4 per cent, are of Asian origin. The picture is, however, geographically varied. In more peripheral regions, including the 'Celtic fringe' of Wales, Scotland and Northern Ireland, ethnic minorities are less than 2.5 per cent of the population, with the Asian share ranging from 1.6 per cent in north-east England to 0.4 per cent in Northern Ireland. In the rest of England the minority proportion is higher, with figures ranging from 4.9 per cent in the East to 28.9 per cent in London. Two main explanations can be discerned. First, old industrial regions have minority populations which have developed from the arrival of replacement labour in earlier decades. Second London, which also attracted migrants in this earlier phase, functions as a 'global city' and consequently continues to attract migrants, both to the upper tiers of the labour market and to the low-paid service jobs which support the lifestyle of the affluent sectors of the metropolitan elite (Sassen, 2001).

Table 2.1 Ethnic groups in Great Britain (%)

Region	White	Indian	Pakistani	Bangladeshi	Chinese	Other Asian	Asian total
United Kingdom	92.1	1.8	1.3	0.5	0.4	0.4	4.4
England	90.0	2.1	1.4	0.6	0.5	0.5	5.0
North East	97.6	0.4	0.6	0.3	0.2	0.1	1.6
North West	94.4	1.1	1.7	0.4	0.4	0.2	3.8
Yorkshire and the Humber	93.5	1.0	3.0	0.3	0.3	0.3	4.7
East Midlands	93.5	2.9	0.7	0.2	0.3	0.3	4.4
West Midlands	88.7	3.4	2.9	0.6	0.3	0.4	7.6
East	95.1	1.0	0.7	0.3	0.4	0.3	2.7
London	71.2	6.1	2.0	2.2	1.1	1.9	13.2
South East	95.1	1.1	0.7	0.2	0.4	0.3	2.8
South West	97.7	0.3	0.1	0.1	0.3	0.1	0.9
Wales	97.9	0.3	0.3	0.2	0.2	0.1	1.1
Scotland	98.0	0.3	0.6	0.0	0.3	0.1	1.4
Northern Ireland	99.5	0.1	0.0	0.0	0.3	0.0	0.4

Sources: National Statistics (2003a, b); NISRA (2002); Registrar General for Scotland (2003).

Note
All figures have been rounded. A degree of approximation by the authors has been involved in making the figures for Scotland and Northern Ireland compatible with those for England and Wales. Any errors in this approximation are likely to be of no greater magnitude than rounding errors.

Within these overall patterns particular Asian minorities display different geographies. The Chinese are the most evenly distributed group. In all regions except London they are present as a small minority of between 0.2 per cent and 0.4 per cent. This wide dispersal is a reflection of their concentration in a single economic activity, the Chinese restaurant or take-away, which was virtually ubiquitous as early as the 1970s (Watson, 1977). Even in Belfast's notoriously segregated districts, the Protestant Shankill and the Catholic Falls, Chinese food outlets provide reminders of the possibility of multicultural life. The picture in London differs. Although Chinese cuisine is as available as elsewhere there is also a concentration of those with professional and technical qualifications. These people have more widespread origins than those associated with the restaurant trade. While people from Hong Kong dominate the latter, the more educated groups include greater proportions from mainland China and from the overseas communities of South East Asia (Cheng, 1996).

The Pakistani pattern contrasts most markedly with that of the Chinese. The group's share in the total population is highly uneven, varying from 3 per cent in Yorkshire and the Humber to less than 0.1 per cent in Northern Ireland. Moreover London is not the principal centre of this community, either proportionally or absolutely. The number of Pakistanis in London is exceeded by the numbers in the West Midlands and in Yorkshire and Humberside. Both these regions recruited Pakistani labour into their formerly dominant industries, engineering in the West Midlands and wool textiles in West Yorkshire (Dahya, 1974).

In contrast London is the principal centre for both Bangladeshis and Indians. The Bangladeshi presence can be traced to nineteenth-century employment in British merchant ships. Galley workers subsequently established shore-based cafés in London's docklands. This eventually led to chain migration into the area from the 1960s onwards as local clothing and leather industries recruited cheap labour in an era of increasing international competition (Rhodes and Nabi, 1992). Chain migration also applies to some of London's Indians, for example the Sikh population found in the Southall area of west London, where a rubber factory was an important magnet in the 1950s. The Indian population, however, also shares with the Chinese the characteristic of possessing many well educated members, attracted to London by its commercial, financial and professional dominance of British society.

Outside London Bangladeshis are far fewer. Like the Chinese their presence in most of these areas is strongly associated with the catering trade, although their restaurants are invariably referred to as 'Indian'. Indians themselves are more numerous than Bangladeshis in regions outside London, especially in the former textile areas of the East Midlands, and, like the Pakistanis, in the engineering districts of the West Midlands.

The 'Other Asian' group also displays a concentration in London. The group is so varied in its origins that one author was moved to describe it as 'the salad bowl'. About 136,000 of its approximately 235,000 current members are to be found in the capital city. Figures from the 2001 census are not published at the time of writing, but in 1991 about 15 per cent of the group originated in Sri Lanka; about 12 per cent came from Japan; about 9 per cent came from the Philippines and about 6 per cent originated in the Asian populations of Mauritius; no other group exceeded 5 per cent of the total. A high proportion of the group possessed higher educational qualifications (Owen, 1996). As in the Chinese and Indian cases, we are seeing the pull of a globally influential city for well qualified, and therefore internationally mobile, individuals.

Labour market data from the 2001 census were not available at the time of writing. Fortunately, however, there are other sources. Prime among these is the Labour Force Survey (LFS), a continuous survey based on household interviews conducted by the Office of National Statistics.[1] The results were published biennially from 1973, then annually until 1991, and have emerged quarterly since. The quarterly sample size is about 59,000 addresses, with about 138,000 individuals, including both the economically active and the economically inactive. The survey incorporates many questions about the labour-market status of individuals, such as occupation, industrial sector, hours of work and income. The latter is an issue which the British census has always avoided, since it is a topic of great sensitivity for many individuals. It is thought that its inclusion in the census would be likely to compromise willingness to co-operate. The LFS also asks about qualifications, training, disability and a range of relevant family and household variables.

Compared with the census the LFS has the advantage of frequency, and is therefore always relatively up-to-date. On the other hand its sample size is such that its results have to be handled carefully. Particularly when information for small groups is considered, sampling error can be considerable. For example for a group constituting only 0.5 per cent of the population, like Bangladeshis in 2001, the number of individuals covered is likely to be of the order of 700. When this is differentiated by age, gender, economic sector, or other variables, all possibility of precision is lost. While the problem can be addressed by combining the data for several waves of interviews this leads to a loss of currency.

Work by Dustmann et al. (2003) based on the LFS is extremely useful for our purposes. Its main focus is on immigrants rather than ethnic minorities. Since, however, there are marked differences in the employment experiences of immigrants and their British-born descendants this is not a weakness. Moreover a number of illuminating inter-generational comparisons are examined. It includes the full range of immigrant groups as well as those which are the subject of the present chapter. What follows is therefore a selection and interpretation of some relevant items.

Table 2.2 provides a basic comparison between non-immigrant whites and the four main Asian immigrant groups for the population of working

Table 2.2 Immigrants and the UK-born compared (working age population, excluding full-time students)

Year	UK-born white	UK-born non-white	Indian	Pakistani	Bangladeshi	Chinese
% of population						
1979	92.2	0.5	1.2	0.4	0.1	0.2
2000	88.3	2.4	1.2	0.7	0.3	0.2
Median age						
1979	36	19	33	31	37	29
2000	39	27	43	37	33	37
Median years in UK						
1983	–	–	14	14	10	8
2000	–	–	25	20	16	14
Percentage with degrees (men)						
1983	10	3	16	8	11	14
2000	16	18	23	12	7	31
Percentage with no qualification (men)						
1983	46	35	42	67	85	47
2000	14	13	16	35	41	21
Percentage with degrees (women)						
1983	4	2	9	4	2	9
2000	12	16	14	6	6	23
Percentage with no qualification (women)						
1983	51	31	57	75	91	47
2000	19	11	28	52	55	15

Source: Dustmann *et al*. (2003: 21).

age. (The source does not use the census's 'Other Asian' category.) The working-age population is that between Britain's minimum school-leaving age of sixteen and the age of eligibility for the basic state retirement pension, which is sixty for women and sixty-five for men. Two years are given for each variable so that the changing characteristics of natives and immigrants can be appreciated. Differences for particular groups between the 2000 figures in this table and those for 2001 in Table 2.1 are mainly the result of the exclusion of those over working age from this table, especially for whites, and the inclusion of the British-born members of ethnic minorities in a separate category from their immigrant forebears. This has effects which are perhaps surprising. Although whites have decreased as a proportion of the labour force between 1979 and 2000, and the number of both Pakistanis and Bangladeshis has increased substantially, the share of Indians and Chinese has remained stable.

Whites and three of the four immigrant groups show an ageing of the labour force since 1979, possibly consequent on a reduction in the rate of immigration, and perhaps on selective return migration. For Bangladeshis, however, the median age of labour-force members has fallen by four years. This is probably a reflection of a higher proportion of Bangladeshis under the age of sixteen at the time of immigration, a feature stimulated by periodic tightening of immigration regulations (Coleman and Salt, 1992: 441) which accelerated family reunion in advance of the restrictions. Immigration controls have also contributed to the increasing length of time which the median member of each immigrant group has spent in the country.

Table 2.2 also compares the qualifications of immigrants with those of indigenous whites. All groups except Bangladeshis have shown a significant increase in the proportion of their male members with degrees, and all groups, including Bangladeshis, have seen a marked decline in their percentage with no qualifications. These trends appear to be in line with the notion of a post-industrial British economy, requiring an ever higher share of the work force to possess advanced qualifications and skills. Both Indian and Chinese men had a higher proportion of degree holders than whites in both 1983 and 2000. Pakistani men had fewer degrees than whites in both years, while Bangladeshi men fell behind whites and behind their own earlier performance. Since most of the 1983 cohort of Bangladeshis are likely also to form part of the 2000 cohort it would appear that this decline is associated with a lower level of university education among recent immigrants. We speculate that this may be related to the community's strong association with the restaurant trade, an economic sector not marked by a demand for graduate labour. At the lower end of the skills hierarchy all the migrant groups have higher percentages without qualifications than UK-born whites. Even for groups with many graduates the polarization Sassen (2001) sees in global cities may apply to immigrants in the entire economy.

For women Table 2.2 shows fewer degree holders than among men for all groups in both 1983 and 2000. However, the situation is improving for both whites and the Asian groups. As in the male case Indians and Chinese women have maintained a lead over whites over the period, and once again Pakistanis and Bangladeshis display lower figures. When the data for women with no qualifications is examined we again see an improving situation for all groups, but with Pakistanis and Bangladeshis trailing other groups. Chinese women, but not Indian women, have somewhat better figures than whites.

The remaining information in Table 2.2 relates to UK-born non-whites. This includes the children and grandchildren of our four Asian migrant groups, but also of other non-white immigrants. The total population in this category has grown rapidly as the migrants of earlier decades have established family life in Britain. The median age of labour force members has increased substantially. In 1979 the figure was only nineteen, indicating that the establishment of UK-born minority populations was quite recent. By 2000 this had increased to twenty-seven, as large numbers of younger co-ethnics joined the twenty-one-year-old cohort in the labour force. The proportion of the group with degrees has increased markedly between 1983 and 2000 for men and women, overtaking the white figure in both cases. Similarly the proportion of UK-born minorities without qualifications has fallen rapidly. The 1991 census showed that the UK-born Chinese of both sexes performed better at degree level than other Asian immigrant groups and whites, while Pakistanis and Bangladeshis lagged in this respect. The comparison between UK-born Indians and whites is less clear: Indian women may have been doing better than white women, but were still behind Indian men and white men, who were quite similar to each other (Blackburn *et al.*, 1997).

Patterns of economic engagement

Our introduction referred to growing flexibility in the British economy. Table 2.3 provides some illustrations of what this has meant for the labour market. The activity rate is the proportion of the population (of the relevant ages, and excluding full-time students) who are either employed, self-employed or available for employment. In 1979 the figure for UK-born white males was 95 per cent, so only 5 per cent of this group were outside the labour market. For UK-born women the activity rate was 66 per cent: an indication that a substantial minority of women were constrained, or chose, to perform purely domestic roles. By 2000 the white male figure had fallen to 88 per cent, but the equivalent female rate had increased to 76 per cent. Falling male activity rates derive substantially from increases in the number of men living on prematurely taken pensions, and on disability-related incomes. Both these devices have been widely used to enable companies, and the public sector, to shed labour in order to remain competitive, or to demonstrate apparent efficiency. The increase in female activity rates is partly social. The reduced incidence of marriage, falling numbers of children, and higher divorce rates, have discouraged economic dependence upon men. However, purely economic considerations also apply. As male jobs have become more insecure the desirability, and perhaps necessity, of female employment has become evident even in stable and highly conventional households.

The remaining figures for UK-born whites in Table 2.3 help to reinforce the picture of flexibility drawn from activity rates. For men, unemployment, which is a percentage of the number economically active, has risen. The number of men taking part-time jobs has multiplied, and their self-employment rate has risen. A major reason for self-employment is insecurity in the employed sector. An additional indication of this insecurity is the availability of a temporary employment figure for 2000, information not reported for 1979. The message is less clear in the figures for women. Self-employment has increased, and part-time working has gone up, but only marginally. The proportion of women on temporary contracts is higher than

Table 2.3 Measures of labour market integration (working age population, excluding full-time students) (%)

Ethnicity	Activity rate	Unemployed	Part-time	Temporary	Self-employed
Men					
UK-born white					
1979	95	4	1	–	9
2000	88	5	4	5	14
Indian					
1979	96	5	0	–	12
2000	84	7	5	7	24
Pakistani					
1979	97	9	1	–	10
2000	76	13	15	8	34
Bangladeshi					
1979	93	1	3	–	13
2000	73	19	21	7	16
Chinese					
1979	99	0	2	–	26
2000	77	5	7	15	25
Women					
UK-born white					
1979	66	6	38	–	3
2000	76	4	39	7	6
Indian					
1979	58	9	16	–	6
2000	61	8	28	10	12
Pakistani					
1979	16	30	24	–	4
2000	24	17	48	17	10
Bangladeshi					
1979	25	35	n/a	–	n.a.
2000	20	34	43	12	1
Chinese					
1979	53	2	23	–	14
2000	59	4	30	11	15

Source: Dustmann *et al.* (2003: 28, 31–2).

Note
Ethnic minority figures exclude UK-born.

the equivalent male figure. Female unemployment has, however, fallen below the male figure. This may show the increasing attractions of a female labour force for employers in an era of global competition. Despite equality legislation, which dictates equal pay for work of equal value, many jobs remain heavily gendered, and women are often still perceived as cheaper and more docile employees than men. Enthusiasts for competitive economic solutions might, however, suggest that this is an over-interpretation of the data, and that the unemployment rate simply shows an improvement in the experience of women.

If the white data on economic engagement illustrate the increased flexibility of the labour market, the information for Asian immigrant minorities shows that flexibility has an uneven impact. The pattern of Chinese and Indian advantage over Bangladeshis and Pakistanis, demonstrated in our analysis of qualifications, is strongly reinforced. In addition it is arguable that even the Chinese and Indians are at a disadvantage compared with whites. The male figures are easier to interpret, since they do not involve the impact of distinct cultural views on the economic position of women. In Asian communities, as well as white men, are seen as possessing responsibility for a primarily economic role, even if some groups also require it of women.

As for whites, activity rates for Asian men fell between 1979 and 2000, but they did so more dramatically. A white decline of seven percentage points compares with twelve for Indians, twenty-one for Pakistanis, twenty for Bangladeshis, and no less than twenty-two for Chinese. Unemployment rates have increased for all Asian groups, but are similar to white levels for Indians and Chinese, and very much higher for Pakistanis and Bangladeshis. Temporary work is above white levels for all Asian groups, especially the Chinese. The self-employment figures are higher for Asians than for whites for all groups in both 1979 and 2000. Moreover the level of self-employment increased for Bangladeshis, increased dramatically for both Indians and Pakistanis, and fell marginally, while remaining at a high level, for the Chinese. Some would interpret high levels of self-employment among Asians as indicative of their clear advantage, based on cultures which value hard work and family solidarity, in an economic environment that encourages entrepreneurialism. If, however, our suggestion that self-employment for white males is partly a result of insecurity in employment is valid, it becomes less reasonable to see Asian entrepreneurialism as demonstrating the economic potency of culture. Instead high self-employment is an adjustment to lack of alternative economic opportunities (Barrett *et al.*, 2001, 2002).

Turning to the female data in Table 2.3 we see an even stronger distinction between Bangladeshis and Pakistanis, on the one hand, and all other groups on the other. Activity rates for the former remain below 25 per cent, and actually fell between 1979 and 2001 for Bangladeshis. This decline may, however, be a temporary phenomenon, since intervening years have seen figures as low as 10 per cent. (Dustmann *et al.*, 2003: 28). Unemployment levels for the two groups are also much higher than for whites, for Indians and for the Chinese; they also exceed the already high equivalent figures for men. There is a higher incidence of part-time work and temporary work for Bangladeshi and Pakistani females. Only the self-employment figures paint a slightly different picture; Bangladeshis remained distinguished by a very low figure, but Pakistani women now exceed the white level, though remaining below the rates for Indians and the Chinese.

In comparison with the differentials with Bangladeshi and Pakistani women the differences between white women and Indian and Chinese women are more muted. Indian and Chinese activity rates are about a fifth down on white levels. Indian women have higher unemployment but whites and the Chinese are now similar on this measure. Whites have the highest level of part-time work among the three groups, with Indians and Chinese closer together. Whites are lowest on temporary employment, and the other groups again similar. The Chinese have the higher levels of self-employment, with Indians next, but all three groups have seen an increase in this figure while remaining below their male equivalents.

Overall the picture of economic engagement is different for men and women. Men's activity rates are much higher, their level of part-time working is lower and the level of entrepreneurship is greater. Unemployment differentials and the scale of temporary employment are not so clearly linked to gender. Within each gender, however, a strong pattern of disadvantage appears for Bangladeshis and Pakistanis, with whites probably retaining an advantage over the Chinese and Indians.

Dustmann *et al.* (2003) attempt to explain their figures for employment (total active labour force minus the unemployed) and activity levels using regression techniques. If an immigrant group is concentrated in regions of economic disadvantage, as for example might be argued for Pakistanis, or has a rather younger age profile, as in the Bangladeshi case, then any disadvantage it has in the labour market may be the result of these characteristics rather than any innate disadvantage it suffers, or of discrimination against it. The authors therefore control the individual characteristics of immigrants in terms of region of residence, family status, age and education. Their results show that, allowing for the control variables, the employment outcomes for Chinese males are very similar for those of UK-born whites. For Indian males there is a slightly lower chance of employment, and for Pakistani and Bangladeshi males it is substantially lower. In the case of women, again allowing for the control variables, Chinese and Indian immigrants have a slightly lower chance of employment than UK-born whites, and Pakistanis and Bangladeshis have a very much lower chance.

Extending the analysis to activity rates they find that, after controls, all four Asian male immigrant groups display lower participation in the economy than UK-born whites. The four equivalent female groups are at an even greater disadvantage to UK-born white women than the male groups are to their white comparators. This is especially so in the Pakistani and Bangladeshi cases. Dustmann and his colleagues also look at the impact of their control variables on gross hourly wage rates. All male Asian immigrant groups are found to be significantly less well rewarded than UK-born white males. Bangladeshi males in particular suffer. Similar patterns apply to female Asian immigrants, although the disadvantage for Bangladeshi females may be less than for Bangladeshi males.

It is tempting to explain part of these patterns by relying on the most obvious commonality of the Pakistani and Bangladeshi communities, the adherence to Islam of most of their members. Compared with the host community in Britain these groups appear to have a protective, and some would say restrictive, attitude to women. Dale and Holdsworth (1998) found for example that women from these groups were more likely than others to leave the labour market on marriage, and particularly upon the birth of children. For religiously conscientious Muslim men it might be that requirements for regular prayer, Friday visits to the mosque, and maybe particular styles of clothing, have an impact on their ability to compete for a full range of jobs. Even in the unlikely absence of employer prejudice it might be that the need to work within easy access of a suitable mosque limits the range of available opportunities.

These issues are further investigated by Lindley (2002). She subjects data from The Fourth National Survey of Ethnic Minorities (Modood *et al.*, 1997) to probit analyses in order to ascertain the relative impact of religion and ethnicity on employment/unemployment and on earnings. She demonstrates that within the male Indian population Hindus have lower unemployment than Sikhs, who in turn do better than Muslims. However, Indian Muslim unemployment figures are better than those for Pakistani and Bangladeshi Muslims. White Christians outperform each of these Asian groups. Within the male Indian population the story for average weekly earnings is similar to that for unemployment, but the comparison with other nationalities is slightly different because Pakistani Muslims who avoid unemployment out-perform Sikhs and their fellow Muslims. The picture for females is different again. Lindley's conclusions allow for a range of socio-economic character-istics in addition to ethnicity and religion. These include the distinction between foreign-born and British-born Asians. Overall she finds that, over and above an ethnic penalty in the labour market, there is also a religious penalty suffered by Sikhs, but to a greater extent by Muslims. For women these penalties are exhibited more for incomes than for unemployment, female unemployment being replaced by absence from the labour market.

Sectoral patterns

As with migrant-origin groups the world over, Britain's Asians are not evenly distributed across the economy. Their differing backgrounds, quali-fications and levels of adaptation to living in the United Kingdom have produced distinct patterns of presence in the various economic sectors. Geographical distribution across the country and time of immigration may well be important factors in contributing to these differences. Table 2.4 shows the relative distribution of Asian minorities and whites across eleven employment categories. Men and women are shown separately, and the immigrant generation is shown separately from those born overseas. The table is drawn from the Labour Force Survey for a single quarter, the winter of 2001–02. Because this is a sample survey, albeit a relatively large one, the sample size becomes quite small when the table's individual columns and cells are considered. In consequence any percentage in the table which is based on a denominator of fewer than fifty respondents has been italicized to indicate its possible unreliability.

Before looking at the details in Table 2.4 let us attempt to assess the overall pattern which it displays. A standard technique for measuring the differences between distributions of data across a set of categories is the dissimilarity index. This measures the percentage of one distribution which would have to move between categories if it were to replicate the pattern of a second distribution. Let us consider the first two columns of figures for men in Table 2.4. These show the distribution across the employment sectors of British-born whites and of whites born overseas. Visual inspection reveals that the two sets of numbers are quite similar. The dissimilarity index gives quantitative form to this qualitative judgement. It is calculated by taking the differences between the percentages for each ethnic group for each industry in turn. Some of the resulting numbers are positive, for example in manufacturing, where the proportion of UK-born whites in the industry is 23.2 per cent, 3.5 per cent greater than the equivalent figure for those born overseas. Other differences emerge as negative, as in the case of hotels and restaurants, where a higher proportion of the overseas-born are in this sector than is the case for the UK-born. The index is calculated by summing all the positive numbers, or all the negative numbers: subject to rounding errors, the two figures are identical. In the case of British-born white males compared with overseas-born white males our calculations produce a figure of 8, implying that this percentage of immigrant whites would have to move sector in order to replicate the industrial pattern of the native-born.

It would be possible to calculate index values between any pair of origins. We choose, however, to concentrate on the differences between UK-born whites and each of the other groups. UK-born whites are much the largest of the groups, and may reasonably be viewed as the 'charter group' in British society. Their distribution across industrial sectors can therefore be taken as a norm against which to measure minority and immigrant difference. Table 2.5 therefore lists the dissimilarity indices between British-born whites and each of the other groups shown in Table 2.4. Figures are shown separately for males and females because gender is clearly a dimension of economic variation which is additional to ethnic variation. As in Table 2.4 the results

Table 2.4 Sectoral specialization by ethnic group 2001/2002 (%)

Sector	White UK-born	White Born elsewhere	Indian UK-born	Indian Born elsewhere	Pakistani UK-born	Pakistani Born elsewhere	Bangladeshi UK-born	Bangladeshi Born elsewhere	Chinese UK-born	Chinese Born elsewhere
Men										
Primary, utilities	3.5	6.0	0.7	1.9	3.3	2.1	0.0	0.0	6.3	0.0
Manufacturing	23.2	19.7	13.9	22.1	15.6	22.8	9.1	10.3	12.5	6.1
Construction	12.8	13.1	5.3	4.4	0.0	1.1	0.0	0.0	6.3	3.0
Distributive	13.7	12.4	24.5	22.4	20.0	16.8	18.2	10.3	12.5	9.1
Hotels and restaurants	2.6	3.7	2.6	2.2	11.1	10.9	9.1	59.0	18.8	47.0
Transport, communication	9.9	8.5	7.9	18.3	22.2	27.7	9.1	3.8	0.0	6.1
Financial, property	16.2	14.3	29.1	14.8	17.8	13.6	18.2	6.4	25.0	15.2
Public administration and defence	6.0	8.0	1.3	3.8	3.3	0.5	0.0	2.6	6.3	6.1
Education	4.1	4.8	6.0	1.6	2.2	1.1	0.0	2.6	0.0	3.0
Health and social work	3.4	4.4	6.6	6.3	3.3	2.7	27.3	0.0	12.5	4.5
Other services	4.8	5.0	2.0	2.2	1.1	0.5	9.1	5.1	0.0	0.0
Women										
Primary, utilities	1.3	1.4	1.4	0.7	0.0	0.0	0.0	0.0	0.0	0.0
Manufacturing	9.5	8.2	18.6	4.8	16.4	5.1	8.7	0.0	5.9	5.1
Construction	1.7	1.3	0.0	0.7	0.0	0.0	0.0	0.0	0.0	0.0
Distributive	17.0	15.2	21.9	28.8	21.3	23.7	17.4	50.0	23.5	8.5
Hotels and restaurants	5.3	5.9	3.6	2.7	4.9	5.1	4.3	8.3	11.8	33.9
Transport, communication	4.0	2.9	5.0	6.8	1.6	1.7	0.0	0.0	0.0	0.0
Financial, property	15.5	14.7	10.0	21.2	11.5	16.9	13.0	16.7	17.6	25.4
Public administration and defence	6.7	10.1	7.2	8.9	6.6	10.2	0.0	16.7	11.8	0.0
Education	13.2	12.7	12.9	8.2	18.0	11.9	21.7	8.3	17.6	10.2
Health and social work	19.0	21.7	15.8	12.3	19.7	16.9	30.4	0.0	5.9	15.3
Other services	6.9	6.1	3.6	4.8	0.0	8.5	4.3	0.0	5.9	1.7

Source: Labour Force Survey, December 2001–February 2002.

Note
Italic figures indicate that the sample size for the ethnic/gender group is below fifty.

Table 2.5 Dissimilarity indices of sectoral distribution

Ethnic group	White born in UK	
	Men	Women
White born overseas	8	7
Indian born overseas	20	23
Indian born in UK	29	15
Pakistani born overseas	29	13
Pakistani born in UK	29	17
Bangladeshi born overseas	56	47
Bangladeshi born in UK	*41*	*20*
Chinese born overseas	46	39
Chinese born in UK	*37*	*25*

Note
Italic figures indicate that the sample size for the minority ethnic/gender group is below fifty.

of any calculation involving a group where the sample size is below fifty have been italicized.

It can be seen that the index value of 8 for the difference between native-born males and immigrant white males is one of the lower numbers in the table. Only the equivalent figure for women is lower, at 7. This shows that Britain's white immigrants have a similar industrial distribution to the native-born. Since white immigrants come predominantly from countries with strong cultural and educational similarities to the United Kingdom, the finding is not unexpected. The comparisons involving the culturally more distinct Asian groups produce substantially higher index values. We may surmise that cultural difference may be feeding into economic experience. If such is the case then the 56 per cent of overseas-born Bangladeshi men who would need to move economic sector in order to match the British-born white men pattern is informative. This group may be interpreted as the most disadvantaged of those we are considering.

Interestingly the dissimilarity figures for men are generally higher than for women, with overseas-born Indians constituting the only exception. Remember, however, that the comparisons are within genders, and that native-born white women are usually recognized as being at an employment disadvantage to their male peers. The figures are therefore saying that the difference between minority women and native white women is not as great as the comparable male figures. British-born white males are at an advantage in the labour market and the positions they have not taken are filled by others. Having not been disadvantaged by gender, Asian-origin males may have more opportunity to be disadvantaged by ethnicity!

Immigration is often seen as a precursor of potential integration, at least in economic terms. It is therefore to be expected that the jobs profile of second and subsequent generations of an incoming group would become

more similar to that of the charter group. Fluency in English, a British education and greater familiarity with societal conventions should ensure that
British-born Asians have better access to the full range of economic
opportunities than their immigrant parents. Table 2.5 confirms that this is
happening for Bangladeshi men, Chinese men, Indian women, Bangladeshi
women and Chinese women. The figures move in the opposite direction for
Indian men and Pakistani women, however, and Pakistani males appear not
to have changed. A fuller understanding of these generational changes may
emerge from a return to a consideration of Table 2.4 and the particular concentrations in economic sectors of the individual groups.

For overseas-born male Indians the major job concentrations are in manufacturing, the distributive trades, transport and communications, and
finance. However, only the distributive and transport and communications
figures are markedly above the figures for UK-born whites. In the UK-born
generations the focus on distributive trades has increased slightly, and in the
finance sector the Indian presence substantially exceeds the white proportion. The higher second (and third) generation presence in these two industries is compensated for by much lower involvement in manufacturing, and
in transport and communications, than is the case for Indian immigrants. In
both sectors the figure is lower than for UK-born whites, markedly so for
manufacturing. The increased dissimilarity index between generations seems
for Indian males to be a move from manual and low-skill sectors into activities based on the possession of educational credentials. This represents economic progress rather than the decline possibly implied in the previous
paragraph. The continuing strong presence in the distributive trades, where
there is a well known concentration of Indian-owned convenience stores, a
sector subject to strong competitive pressures, may be a cause for future
concern (Barrett *et al.*, 2001). For Indian women, however, some contrary
trends emerge. The concentration in distribution is lower in the British-
born cohorts than in the immigrant generation, and the figure for the
finance sector is lower for the UK-born. In manufacturing, however, the percentage is higher for the UK-born.

For overseas-born Pakistani men the sectors where the UK-born white
figure is significantly exceeded are distribution, hotels and restaurants, and
transport and communications. The concentration in transport relates
particularly to a very high proportion of self-employed taxi drivers. As Kalra
(2000) shows, this is not a niche comprising truly independent entrepreneurship. Rather it is a form of subcontracting in which the owner-driver is
subordinate to the organizer of his radio service, but has to make his own tax
and social security arrangements, has no paid holidays, and is subject to high
risk of theft and violence. The transport focus falls in the British-born generations, but is still well above the UK-born white figure. Meanwhile their
dependence upon the distributive trades is higher than for their immigrant
forebears. Reliance on manufacturing is, however, less in successor generations, and the UK-born have overtaken their white equivalents in the

growing financial sector. Overseas-born Pakistani women are more similar to their UK-born white comparators than any other Asian group, but UK-born Pakistani women are more dependent on manufacturing than the first generation. The increased presence in the growing education sector in later generations is perhaps a hopeful sign.

Generational comparisons are invalidated for Bangladeshis and the Chinese by the small size of the British-born samples. It is, however, clear that both communities rely heavily on the hotel and restaurant sector. As already indicated this reflects the dominance of restaurant and take-away employment in the two communities, especially among men. The high figure for UK-born Chinese women reflects the importance of family-run take-away food establishments within the restaurant category. In much of Britain this has involved the acquisition of traditional fish-and-chip shops as former white owners withdraw from the trade, perhaps on age grounds, but also because of inconvenient working hours and limited profits. Bangladeshis are more involved with table-service establishments, but many of these now also offer take-away facilities. Both Bangladeshis and Chinese are operating in extremely competitive contexts. Not only is there strong competition within and between the cuisines the two communities offer, but there is also great pressure from other food formats, much of it organized in large chains such as McDonald's and Pizza Hut (Ball, 1999).

Conclusion

We began by referring to a popular belief in Asian success in the British economy, even as the labour markets with which they have had to cope have been subject to state-sponsored restructuring. How far have the data we have presented supported this impression? It seems clear that the question cannot be answered in these simple terms because of divergences between the experiences of different Asian groups. Moreover examining the sectors in which Asians are employed complicates the analysis. Most groups show increasing similarity across generations with the distribution of jobs held by UK-born whites, but there are exceptions. Moreover the strong Chinese presence in the restaurant business makes their industrial profile less similar to whites than that of Pakistanis, who otherwise underperform them. The sectoral detail contains a confusing mix of progress and continued or growing reliance on industries in difficulty.

It is true that Asian males appear to outperform native whites on self-employment, but this relies upon the conviction that higher figures here constitute greater achievement, which is disputed. If self-employment always connotes success then we face the picture of Pakistani men doing better than Indians and the Chinese, which is inconsistent with our other evidence. We need to recognize that the national rise in self-employment is not independent of the policy agenda which has transformed other aspects of the economy. Much self-employment may be a desperate attempt to cope with the increased flexibility which has been designed into the labour market since 1979. As marginal groups in the economy ethnic minorities are particularly vulnerable to insecurity arising from this process. Asian self-employment, so often concentrated in sectors involving antisocial hours, unpleasant conditions and dubious rewards, may represent self-exploitation rather than self-reliance.

As we have seen Dustmann *et al.* (2003) and Lindley (2002) have shown that in employment disadvantage is not merely the result of minorities having a set of non-labour characteristics which handicap them in employment. Asians have worse jobs than one would expect, even allowing for their education, age, place of residence and other variables. On a pessimistic interpretation this may mean that Asian ethnic minorities are continuing to suffer the disadvantages of racism in a white-dominated labour market. It could be argued that, in an economy in which government has deliberately rearranged the rules of the labour market for the benefit of employers, groups of workers are inevitably played off against one another. The feminization of the labour force is one manifestation of this. Any numerical progress by Asian minorities is simply a further device to ensure the continued fragmentation of labour power.

A more sanguine interpretation would be that economic equality between ethnic groups does not emerge in any short period; it occurs over generations. Individual prosperity is a product of qualifications and other forms of human capital. Increasing Asian presence in some more buoyant parts of the labour market is evidence that individuals from ethnic minorities are not stigmatized, but can obtain their individual deserts. The continued presence of many Asians in industries with strong downside potential is an experience shared with many whites. The lower success rates of Bangladeshis and Pakistanis may be associated with cultural characteristics rather than specific discrimination. The inability of many British whites to distinguish between Indians, Bangladeshis and Pakistanis means that prejudice applies to a relatively successful group as much as to the less successful. Moreover the relative performance of religious groups within the Indian population also shows that culture, not discrimination, is at work.

Perhaps we can conclude by acknowledging that in a liberalized economy not all can win prizes. As in any race the winners and their supporters believe that this is a just outcome. As promoters of the liberalization, governments belong in this camp, even as they berate their political rivals for advocating similar strategies. The losers and their sympathizers suspect, however, that the rules of the game are stacked against them.

Note

1 The Labour Force Survey data were obtained from the UK Data Archive at Essex University. Neither the archive nor ONS bears any responsibility for the interpretations presented here.

References

Anwar, M. (1979) *The Myth of Return. Pakistanis in Britain.* London: Heinemann.

Ball, S. (1999) 'Whither the small independent take-away?' *British Food Journal*, 101 (9): 715–23.

Barrett, G.A., Jones, T.P. and McEvoy, D. (2001) 'Socio-economic and policy dimensions of the mixed embeddedness of ethnic minority business in Britain', *Journal of Ethnic and Migration Studies*, 27 (2): 241–58.

Barrett, G., Jones, T., McEvoy, D. and McGoldrick, C. (2002) 'The economic embeddedness of immigrant enterprise in Britain', *International Journal of Entrepreneurial Behaviour and Research*, 8 (1–2): 11–31.

Blackburn, R.M., Dale, A. and Jarman, J. (1997) 'Ethnic differences in attainment in education, occupation and life-style', in V. Karn (ed.) *Ethnicity in the 1991 Census. IV. Employment, Education and Housing among the Ethnic Minority Populations of Britain.* London: Stationery Office, 242–64.

Bradford Heritage Recording Unit (1994) *Here to stay. Bradford's South Asian Communities.* Bradford: City of Bradford Metropolitan Council, Arts, Museums and Libraries Division.

Cheng, Y. (1996) 'The Chinese: upwardly mobile', in C. Peach (ed.) *Ethnicity in the 1991 Census. II. The Ethnic Minority Populations of Great Britain.* London: HMSO, 161–80.

Coleman, D. and Salt, J. (1992) *The British Population. Patterns, Trends and Processes.* Oxford: Oxford University Press.

Dahya, B. (1974) 'Pakistani ethnicity in industrial cities in England', in A. Cohen (ed.) *Urban Ethnicity.* London: Tavistock, 77–113.

Dale, A. and Holdsworth, C. (1998) 'Why don't ethnic minority women work part-time', in J. O'Reilly and C. Fagan (eds) *Part-time Prospects. An International Comparison of Part-time Work in Europe, North America and the Pacific Rim.* London: Routledge, 77–95.

Dustmann, C, Fabbri, F., Preston, I. and Wadsworth, J. (2003) Labour Market Performance of Immigrants in the UK Labour Market, Home Office online report 05/03, London: Home Office (www.homeoffice.gov.uk/rds/pdfs2/rdsolr0503.pdf. Last accessed 31 March 2003).

Kalra, V.S. (2000) *From Textile Mills to Taxi Ranks.* Aldershot: Ashgate.

Lindley, J. (2002) 'Race or religion? The impact of religion on the employment and earnings of Britain's ethnic communities', *Journal of Ethnic and Migration Studies*, 28 (3): 427–42.

Modood, T., Berthoud, R., Lakey, J., Nazroo, J., Smith, P., Virdee, S. and Beishon, S. (1997) *Ethnic Minorities in Britain. Diversity and Disadvantage.* London: Policy Studies Institute.

National Statistics (2003a) *Neighbourhood Statistics. 2001 Census. Key Statistics.* Table KS06P, Ethnic group, 2001 percentages, Crown copyright (http://www.neighbourhood.statistics.gov.uk/Reports/eng/TableViewer/wdsview/dispviewp.asp?dsid=1519. Last accessed 30 July 2003).

National Statistics (2003b) *Neighbourhood Statistics. 2001 Census. Key Statistics.* Table KS06AP, Ethnic group and identification as Welsh, 2001 percentages, Crown copyright (http://www.neighbourhood.statistics.gov.uk/Reports/eng/TableViewer/wdsview/dispviewp.asp. Last accessed 30 July 2003).

NISRA (2002) *Northern Ireland Census 2001: Key Statistics.* Belfast: Northern Ireland Statistics and Research Agency. Crown copyright.

NISRA (2003) *Census Statistics 2001. Evaluation.* Part 4, Data classifications, Ethnic group ETHPUK. Belfast: Northern Ireland Statistics and Research Agency. Crown copyright (www.nisra.gov.uk/census/metadata/ETHPUK.html. Last accessed 20 March 2003).

Owen, D. (1996) 'The other Asians: the salad bowl, in C. Peach (ed.) *Ethnicity in the 1991 Census.* II. *The Ethnic Minority Populations of Great Britain.* London: HMSO, 181–205.

Registrar General for Scotland (2003) *Scotland's Census 2001. The Registrar General's 2001 Census Report to the Scottish Parliament.* Edinburgh: General Register Office for Scotland. Crown copyright.

Rhodes, C. and Nabi, N. (1992) 'Brick Lane: a village economy in the shadow of the city?' in L. Budd and S. Whimster (eds) *Global Finance and Urban Living.* London: Routledge, 333–52.

Sassen, S. (2001) *The Global City.: New York, London, Tokyo,* second edition. Oxford: Princeton University Press.

Watson, J.L. (1977) 'The Chinese: Hong Kong villagers in the British catering trade', in J.L. Watson (ed.) Between two cultures. Migrants and minorities in Britain, Oxford: Blackwell, 181–213.

3 Chinese immigration and the (ethnic) labour market in France

Emmanuel Ma Mung

Unlike other European nations, France has a long history of labour immigration. Indeed, it started using foreign labour as early as the nineteenth century, first recruiting it from countries along its northern and eastern border (Germany, Belgium and Switzerland), then from Italy and Poland in the 1920s and 1930s, and Spain and Portugal in the 1960s and 1970s. It was also during these last two decades that it began to tap the labour force of its former colonies, Algeria, then Tunisia, Morocco and West Africa. It was against this backdrop that Asian immigration developed. Unlike the previous waves, however, it is made up not of a work force matching the needs of French companies, but mainly of refugees who arrived in France in the aftermath of conflicts in South East Asia, together with an influx of ethnic Chinese linked with the migratory networks that were set up several decades before.

The Chinese immigration that can be observed in France is a local manifestation of a global phenomenon, the Chinese diaspora (Ma Mung, 2000). As a transnational social body, it feeds upon its own impetus through the networks it generates, which constitute real migratory resources, facilitating personal mobility by shaping traffic channels, identifying potential destinations and permanent residence points, providing traffic logistics (means of transport and of entry into France) as well as means of settlement (employment and housing opportunities). The diaspora can thus achieve its own migratory autonomy: as a migration-encouraging structure, it generates the mobility that feeds its own movement. The phenomenon affects not only those populations already settled in the various poles of the diaspora, but also those migrants coming direct from China, as is revealed by the fact that migratory flows tend to head directly for those countries which already do host a diaspora.

Like other migrations, it has generated over time, in the various host countries, an *ethnic economy* that is similar to those set up by other migrant populations (Portes and Bach, 1985; Light and Gold, 2000; Waldinger *et al.*, 1990). This economy, however, is indeed transnational – a diaspora economy – since the relations between its economic agents tend to be established not only on a local scale, but also on a higher, global level (Ma Mung,

1992). This transnational ethnic economy can be characterized by the existence of local economic configurations (Fr. *dispositif économique*) linked together at the global level, and increasingly linked with mainland China since the country began to open up economically to the rest of the world in the 1980s. The development of this economy generates, at the local level, a demand for work and labour on an *ethnic labour market*. Should the local work supply be too low, the demand will be met by international migrations that will thus help further reinforce these economic arrangements. This is what this chapter will try to highlight.

An assessment of the Chinese population in France

In this chapter, what we mean by 'Chinese migrations' is not only the inflow coming directly from China, but also those from the other poles of the Chinese diaspora. They may comprise individuals belonging to populations that settled generations ago in those poles. Thus the Chinese population in France includes more than twenty different national origins, mainly Vietnam, Laos, Cambodia and China, but also including Thailand, Singapore, Malaysia, as well as the United Kingdom, the Netherlands, Italy and Spain. Even though we don't need to go into the details of the history of Chinese immigration in France (Guillon, 2003; Ma Mung, 2000), suffice it to say that the Chinese population in France can be broken down into two main groups. The first is made up of people from diaspora host countries, for example, mostly nationals of ex-Indochina (Vietnam, Laos, Cambodia), that population itself being made up of ethnic Chinese groups from any number of places, but all originally from southern China (Chaozhou, Hakka, Hokkien, Kwantung, etc.). The second group is made up of migrants coming direct from China, mainly from the province of Zhejiang[1] (specifically from the Wenzhou and Qingtian communities), and more recently from the large cities and provinces of north China.

China's opening-up resulted in easier access to, but also easier exit from, the country. Even though complete freedom of movement is still a remote goal, steps were taken in the 1980s and 1990s to facilitate the obtaining of passports and permits to leave the country (Guerassimoff, 1997). In the past two decades this resulted in the migration of hundreds of thousands,[2] maybe even millions. Population flows from China must have sharply outnumbered intra-diasporic migrations. Traditional emigration areas (southern and middle Chinese provinces, Guandong, Fujian, Zhejiang) still remain so, but other departure areas have emerged, as will be discussed later. The Western world, in a broad sense, displaced South East Asia (Ma Mung, 2000) as the main destination in the 1990s. Although North America remains the preferred destination,[3] emigrant flows also head for Australia, New Zealand, Japan and Western Europe, especially France, Spain and Italy for Zhejiang Chinese. There are also tens of thousands of overseas Chinese in ex-Soviet bloc countries like Russia, Romania, Hungary and the Czech Republic.

What is new is that it is no longer only the unskilled workers, but also students, artists,[4] skilled and highly skilled workers, executives, technicians and entrepreneurs who are leaving (see below). There is a long tradition of migration of Chinese political elites to France (Dirlik, 1991; Kriegel, 1968), e.g. Zhou-En-lai or Deng Xiao-ping in the 1920s and 1930s,[5] and this tradition has continued with the arrival of refugees in the aftermath of the 1989 Beijing spring, but remains numerically limited.

Why ethnic Chinese numbers are difficult to assess

In the French census, individuals are classified by nationalities when they are aliens and by nationalities of origin when they are naturalized French citizens. But, unlike, say, North American censuses (that do have a question about which ethnic group respondents claim to belong to), they are not broken down into ethnic groups. Since 1999 the census also categorizes immigrants according to INSEE[6] standards, for example, aliens and naturalized French individuals born outside French territory.[7] Thus the figures for both Chinese (People's Republic of China) nationals and French-naturalized ex-Chinese nationals born in China are available. However, it is impossible to assess the numbers of ethnic Chinese immigrants born in Cambodia, Laos or Vietnam, even though it is well known that the proportion of ethnic Chinese was indeed quite high among South East Asian refugees. A survey carried out in the 1980s put that proportion as high as 70 per cent among South East Asian refugees in France. Li (1999) puts it at 60 per cent to 70 per cent of the 1.5 million to two million refugees who fled Vietnam.[8] It should therefore always be borne in mind that our numbers, which also include 'non-Chinese', are higher than the numbers of ethnic Chinese from those countries.

The 2002 publication of the 1999 census details gives a better view of the weight of ethnic Chinese in France. The figures show 91,197 *foreigners*, for example, citizens of Cambodia, Laos, Vietnam and the People's Republic of China, along with 138,967 *naturalized French nationals*, ex-citizens of one of those countries, 230,164 on aggregate (see Table 3.1), against 206,459 back in 1990, a 24,000 increase. Roughly speaking, *registered* ethnic Chinese can be estimated to number 170,000 to 200,000, a figure that does not, obviously, include those illegal immigrants who have not registered in the census. From 1990 to 1999 the number of foreigners and naturalized French citizens from the four countries did increase, which is mainly due to the Vietnamese (increased 12,918) and Chinese (increase of 18,947) groups, whereas the numbers of Cambodians and Laotians declined (by 3,381 and 4,329 respectively).[9] The population coming directly from China doubled over ten years, acquiring more weight relative to the overall population: 15.9 per cent in 1999, compared with 8.7 per cent in 1990. It still remains a small minority in that overall population, even though its increase was the sharpest of all.

Table 3.1 South East Asian and Chinese populations in France

Population	Nationality and former nationality				
	Cambodia	Laos	Vietnam	China (PRC)	Total
Foreigners					
1990	47,369	31,803	33,743	14,051	126,966
1999	25,969	16,240	21,162	27,826	91,197
Naturalized French					
1990	19,296	17,758	38,435	4,004	79,493
1999	37,315	28,992	63,934	8,726	138,967
Aggregate					
1990	66,665	49,561	72,178	18,055	206,459
1999	63,284	45,232	85,096	36,552	230,164

Source: Adapted from INSEE, recensement de la population 1999.

As far as *immigrants* are concerned, there are 159,750 people born in Cambodia, Laos and Vietnam, and 30,932 in China, for example, 190,682 on aggregate (see Table 3.2), representing 4.4 per cent of the local immigrant population and 0.3 per cent of the total population. Immigrant flows for the 1990–99 period mainly came from China and Vietnam. Indeed, the remarkably low numbers from Cambodia and even lower numbers from Laos reflect a considerable change in the migration structure of those two countries, for which France is no longer a major destination, as was the case in previous

Table 3.2 South East Asian and Chinese immigration in France

Year	Cambodia	Laos	Vietnam	China (PRC)	Total
Total including	50,675	36,838	72,237	30,932	190,682
1990	548	118	2,712	1,739	5,117
1991	193	86	1,596	2,067	3,942
1992	184	85	1,109	1,924	3,302
1993	282	62	896	1,182	2,422
1994	282	34	1,038	1,457	2,811
1995	164	55	1,100	1,246	2,565
1996	203	50	1,340	1,283	2,876
1997	289	49	1,402	1,367	3,107
1998	298	93	1,751	1,526	3,668
1999[a]	48	28	297	400	773
1990–99	2,491	660	13,241	14,191	30,583

Source: Adapted from INSEE, recensement de la population 1999.

Note
a The census was carried out in March 1999; hence the very low numbers recorded for that year.

decades. Almost half the immigration from China took place during that period, and that from Vietnam only inflated an already large population, thus making this group the most important in Asian immigration.

Most of the immigrants from South East Asia and China are to be found in the Paris area, which hosts 17 per cent of the total French population. There are, however, significant discrepancies: 42 per cent of the immigrants from Laos live in that area, as against 48.8 per cent of those from Vietnam and 56.9 per cent of those from Cambodia. But 82.5 per cent of the immigrants from China are to be found in the Paris area, which therefore seems to be particularly attractive to this specific group.

Census figures about the China-born population, however, leave observers doubtful. In-depth studies of the flows from those countries, based on alternative sources, generate quite a different picture. Cattelain *et al.* (2002) analyse in great detail data from various organizations, especially those in charge of foreign labour entry and asylum claims.[10] The authors' treatment of their data can be argued to be debatable on some issues, as there must be cases of double counting. Still, they interestingly count 43,481 immigrants from China compared with the 14,191 officially registered (Cattelain *et al.*, 2002: 48), almost 30,000 more than census figures. The study also provides invaluable details on Chinese inflows over the 1990s, showing their diversity. Thus, the 43,500 entries break down as follows: 19,000 asylum seekers, 7,700 legalized ex-illegal immigrants, 5,000 students, 3,800 spouses or parents of legal residents, 3,600 permanent workers, 1,600 temporary workers, 1,600 spouses or parents of French nationals, and 1,300 visitors, including legal as well as illegal or semi-legal immigrants. Its interest also lies in the characteristics it reveals about the inflows. Women are numerous (over 50 per cent), and their proportion may top 60 per cent to 70 per cent among those from the northern provinces (Liaoning, Shandong, Heilongjiang). Moreover, the latter population group is on average more skilled than that from Zhejiang. This immigration from north-eastern China constitutes a new phenomenon in Chinese migration to France. It is more often than not a little older (thirty-five to forty years old) than the immigrant average, and made up of junior executives or skilled workers made redundant by the closing down of large state companies in their home area. Unlike the Zhejiang or South East Asian Chinese, they do not come from a region with a strong migratory tradition, and therefore cannot benefit from the mutual help of family or community networks. But most of them do share a common characteristic with the other Chinese migrants: their ability to find work for other, earlier Chinese immigrants, as workshop or restaurant employees, or servants for private households (Cattelain *et al.*, 2002: 133 ff.).

Ethnic labour market reinforcement and the strengthening of the Chinese economic configuration in France

The ethnic Chinese on the job market

For the ethnic Chinese, the labour market in France is, first and foremost, an *ethnic* labour market (Ma Mung, 1991, 2002),[11] where labour supply and demand adjust themselves chiefly on an ethnic basis in the choice of economic partners. Ethnic Chinese companies mainly employ ethnic Chinese labour, and in turn ethnic Chinese labour mainly works for ethnic Chinese companies. The sharp increase in the number of such companies generates a labour demand that is met by immigrant workers whenever the local ethnic Chinese labour is not sufficient. Employment of non-ethnic Chinese labour is at best marginal. The Chinese economic configuration creates a true ethnic labour market by generating supply and demand for labour: the demand from ethnic Chinese employers seeking ethnic Chinese workers, and supply from ethnic Chinese workers looking for work in an ethnic Chinese business.

Qualitative surveys (Cattelain *et al.*, 2002; Ma Mung, 1991) show that few employees leave the ethnic labour market, for example to work in French businesses. The pool of job seekers is fed by waves of new immigrants, including many illegal aliens. However, whether these are economic migrants who do not hold a French residence permit or political refugees unable to obtain official status as such, nearly all of them rely on the ethnic labour market for their survival. The pool of employers is fed both by the ongoing creation of businesses, which stimulates the demand for workers, and by already existing businesses with a high rate of employee turnover. Compared with the French average, working conditions are poor. Career perspectives are more or less nil, because these are usually small businesses that offer no chances of promotion. As for job security, workers are usually in a precarious position. The most striking feature of the ethnic labour market is its great fluidity, a result of the flexibility of the work force because hiring and firing are so casual. Many jobs are short-lived (a matter of weeks or months), and the workers themselves are quite mobile owing to the low pay and the absence of career perspectives within small businesses.

As underlined above the majority of the workers remain on the ethnic labour market. One reason is the shortage of jobs on the general labour market. But the will to remain on the ethnic labour market can frequently be observed. From the workers' viewpoint, this tendency can be interpreted as a desire to remain within the Chinese economic circuit. The desire is probably due to the language barrier. Likewise, a worker's 'connections' or social networks are usually more solidly established within the ethnic job market than within the general one. But the main reason is that remaining in the Chinese economic circuit offers opportunities for social mobility which do not exist elsewhere. Despite the fact that working conditions and pay are less favourable within overseas Chinese businesses than within French-owned ones, the fact of working for a Chinese employer enables the migrant to remain within a system of social solidarity networks he intends to draw upon if and when he opens his own small business. The migrants' expectations are continually confirmed by the large number of small independent businesses they are able to observe starting up around them. The workers who were interviewed almost always cited this reason. Moreover, aware of the type of jobs available to migrant workers in non-Chinese businesses (nearly always unskilled, low-paid work), they have no hope of upward social mobility. This encourages them to remain on the ethnic labour market: the only avenue, in their eyes, to social success – epitomized by ownership of an independent business.

Census data do back up field observations. The ethnic Chinese labour is mainly employed in ethnic Chinese companies, which in turn mainly employ ethnic Chinese labour. The observation is confirmed (albeit indirectly, since data do not indicate ethnic groups) by several of the census details. Indeed, immigrants from South East Asia and China are overrepresented in three areas:

1   The socio-professional categories that are likely to be employed by ethnic Chinese companies (see Table 3.3).
2   The economic sectors in which ethnic Chinese companies are mainly to be found (see Table 3.4).
3   The socio-professional categories entrepreneurs belong to (see Table 3.3).

Moreover, the numbers of such entrepreneurs can be seen to be growing very fast (see Table 3.5).

Workers, employees and entrepreneurs are proportionally more numerous than the national average

Workers (skilled and unskilled) represent 37.8 per cent of the target population, which is significantly higher, irrespective of the group considered, as compared with their proportion in the French working population (25.5 per cent); the proportion can even be as high as 46.5 per cent for migrants from Laos (see Table 3.3). Shop employees are also proportionally more numerous than the national average (4.2 per cent), whatever their national origin. The 'direct services to individuals' category includes restaurant and catering employees, and the proportion of immigrants is also higher than average (6.2 per cent), with the exception of those from Laos (6.1 per cent). It can be significantly higher, e.g. for immigrants born in Vietnam (10.4 per cent) or even 12.8 per cent, more than twice the average, for those born in China. Those categories normally cover those wage earners that are likely to work for ethnic Chinese companies, but of course, the aggregate figure is higher

Table 3.3 A breakdown of working immigrants by socio-professional category and country of birth (%)

Socio-professional category	Total working population	1 Cambodia	2 Laos	3 Vietnam	4 China (RPC)	Total 1–4
Numbers	26,537,436.0	36,461.0	27,525.0	40,627.0	19,273.0	123,886.0
% total	100.0	100.0	100.0	100.0	100.0	100.0
Farmers	2.4	0.2	0.7	0.0	0.0	0.2
Craftsmen	2.9	4.7	2.2	2.3	2.8	3.0
Shopkeepers	2.7	6.3	4.1	6.0	10.6	6.4
Managers of ten employees or more	0.7	0.5	0.3	0.3	0.5	0.4
Professionals	1.3	1.5	0.5	3.5	0.8	1.8
Civil servants, intellectual and artistic professions	4.8	2.0	1.0	4.3	5.3	3.0
Executives	5.8	4.6	2.9	7.4	5.9	5.4
Intermediate professions: education, health, civil service and similar	9.0	2.6	3.3	4.5	2.8	3.4
Intermediate professions: administrative and commercial company departments	6.9	5.2	5.4	4.6	5.0	5.0
Technicians	3.6	4.2	4.9	5.4	1.2	4.3
Foremen	2.2	1.2	1.2	1.0	1.2	1.2
Civil service employees	10.8	4.6	4.9	5.2	1.6	4.4
Company administrative employees	8.2	6.2	6.5	6.3	3.1	5.8
Commercial employees	4.2	6.7	5.3	5.1	5.9	5.7
Directs services to individuals	6.2	10.4	6.1	7.9	12.8	9.0
Skilled workers	15.7	19.8	23.4	17.9	18.8	19.8
Unskilled workers	9.8	16.9	23.1	15.2	19.0	18.0
Farmhands	1.1	0.4	2.1	0.3	–	0.7
Jobless individuals who have never worked	1.3	1.7	1.9	2.7	2.6	2.2
Other (including military service)	0.3	0.1	0.2	0.2	–	0.1

Source: 1999 population census, additional data.

than ethnic Chinese potential work supply, since, as already noted, those immigrants are not all ethnic Chinese.

If we now turn to work demand, the proportion of Asian immigrants can be observed to be higher, sometimes even significantly higher, among entrepreneurs (craftsmen, shopkeepers, managers of ten or more employees) than the national average (6.3 per cent). That is especially true of shopkeepers, who represent 6 per cent of the working population coming from Vietnam (compared with a 2.7 per cent national average), 6.3 per cent of those coming from Cambodia, and as much as 10.6 per cent for those from China. The picture is less clear for the craftsmen category, where the proportion is significantly higher than average for immigrants from Cambodia, but lower for the others. As for managers of ten or more employees, it is lower than average, which again shows that the small company is indeed the predominant structure for Asians.

The group including the highest proportion of entrepreneurs is that of immigrants from China (13.6 per cent), and that group also displays the most skewed socio-professional structure towards workers likely to be employed by Chinese companies, and entrepreneurs. Those two categories represent 69.9 per cent of working Chinese immigrants (compared with 42.2 per cent for the overall working population). Immigrants from China are thus the group whose socio-professional profile best fits an ethnic labour market-centred economic organization.

The socio-professional structure of immigrants from Cambodia and Vietnam is less skewed, but still close to the structure of those from China. Those from Laos, however, have quite a different structure, with a higher proportion of workers and a lower proportion of entrepreneurs. One possible explanation is a lower proportion of ethnic Chinese than in other groups, but again, we lack the statistical elements to confirm this view.

Concentration of the working population in specific activity niches

The main feature of the immigrant working (employed[12]) population from Cambodia, Laos, Vietnam and China (see Table 3.4) is its overrepresentation in activity sectors that match those where Chinese companies are numerous: the consumer goods industry, trade and services to individuals.[13] Only 25.9 per cent of the total working population is to be found in these three sectors, as against almost half (48.9 per cent) of the Asian immigrant population: 19.5 per cent in services to individuals (including catering), 19.3 per cent in trade and 10.1 per cent in the consumer goods sector (including clothing and leather industries). Moreover, they are also quite numerous in the 'services to companies' category. Yet there are noticeable differences between groups, which may help fine-tune the previous findings on the specific profile of China-born migrants. Their proportion in the three sectors is significantly higher than that of the others: 76.7 per cent are employed

Table 3.4 A breakdown of working (employed) immigrants by socio-professional category and country of birth (%)

Economic sector	Total working population	1 Cambodia	2 Laos	3 Vietnam	4 China	Total 1–4
Numbers	20,913,743.0	29,224.0	21,442.0	32,222.0	16,770.0	99,658.0
Farming, forestry, fishing	4.3	0.5	2.1	0.4	0.1	0.7
Food	2.8	3.5	3.5	2.8	1.0	2.9
Consumer goods[a]	3.3	8.4	8.6	6.2	22.7	10.1
Automotive	1.2	2.9	2.7	1.5	0.4	2.0
Capital goods	3.7	3.1	6.1	4.3	1.2	3.8
Intermediate goods	6.3	6.8	11.9	7.1	1.3	7.1
Energy	1.1	0.2	0.4	0.6	0.3	0.4
Construction	5.3	1.5	2.7	2.0	1.0	1.9
Commerce	13.3	19.8	18.6	17.1	23.8	19.3
Transport	4.4	5.2	4.4	3.2	1.4	3.7
Financial	3.1	2.0	1.5	2.6	1.2	2.0
Real estate	1.1	0.5	0.8	0.6	0.5	0.6
Services to companies	12.1	13.0	13.6	15.2	7.4	12.9
Services to individuals[b]	6.9	21.5	11.6	17.4	30.2	19.5
Education, health, social work	19.4	7.6	7.7	13.7	5.3	9.2
Administration	11.8	3.5	3.8	5.3	2.3	3.9

Source: INSEE, 1999 population census.

Notes
a Including clothing and leather industries.
b Including hotels, restaurants and personal and housework services.

there, especially in the catering (30.2 per cent) and trade (23.8 per cent) sectors. They are strikingly underrepresented in other activity sectors, which sets their group profile in sharp contrast to the other groups, and, albeit indirectly, emphasizes the importance of the ethnic labour market for the group. The profile of those immigrants coming from Cambodia, Laos and Vietnam is less polarized, as they are overrepresented in each of the three sectors under scrutiny, but are also present in other sectors, like 'services to companies'. Their lower overrepresentation in sectors mainly cornered by ethnic Chinese companies make them probably less dependent on the ethnic labour market than their China-born counterparts.

A sharp increase in labour demand

The increase in the number of Asian entrepreneurs from 1990 to 1999 is impossible to assess, for we do not have any comparable data in the two censuses. The number of Chinese entrepreneurs, whether naturalized French nationals or not, was unavailable in 1990, nor is the number of naturalized French entrepreneurs who used to be citizens of Cambodia, Laos, Vietnam or China in 1990. However, it is still possible to assess over a longer period, 1982 to 1999, if we restrict ourselves to Cambodians, Laotians or Vietnamese, whether naturalized or not (Table 3.5). These data do illustrate a more general trend.

The number of entrepreneurs soared threefold between those two years, from 2,724 to 9,587, an increase that is mainly due to shopkeepers and merchants (an increase of 3,749), but the progression of craftsmen – including the small entrepreneurs in the clothing industry – is remarkable (+2,773, a 6.5 times increase). As entrepreneurs multiply, obviously so do companies, resulting in a sharp rise in labour demand. The demand mainly turns to the ethnic labour market, further reinforcing it.

Table 3.5 Foreign and naturalized French entrepreneurs from Cambodia, Laos and Vietnam, 1982 and 1999

Economic sector	1982	1999	Progression	Multiplying coefficient
Craftsmen	504	3,277	+2,773	6.5
Shopkeepers and merchants	2,152	5,901	+3,749	2.7
Managers of ten or more employees	68	409	+341	6.0
Total	2,724	9,587	+6,863	3.5

Sources: Population censuses 1982 and 1999.

Conclusion

Though the ethnic Chinese immigrant population in France is one of the most numerous in Europe, it represents only a tiny proportion (4.4 per cent) of the overall immigrant population in this country, and a minute proportion (0.3 per cent) of the total population. It does, however, have high visibility in large cities and increasingly in small and medium-sized towns. Its presence is mainly associated, for the French population, with commercial activities, and even more so with the shop signs advertising those activities. Those are symbols of the ethnic Chinese presence, and, in the final analysis, an adequate representation of its reality, that presence being, first and foremost, entrepreneurial, in the sense of being linked with an economic arrangement of companies.

The number of small entrepreneurs is proportionally higher than the national average, and so is the number of workers at the bottom of the social ladder. The labour demand increase fostered by the development of the Chinese economic configuration in France, together with the labour supply increase linked with the development of immigration, both contribute to the strengthening of a labour market that is mainly an ethnic labour market. Work conditions are harsh and wages are low, and yet very few workers leave it to join the mainstream labour market. Maybe work conditions and wages are hardly better there, but the main reason is that, according to them, staying in an ethnic labour market is tantamount to staying in a social network that will eventually help them, in turn, to become bosses. For social mobility, in their view, is embodied by a single figure, that of the entrepreneur.

Notes

1 Migrations from Zhejiang, a coastal province of middle China north of Fujian, are different from those from the southern provinces: they are not originally linked with the coolie trade, and are mostly (75 per cent) Europe-oriented, whereas the latter mostly focus on South East Asia and North and South America.

2 Almost one million Chinese citizens are estimated to have obtained passports and visas in a single decade (1980–90).

3 The number of overseas Chinese in the United States went up by almost 800,000 between 1990 and 2000.

4 E.g. Gao Zingjian, who has lived in exile since 1988 and was awarded the Nobel Prize for Literature in 2000.

5 France has hosted other famous Asian communist cadres, such as Ho Chi Minh in the 1920s or Pol Pot in the 1950s.

6 Institut National de la Statistique et des Études Économiques.

7 A foreign-born French citizen is not registered as an immigrant, nor is a foreign national born on French territory.

8 It should be borne in mind that ethnic Chinese were very numerous in ex-Indochina countries. In the early 1970s their numbers were estimated at around four million in Vietnam (Tran, 1993), 400,000 to 500,000 in Cambodia and 130,000 to 150,000 in Laos (Pan, 1999).

9 No plausible explanation for the decrease can be offered at the moment; we might be witnessing returns to Cambodia or Laos, but also re-emigration to other countries.

10 Namely the OMI (Office des Migrations Internationales, Ministère de l'Emploi et de la Solidarité) and OFPRA (Office Français de Protection des Réfugiés et Apatrides).

11 See, *inter alia*, Paul Ong (1984) and Don Mar (1991) on the analysis of the Chinese ethnic labour market in the United States.

12 Data on economic activity sectors take into account employed workers, while those on socio-professional categories cover the whole working population (including the unemployed).

13 With the exception of the trade sector, their presence can be appraised only indirectly, since INSEE's published data are not detailed enough to assess numbers in the catering and clothing and leather sectors, where a special exploitation of data would be needed.

References

Cattelain, C., Moussaoui, A., Lieber, M., Ngugen, S., Poisson, V., Saillard, C. and Ta, A. (2002) *Les Modalités d'entrée des ressortissants chinois en France*. Paris: Ministère des Affaires Sociales, du Travail et de la Solidarité, Direction de la Population et des Migrations, 183.

Dirlik, A. (1991) *Anarchism in the Chinese revolution*. Berkeley CA and Los Angeles: University of California Press, 326.

Guerassimoff, C. (1997) *L'Etat chinois et les communautés chinoises d'Outre-Mer*. Paris: Harmattan, 343.

Guillon, M. (2003) 'Les Chinois de France: anciennes et nouvelles migrations', *Historiens et Géographes*, 383: 373–92.

INSEE Institut National de la Statistique et des Études Économiques (1999) Recensement de la population, mars 1999. Paris: INSEE.

Kriegel, A. (1968) 'Aux origines françaises du communisme chinois', *Preuves*, 209–10: 24–41.

Li, K. (1999), 'Vietnam', in Lynn Pan (ed.) *The Encyclopedia of the Chinese Overseas*. Richmond: Curzon, 275–83.

Light, I. and Gold, Steven G.J. (2000) *Ethnic Economies*. San Diego, CA: Academic Press, xiii, 302.

Ma Mung, E. (1991) 'Logiques du travail clandestin des Chinois', in S. Montagné-Villette (ed.) *Espaces et travail clandestins/sous*. Paris: Masson, 99–106.

Ma Mung, E. (1992) 'Dispositif économique et ressources spatiales: éléments d'une économie de diaspora', *Revue européenne des migrations internationales*, 8 (3): 175–94.

Ma Mung, E. (2000) *La Diaspora chinoise. Géographie d'une migration*. Paris: Ophrys, 175.

Ma Mung, E. (2002) 'Migratory and economic networks of the Chinese diaspora in southern Europe', in Tsun-Wu Chang and Shi-Yeoung Tang (eds) *Essays on Ethnic Chinese Abroad*. Taiwan: Academia Sinica Press.

Mar, D. (1991) 'Another look at the enclave thesis: Chinese immigrants in the ethnic labour market', *Amerasia Journal*, 17 (2): 5–21.

Ong, P. (1984) 'Chinatown unemployment and the ethnic labour market', *Amerasia Journal*, 11 (1): 35–54.

Pan, L. (ed.) (1999) *The Encyclopedia of the Chinese Overseas*. Richmond: Curzon, 399.

Portes, A. and Bach, R. (1985) *Latin Journey: Cuban and Mexican Immigrants in the United States*. Berkeley CA and Los Angeles: University of California Press.

Tran, K. (1993) *The Ethnic Chinese and Economic Development in Vietnam*. Singapore: Institute of Southeast Asian Studies, 127.

Waldinger, R., Aldrich, H. and Ward, R. (eds) (1990) *Ethnic Entrepreneurs*. New York: Sage.

4 On nurses and news vendors

Asian immigrants on the Vienna labour market

Christiane Hintermann and Ursula Reeger

The paths of immigrants into urban labour markets are generally deter-mined by three factors: legal regulations concerning access to the labour market, the structural conditions of the national and urban labour market and the networks of the individual migrant. Looking at the Austrian case, the first two conditions are rather unfavourable to migrant workers. Access to the labour market is strictly regulated and the ethnic segmentation of the labour market is still very high. Migrants from the traditional 'guest worker' countries – such as the former Yugoslavia and Turkey – have traditionally been channelled into special branches (niches) of the labour market that faced a labour shortage, most of all jobs in industry in the secondary sector.

With the exception of Philippine nurses, migrants from South East Asia have not been part of a migration process based on direct recruiting on the part of Austria. There was no pre-structured path into the Austrian labour market but opportunities that have been spotted and used. It can be argued that this 'wealth of invention' may be one of the reasons why these small groups of migrants are employed in very different branches compared with traditional 'guest workers'. Nevertheless, this does not answer the question of the chicken and the egg. Was it the immigrant group that created the niche or did the niche exist in the first place? In this chapter we look at such different paths of Asian migrants into the labour market.

The importance of existing migrant networks as a third decisive factor for entry to the labour market is empirically evident. For getting a job, it is very important to know the 'right people', for example, someone who is informed about the situation on the labour market of the target city. As Cross and Waldinger (1997) note, 'Because getting a job remains very much a matter of whom one knows, immigrants and members of ethnic minorities get hired through networks; the repeated action of network recruitment leads to ethnic employment concentrations, or "ethnic niches" as these have been termed.'

Some empirical studies indicate that different groups with varying educa-tional, occupational and social backgrounds choose and use different kinds of networks. While high occupational groups rely more on networks of col-leagues or organizations, workers tend to use networks of friends and relatives (Vertovec, 2000: 4), which once more channels them into different kinds of niches. Furthermore, some authors argue that ethnic networking does not only bring along positive, but also negative effects on the immigrants, such as lack of mobility chances or a certain amount of isolation and lacking integration (Gurak and Caces, 1992). There is the danger of being 'stuck in the network'.

In the case of Asian migrants the incorporation of a gender perspective is unavoidable if one wants to tell the whole story about South East Asian immigration to Vienna. Two groups – Philippine nurses and nurses from Kerala in India – not only opened a then new employment branch for migrants in Austria in the 1970s but also changed a commonly accepted migration pattern. They did not migrate as family members or as dependants but took the lead in the migration process and brought their husband and family to the country afterwards.

In order to get an idea of the variety of paths into and positions within the Vienna employment system, this chapter merges quantitative with qualitative data. The quantitative research is based on an analysis of the official employment statistics on immigrants from seven South and East Asian countries: Bangladesh, the People's Republic of China (in the following, 'China'), India, Japan, the Philippines, Sri Lanka and Vietnam.[1]

In-depth qualitative information showing some features official statistics could never provide are included for the immigrant group from India that represents the most important group regarding South East Asian migrants in Vienna. If not cited otherwise this qualitative information is based on half-standardized interviews conducted by one of the authors.[2]

For the reader it may be important to understand that migration to Austria is to a great extent a 'Vienna phenomenon'. More than 40 per cent of all foreigners registered in Austria live in its capital, with the metropolitan services sector on the one hand and ethnic networks on the other offering the necessary opportunities for newcomers. Therefore, this chapter focuses on Vienna and is conceptualized as an example for an urban setting within the European Union. The basic questions addressed here are as follows:

1 Which are the main traits of immigration from selected South and East Asian countries to Vienna? What are the demographic features and what about the importance of naturalization in Austria?
2 Which are the relevant legal and structural conditions of the Vienna labour market?
3 How strong is the labour market participation of the groups under consideration, and what are the differences between the various groups?
4 Which special patterns are there regarding actual labour market positions of gainfully employed immigrants?
5 What about immigrant self-employment? Do migrants from South and East Asia participate in that economic field and what are the characteristics of their presence?

Our statistical and empirical investigation, presented here, indicates that there is some sort of new pattern in the labour market integration of Asian migrants: Asian newsvendors, for instance, end up stuck in pseudo-self-employment and seem to embody a new quality of discrimination. Within the range of the different modes of labour market integration of Asian migrants the newsvendors represent nearly the opposite case to the institutionalized and well established integration of nurses. The chapter is structured as follows: first we give an overview on the 'statistical' dimension of Asian immigration, then we focus on the gendered labour market insertion and, third, we concentrate on self-employment.

Overall structure of the immigrant population

Immigration has always played an important role in Austria's and Vienna's history, especially during the Founders' period (the late nineteenth century), when thousands of people from within the Habsburg Empire came to Vienna to live and work. Nowadays, the city is a place of residence for people from almost all over the world. In 1999, 284,691 persons with foreign citizenship (including the European Union) were living in Vienna, their proportion in the total population was 17.7 per cent. Most of them (about 59 per cent) have their roots in the 'classic' sending countries of guest-worker migration, the former Yugoslavia and Turkey. This type of migration has been the most important source of immigration in Austria since the 1960s, when the recruitment agreements were signed (see Table 4.1). The picture of the foreign population in the Austrian public is still very much determined by this type of immigration: in a survey concerning attitudes to foreigners in Austria undertaken in 1998, 96.4 per cent of the interviewees answered 'former Yugoslavia' to the question 'If you think of foreigners living in Austria which countries of origin play a role?' and 93 per cent named Turkey (Lebhart and Münz, 1999: 19).

But this uniform perception does not reflect the real situation any more. The migratory landscape in Vienna has become much more diverse. During the last twenty years and especially since the fall of the Iron Curtain the countries of origin of migrants as well as the migration patterns and structural characteristics of migrants have changed considerably. Nowadays, migrants from East and East Central Europe account for 14 per cent of the total foreign population, with Poland representing the most important sending country. Immigrants from EU countries, mostly of German origin, make up 9.3 per cent of the total foreign population.

Migrants from South and East Asian countries in Vienna

Clarke *et al.* (1990: 1) pointed out that the number of South Asians living abroad is relatively low compared with the home-based population and compared with overseas communities of other origins. But they also indicated

Table 4.1 Composition of the Vienna population, 1993 and 1999

Category	1993			1999		
	No.	As % age of population		No.	As % age of population	
		Foreign	Total		Foreign	Total
'Guest workers' (former Yugoslavia, Turkey)	174,258	59.4	10.6	167,082	58.7	10.4
'New immigration' (Central and East Central Europe)	41,958	14.3	2.6	39,743	14.0	2.5
European Union	21,296	7.3	1.3	26,458	9.3	1.6
Other	55,979	19.1	3.4	51,408	18.1	3.2
Among them: Selected Asian groups	13,865	4.7	0.8	13,113	4.6	0.8
Foreigners total	293,491	100.0	17.9	284,691	100.0	17.7
Austrians	1,348,900	–	82.1	1,321,453	–	82.3
Total population	1,642,391	–	100.0	1,606,144	–	100.0

Source: own calculations based on the Statistical Yearbook of the City of Vienna.

that South Asian migrants 'are more dramatically spread around the world' (*ibid.*). Asian migration to Austria and Vienna is a rather marginal phenomenon that happened and happens apart from the bigger migration flows. As Austria does not have a 'colonial history' Asian migration never played an important role in quantitative terms as we find it for Great Britain for example (Hillmann, 2003). Although the numbers of Asian immigrants are still relatively small compared with other nationalities, the respective groups have grown substantially, most of all at the end of the 1980s and beginning of the 1990s. This fact is of particular interest for two reasons. First, these immigrant groups differ considerably from 'classical guest workers' in certain respects like their position on the Vienna labour market. Second, they can be characterized by a strong heterogeneity within the group regarding region of origin, religion, and educational level and labour market participation, which holds especially true for immigrants from India.

Regarding the countries of origin under consideration, migrants from India form the biggest group, with 4,331 persons living in Vienna in 1999. Migrants from the Philippines and China[3] are present in almost equal numbers (2,685 and 2,883 persons). The smallest Asian communities under consideration come from Sri Lanka and Vietnam (381 and 319 persons only). Altogether they make up 4.6 per cent of the total foreign population living in Vienna in 1999.

If we look at the development over the past fifteen years, there are some groups that grew very fast but started from a very low quantitative level (like Bangladesh and China) while other groups are relatively stable in size (like the Philippines and migrants from Japan). Philippine migration to Austria started in 1973, when the Council of Vienna signed a bilateral contract with the Philippine Ministry of Labour because of a shortage of nurses in the Vienna hospitals. In the course of the years, hundreds of nurses from the Philippines came to Vienna. Many of them stayed, fetched their families and eventually became naturalized citizens. However, many of them went back to their country of origin.

Another important feature is the break in 1993, when a clear stagnation phase in immigration started (see Table 4.2). The main reasons for this stagnation were changes in the political situation in Austria. As in other European countries, immigration regulations got markedly more restrictive. The asylum procedure was tightened, upper limits were set for the recruitment of new workers and for the number of dependants allowed to join these workers, and quotas were set for the immigration of foreigners.[4] As a matter of fact, it has been almost impossible to immigrate into Austria officially from outside the European Union since then. In 1998, Austria had – together with Germany – the lowest immigration rate within the EU (0.06 per cent). On average, official immigration was twice as high in the rest of Europe (see Fronek, 2000: 93).

The number of asylum seekers from the countries under consideration has varied much over the last twenty years. Between 1980 and 1997, Vietnam

Table 4.2 Migrants from selected South and East Asian countries officially recorded in Vienna, 1986–99 (absolute and index 1986 = 100)

Year	India		China		Philippines		Japan		Bangladesh		Sri Lanka		Vietnam	
	No.	Index	No.	Index	No.	Index	No.	Index	No.	Index	No.	Index	No.	Index
1987	2,242	107	443	112	2,216	105	1,128	104	244	112	278	105	83	96
1989	2,576	123	496	125	2,361	113	1,007	93	591	272	275	104	63	73
1991	3,511	167	618	156	3,038	111	1,380	128	680	313	330	125	54	63
1993	4,667	222	2,426	611	3,520	106	1,644	152	978	451	405	153	225	262
1995	4,324	206	2,660	670	3,014	92	1,718	159	878	405	391	148	260	302
1997	4,119	196	2,782	701	2,658	93	1,555	143	903	416	388	147	285	331
1999	4,331	206	2,883	726	2,685	101	1,510	140	1,004	463	381	144	319	371

Source: City Council, MA 66 (population register). The population register is the duty of the registration authority and thus part of the Ministries of Internal Affairs. It records all 'officially' present persons in Vienna, including their nationality. Thus anyone who stays in Austria and is not registered by the police or the competent authority respectively does not appear in the population register either. As long as such persons are not conspicuous, not taken up by a police patrol or being caught as fare dodgers in public means of transport, they can live in Austria for a very long period of time (see Fassmann and Reeger, 1997).

showed the biggest number of asylum seekers in Austria (2,616); the peak was reached in 1990 and 1991 with more than 300 Vietnamese applying for asylum in these two years.[5]

Table 4.3 illustrates well that the ethnic communities must be bigger than was thought, because the numbers of naturalizations were rather high in comparison with the groups' sizes during the last twenty years. As soon as people are naturalized, they are no longer included in the Austrian statistical sources concerning residence and foreign employment, the place of birth is not of interest there. Generally speaking, the Vienna administration is willing to grant citizenship more readily than other European cities.[6] This reflects the different basis of identity for an Austrian: nationality is not based on ethnicity. In that sense the Austrian concept of nationality is similar to the French model of citizenship. In Austria, immigrants see naturalization as a kind of ticket to faster integration (see Fassmann and Reeger, 2001).

Adding to this high propensity to grant citizenship, Asian migrants show a much higher tendency to take on the Austrian nationality in comparison with other immigrant groups. In 1998 and 1999 the overall proportion of foreigners being naturalized was around 3–4 per cent. This is small compared with 21 per cent of all migrants from Sri Lanka being naturalized in 1999 or 13.5 per cent of all Philippines in 1997, just to give some examples. In interviews with Chinese immigrants, some reasons for naturalization that go beyond the simple view that immigrants only want to improve their legal situation on the labour or housing market in Vienna were found (Schnetzer, 1994: 17). The reasons mentioned were the freedom to travel in Europe without any restrictions on the one hand and the improvement of business connections with China by being a 'foreigner'. These reasons give a picture of the high mobility of Chinese migrants and to some extent let fade the old idea of emigration – immigration – permanent settlement. Similar arguments were found for Indians. The possibility of free movement within Europe seems to be a good reason to take up Austrian citizenship, because then they do not need visa for other European countries any more (Hintermann, 1995).

Table 4.3 Naturalizations of migrants from selected South and East Asian countries, 1981–99

Period	India	China	Philippines	Japan	Bangladesh	Sri Lanka	Vietnam
1981–89	524	323	1,505	46	13	46	333
1990–94	921	787	1,749	14	30	78	136
1995–99	1,565	701	1,247	24	137	180	151
Total 1981–99	3,010	1,811	4,501	84	180	304	620

Source: City Council, MA 66.

With the exception of Bangladesh, gender proportions only changed a little in the course of the 1990s (see Table 4.4). The high proportion of women within the Philippine group still is the result of their 'port of entry' to Vienna as nurses in the health system, but male immigration gained importance in the course of the 1990s. There are markedly more females than males from Japan living in Vienna. The lowest share of women is found in the group from Bangladesh. Gender proportions are relatively balanced in the case of China, but also in the groups from Sri Lanka and Vietnam. At the very beginning of the migration process from India to Vienna the sex proportion was rather balanced. Out of fifty-one migrants counted in the census 1971, twenty-nine were male and twenty-two female. All over the last four decades male migrants were in the majority, which is astonishing, taking into account that the migration from the Indian state of Kerala of the two most important regions of origin was at least at the beginning dominated by women. During the last twenty years the share of female migrants even decreased continuously from more than 40 per cent to just over 30 per cent. This decrease is the result of two simultaneous developments: family reunification on the part of female migrants from Kerala who brought their husbands to Austria and the increased flow of male migrants mainly from Punjab, many of them still living in Vienna without their families. Male migrants are in the majority in all age groups except those aged sixty-four years and more. Especially strong is the male dominance in the age groups between thirty and forty-nine, where they constitute about three-quarters of the Indian population.

Educational level or marital status are not recorded by the official data base. Here we rely on qualitative data on Indian migrants which indicate that it is necessary to distinguish between the two main migrant groups: the nurses from Kerala and the migrants from Punjab, many of them working as

Table 4.4 Migrants from selected South and East Asian countries in Vienna, by sex, 1993 and 1999

Country	1993			1999		
	Total	*Women*	*% women*	*Total*	*Women*	*% women*
India	4,667	1,665	35.7	4,331	1,403	32.4
China	2,426	1,169	48.2	2,883	1,495	51.9
Philippines	3,520	2,152	61.1	2,685	1,521	56.6
Japan	1,644	992	60.3	1,510	947	62.7
Bangladesh	978	101	10.3	1,004	187	18.6
Sri Lanka	405	189	46.7	381	190	49.9
Vietnam	225	95	42.2	319	146	45.8
Total	13,865	6,363	45.9	13,113	5,889	44.9

Source: City Council, MA 66.

newsvendors in the streets of Vienna. Most of the nurses have a high or very high educational level. Some of them and also their husbands even hold a university degree. Their migration thus led to a brain drain to the debit of India and the benefit of Austria. Compared with the nurses the educational level of the mostly male migrants from Punjab is lower. Their knowledge of English, which is rather poor, may be taken as an indicator. A former embassy counsellor of the Indian embassy in Vienna tried to sum up the educational level of the newsvendors as follows:

> The average boy probably would have done high school. But in India that's nothing. You have to go beyond that to get into a profession. In India it's twelve years of school. If you don't go beyond that, you don't get a job – an office job. And office jobs pay. So I think most of the boys have finished high school but I don't think many of them went beyond that.

Clarke *et al.* referred to the observation that 'fundamental bases of differentiation within any South Asian population will include (a) *religion*...; (b) *language*...; (c) *region of origin*...; *caste*...; (d) *degree of "cultural homogenization"*...' (1990: 6). At least three of these items can explicitly be duplicated in the case of the Indian migrants in Vienna: region of origin and, connected with that, religion and language. Migrants from Kerala are almost exclusively Christians, most of them Catholics, some Protestants. Their regional language is Malayalam, which is not comparable with Hindi, for example. Some of those interviewed also mentioned other points of differentiation between Malayalees (from Kerala) and people from north India, like dress or diet. Migrants in Austria from Punjab are mainly Sikhs, who speak Punjabi. In addition, there are some other smaller groups with their own regional languages. Migrants from north India sometimes also speak Hindi, but only a few of the Malayalees. The third biggest religious groups of Indians in Vienna are Hindus from different parts of the country.

The way to Austria: chain migration and networking systems: the Indian example

Using the qualitative data on Indian migrants, it is possible to sketch and to analyse the development of a migration system between regions of two countries without obvious links and bonds. The migration of Indians to Vienna can best be described as chain migration.[7] Early migrants serve as source of information about the country of destination for family members, relatives and friends at home, they build up the image of the destination country and attract other members of the home society to follow in their footsteps. As many examples show, early migrants don't only inform and attract, they also give help to the newcomers in the destination country: financial help, help in finding a job and a flat but also emotional support in the new environment. This is the case also with the Indian migrants in Vienna. Today hardly any migrant lives there without any family member, friend or relative. Many migrants were actively involved in bringing friends or relatives, as interview passages from 1995 show:[8]

> Most of them are on a family network. So supposed I am here all my brothers and cousins would try to come here too.

> All of them who are coming now have relatives or friends here. I brought some of my relatives as well.

> The aunt of my husband is here as well. She has been living in Austria for 17 years and also brought my husband to Vienna.

> My three brothers were already here when I came. I share a room with them.

The qualitative data allow us to distinguish five characteristic patterns in the migration history of Indians to Vienna that did not directly succeed one another but sometimes overlapped.

1   The first Indian migrants who came to Austria were mainly students, academics, employees of international organizations and diplomats. It was a migration of highly educated individuals who came to Austria almost by chance.
2   At the beginning of the 1970s the introduction of Africanization programmes in Uganda had the effect that Indian migrants came to Austria. In August 1972 the Ugandan dictator Idi Amin expelled all people of Asian origin. Austria temporarily sheltered about 1,500 Indians from Uganda, and fifty families (about 200 people) were given permanent asylum (Stanek, 1985: 98 ff.).
3   The migration of Catholic nurses from Kerala goes back to the beginning of the 1970s and is very much connected with the activities of individuals but at the same time embedded in the network of the Catholic Church. Kerala has the highest share of Catholics in India, a relatively well established health system and a high number of educated nurses compared with other Indian provinces. Nurses from Kerala not only migrate to Europe but also to other parts of India as well as to Saudi Arabia or the United States.

    Talking about the importance of individual migrants to foster migration processes the initiative of a migrant from Kerala, who came to Vienna in 1973 as a student has to be mentioned especially. He spotted the chance for Indian nurses to find work there and contacted the Austrian and Vienna authorities as well as the Indian embassy in Vienna to look for possibilities to officially recruit Indian nurses. Although all the efforts did not succeed in an official recruitment contract between Austria and India, Mr Kizhakkekara got at least a 'letter of good will' signed by the former Vienna Town Councillor for Public Health declaring that the city of Vienna 'will employ Indian nurses as far as possible'.

An important source for the migration of Indian Catholics to Vienna was the 'Queen of the Apostles' order, founded in Vienna in 1923 for the missionizing of India. During the eighty years of its existence the order established numerous missions in India. Due to the initiative of people from the order ten Indian women came to a Catholic hospital in Lower Austria (the province next to Vienna) and twelve to a Catholic hospital in Vienna in 1972/73. Some of them still live in Vienna, together with their husbands and children.

The migration of nurses from Kerala to Austria is, apart from Philippine nurses, an exceptional case compared with all other migrant groups in Austria. Young female individual migrants took the lead in the migration process and therewith reversed a widely observed migration pattern.

4   Parallel to the migration of nurses from Kerala the second important migration branch from India to Austria developed during the last thirty years: the migration of Punjabis. Compared with the migration from Kerala it corresponds to the classical migration pattern that young male individuals migrate first and bring their wives and children to the destination country after a certain period of consolidation there (Fuchs, 1997: 149). It was the migrants from Punjab who made the presence of people from India in Vienna more visible. Many of them work as newsvendors in the streets or as market vendors at open markets all over Austria. Newsvendors have a special legal status in Austria: compared with other migrant workers they do not need a work permit, as they are legally treated as self-employed. Fuchs (1997: 153) reports that Austrian newspaper publishers displayed info-sheets for newsvendors in Austrian embassies and missions like Islamabad and that travel agencies act as middlemen. They are specialists in immigration questions and have a detailed knowledge of immigration regulations of many countries of destination.

5   The so far last step in the migration process from India to Austria is family reunification. Newsvendors had greater difficulties in fetching their families than other groups. Their income is often too small and their housing situation is not adequate. It is quite common that newsvendors share a flat or even a room. Especially at the beginning of their life in Austria they often live in a 'men's world' with hardly any contact with women apart from customers and visits to the religious centres.

Labour market integration

Legal and structural conditions of the Vienna labour market

The occupational status of immigrants in the receiving countries does not only depend on their age, educational status or a good command of the foreign language, but also on structural factors of the receiving labour market. Access to the Austrian labour market for non-EU citizens is directed by the state. The proportion of foreigners from these countries who are allowed to work in Austria is regulated by an annual quota (about 8 per cent of the total work force and always too low for all foreigners to gain access to the official labour market). Unlike other European countries, Austria distinguishes between residence permits and work permits, which means that not everybody who is allowed to stay in Austria also has the right to work there. It is especially women desiring to (re)enter the labour market who are affected by this regulation. Some estimations put the number of immigrants legally present but excluded from 'official' jobs at a high 35,000 persons.

The Austrian labour market turns out to be ethnically segmented. The limited access to the labour market and the concentration of foreign workers in some branches did not change much since the beginning of labour migration to Austria in the 1960s. The majority of migrants – especially 'guest workers' – are still employed in sectors with low social status, high fluctuation, low wages and hardly any promotion opportunities. In 1998 about 43 per cent of all gainfully employed citizens of the former Yugoslavia and even about 59 per cent of all employed Turkish migrants in Vienna worked in the 'industry and crafts' sector. Upward professional mobility is hardly discernible, even among migrants who have been living in Austria for the last two or three decades (Fassmann *et al.*, 1999: 106 ff.).

Another special feature of the Austrian labour market for immigrants is the fact that a longer period of unemployment can lead to loss of the right to stay in Austria. The right to emergency social benefits in addition to unemployment benefit does not, under the existing administrative system, provide the subsistence minimum required to retain the right to residence. Therefore unemployed foreigners in Vienna are under great pressure to accept a new job, because they are likely to have to return home, as they lose the legal basis to stay when they cannot find a new one. This fact also explains, why foreigners' unemployment is quite low compared with other European countries and metropoles.

Studies show that the Vienna labour market can be characterized as being rather discriminating against immigrants. They are still to be found mainly in those branches that are economically risky and display unfavourable working conditions. The step into self-employment or into very special niches seems to offer the only way out of this dilemma for some of them.

Gainfully employed SEA immigrants: participation and labour market positions

According to the official data, the labour market participation of Asian immigrants is rather low compared with other immigrant groups. Among the male Asian migrants, those from China show the highest labour market participation by far: more than one in three male Chinese resident in Vienna is gainfully employed. All the other groups under consideration show much lower values. Concerning women, females from the Philippines are most likely to hold a gainfully employed position (25.4 per cent, followed by women from China, India and Sri Lanka). There are several reasons for this low participation rate:

1   Family members who follow their spouses or other family members are not entitled to take up a job immediately after their arrival, but they have to wait for up to five years till they are allowed to enter the labour market.

2   Migrants, who are working for international organizations (UN, OPEC, etc., 'elite migration') are not included in the statistics of the gainfully employed labour force in Vienna.

3   Foreign students from outside the European Union have not been allowed to take up a job in Austria until the end of 2002. With the latest amendment of the aliens law they now have the opportunity to work for three months a year.[9]

4   Many of them are self-employed. For example, male Asian migrants, most of all from India and Bangladesh, work as newsvendors. These positions are considered as self-employment. Hintermann (1995: 54) assesses the number of Indian newsvendors at about 400; some 100 trade on open markets in the city.

According to what has already been said about the special features of the Vienna labour market, the occupational positions of the Asian work force show strong concentrations, but with varying patterns from group to group.

Within the Indian community, 41.8 per cent of all gainfully employed hold jobs in the health sector and another 22.8 per cent in tourism (most of all in restaurants as cooks and assistants). Only one in ten Indian immigrants is employed in industry and crafts. According to Hintermann (1995), most of the Indians who are gainfully employed in the catering trade work in restaurants and firms of compatriots. The same holds true of all other immigrants from South and East Asia who are gainfully employed in the restaurant sector in Vienna (Table 4.5). As expected, there is a clear dominance of the catering trade within the Chinese group with a share of more than 65 per cent of all gainfully employed. One in five gainfully employed Chinese works in the health sector. The Chinese show the greatest amount of specialization within the Vienna labour market. Jobs in industry are not

Table 4.5 Occupational positions of gainfully employed migrants from selected South and East Asian countries in Vienna, 1999 (%)

Sector	India	China	Philippines	Japan	Bangladesh	Sri Lanka	Vietnam	Ex-Jugo	Turkey
Agriculture and forestry	0.5	0.2	0.0	0.0	1.7	1.9	0.0	3.7	3.7
Industry and crafts	10.1	1.8	10.3	0.6	17.1	16.7	61.8	42.6	60.8
Building trade	0.8	0.0	0.3	0.0	5.0	1.9	5.9	14.8	22.9
Metal working, electrics	1.9	0.4	2.8	0.0	1.1	3.7	27.9	11.9	13.7
Wood, leather, textile	0.5	0.2	0.0	0.0	0.0	3.7	11.8	1.6	2.1
Assistant workers	5.9	0.7	3.7	0.6	6.6	3.7	7.4	6.8	11.6
Trade and transport	9.0	2.6	1.8	26.0	17.7	13.0	4.4	9.4	6.7
Trade	5.0	2.2	1.2	11.2	2.8	0.0	4.4	3.9	4.0
Transport	4.1	0.5	0.7	14.8	14.9	13.0	0.0	5.5	2.7
Services	27.8	65.8	56.8	15.4	59.1	59.3	32.4	37.3	24.0
Tourism	22.8	63.7	28.5	11.8	55.2	40.7	23.5	11.7	7.6
Domestic help, cleaning	4.5	1.7	28.2	1.2	3.9	18.5	8.8	23.5	15.3
Technicians	3.7	2.6	0.3	0.0	0.6	1.9	1.5	1.0	0.5
Administration, office	3.3	3.5	2.5	22.5	0.0	3.7	0.0	2.4	2.2
Office jobs	3.3	3.4	2.5	21.9	0.0	3.7	0.0	2.4	2.2
Health, teaching	44.8	21.5	28.2	35.5	3.3	1.9	0.0	2.6	1.0
Health	41.8	20.0	27.7	4.1	1.7	1.9	0.0	1.9	0.5
Teaching, culture, etc.	3.0	1.4	0.5	31.4	1.7	0.0	0.0	0.7	0.6
Unknown	0.8	1.0	0.0	0.0	0.6	1.9	0.0	1.0	1.0
	100.0	100.0	100.0	100.0	100.0	100.0	100.0	100.0	100.0
No.	641	833	600	169	181	54	68	48,014	13,649

relevant in the case of gainfully employed Asians working in Vienna. (The only exception are Vietnamese immigrants, but the small size of the group makes it difficult to evaluate this result.)

Only 27.7 per cent of the members of the Philippine community who are gainfully employed in Vienna still work in the health services, which is a sign of diversification during recent years. Nowadays they are to be found in the catering trade (28.5 per cent) and as domestic helpers (28.2 per cent).

Unlike all other immigrant groups from South and East Asia, Japanese immigrants are most likely to be found in the teaching sector, in administration as well as in trade and transport. One in ten Japanese gainfully employed holds a job in the catering trade, which is by far the lowest share of all countries under consideration. Anyway, they show a completely different pattern of employment compared with other Asian groups. Bangladesh employees are concentrated in the service sector as well as in transport and industry, the health services are not important for them. The same holds true for the gainfully employed immigrants from Sri Lanka.

The comparison between classic guest workers from the former Yugoslavia and Turkey and Asian immigrants also shows clear differences in their occupational positions. Guest-worker immigration was launched because of a shortage of manpower in industrial jobs and in the metropolitan services sector. They mainly got the jobs the Austrians did not want to take on any more ('3-D' positions: dirty, dreadful and dangerous). This understratification is still existing: in the case of people from former Yugoslavia, two-thirds are still employed in the two sectors mentioned (jobs in industry: 42.6 per cent, service jobs: 37.3 per cent), with all other fields playing a subordinate role. The respective shares in the Turkish group show an even higher tendency towards concentration (industry: 60.8 per cent, services: 24.0 per cent).

Gender differentiation of labour market positions

As shown above, women who were recruited for clearly defined positions in the health system initiated part of the labour migration from South and East Asia to Austria. The question is whether the health system still is the main provider of jobs for these women or whether they had to find their way into other fields of the urban economy.

Immigrant women very often suffer from even higher discrimination on the labour market than males, simply by being 'female' and a 'foreigner' (Wilpert and Morokvasic, 1983). On the Vienna labour market, the majority of immigrant women are concentrated in only a few occupational segments, most of all in the services sector as domestic helpers and in cleaning jobs (see Figure 4.1, former Yugoslavia and Turkey). The boom in domestic helpers began in the late 1980s, when local households earning more money could afford to pay somebody to do the housework for them (compare Sassen, 2000: 93 f.).

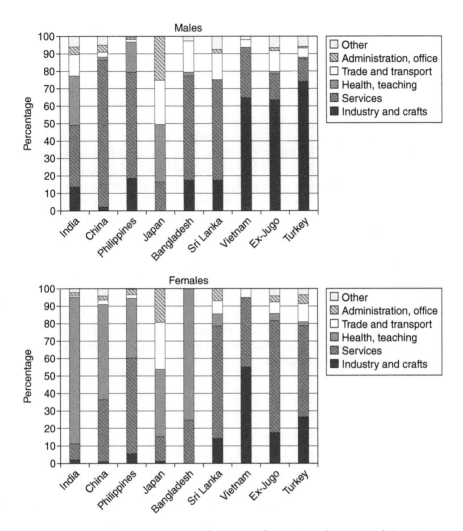

Figure 4.1 Occupational positions of migrants from selected South and East Asian countries in Vienna, by gender, 1999 (source: AMS, own calculations).

In the case of gainfully employed women from India it is still the health sector that provides most of the jobs. More than 80 per cent of them are employed as nurses in the Vienna hospitals, another 6 per cent work as domestic helpers and in cleaning jobs. The respective share in the catering trade is infinitely small (3.7 per cent). In comparison, Indian males show a much bigger diversification of occupational positions: 30.9 per cent found jobs in the catering trade, but they also work in hospitals (24.2 per cent) and in the field of trade and transport. About 10 per cent work as technicians and in office positions and another 13.6 per cent in industry. This pattern is a result of the occupational background of the male migrants. While many female immigrants are trained nurses, males did not work in hospitals in India but in many other professions.

The already described concentration of Chinese immigrants in the catering trade even gets more marked from a gendered perspective: 83.4 per cent of the male Chinese gainfully employed work in such positions, the respective share for women is 31.2 per cent only. Once more, the health system is the main provider of jobs for females, with more than half of the female Chinese gainfully employed working in a hospital.

The concentration of Philippine women in the health services sector decreased during the 1990s. In 1999 exactly one-third were employed there, but almost the same share work as domestic helpers and in cleaning jobs in private households; one in five women from the Philippines is employed in the catering trade. Thus the service sector is now more important for them than the health system. Males are concentrated in the catering trade; one-fifth found a job in industry.

For the immigrants from Japan, the differences between males and females regarding occupational positions are not that marked. They both work in the fields of teaching and culture, in trade and transport and in the catering trade and thus show a completely different pattern. Concerning Bangladesh, Sri Lanka and Vietnam, the small sizes of the groups leave the interpretation kind of vague. Anyway, the service sector remains dominant, also from a gendered perspective.

Self-employment

General features and developments

Wandering around in Vienna nowadays, it is not hard to enjoy sushi in one of the many sushi bars around 'Naschmarkt'[10] and in other parts of the city, to buy an ice cream at an Italian ice-cream maker or to eat a *döner kebab* in one of the market places. In the past few years, one could observe a growing presence of shops and restaurants run by foreigners – also originating from Asia – when walking through some parts of the city, especially those where many immigrants live. Anyway, immigrant entrepreneurship is a rather new phenomenon in Vienna. In some parts of the city, shops and restaurants provide special goods and services, with Chinese restaurants clearly dominating the scene. A growing dynamic can easily be observed, but foreigners' self-employment does not yet play such an important role in Vienna as in many other European cities, like London, Berlin, Paris or Amsterdam.

There are two reasons for this. First of all, Asian migrants who dominate the scene in other cities, like London or Paris, are not that numerous in Vienna. Furthermore, there are many constraints and barriers to ethnic self-employment in Vienna. Citizenship plays a decisive role: it is hard to set up a business without being an Austrian national. Basic requirements to get a trade licence are not easily fulfilled. Thus many foreigners who want to set up in business cannot help but have an Austrian partner within the company. This partner, who figures as a front man, then undertakes all official transactions. The cost of such a partner is often very high and represents a big burden on the entrepreneur.

In Austria, particular restrictive political regulation applies to access to the trade and service sector. The definition of 'free' and 'tied' businesses, deeply anchored in the tradition of the guild, leads to an ethnic structure, which the food sector clearly exemplifies. Foreign entrepreneurs are to be found in so-called free trade, in food retailing and in restaurants.

In their analysis of self-employed guest workers, Haberfellner and Böse (1999) describe some more barriers to ethnic self-employment in Vienna that hold true for all immigrant groups. First of all, the business authority checks, in all cases, whether the business to be set up is of 'economic interest' and does not act 'contrary to other public interests'. Furthermore, the change into self-employment also brings along a change in residence purpose, which means that the residence permit has to be granted anew. Very often, these residence permits are then limited in time, which brings up problems for example at the bank, because nobody likes to give a credit to someone whose stay is limited in time. These two remarks only represent examples of all the disadvantages that come along with immigrant entrepreneurship in Vienna. But examples show also that migrants from East and South East Asia are economically active in Vienna, despite the barriers they are facing.

Some estimations assume that there were around 300 Chinese restaurants in Vienna in the middle of the 1990s. Exact numbers are difficult to find owing to the following problems: owners often change; the owners' nationalities are not registered by the Chamber of Commerce; a survey based on restaurant names remains vague, as the restaurants could also be run by people from Taiwan or other countries outside China (see Schnetzer, 1994: 21).

A special niche concerning South and East Asian self-employment in Vienna is that of newsvendors. From the 1970s onwards news-vending in the streets and public places developed to a 'foreigner's job' in Austria. In the beginning it was almost exclusively migrants from Egypt who worked as newsvendors. Since then the structure has changed considerably. During the 1980s and 1990s more and more vendors from South Asian countries, especially from India but also from Bangladesh or Pakistan, started the 'newsvending business'.

As mentioned before newsvendors have a special legal status in Austria concerning their residence permit ('Z visa') as well as their professional status. The 'Z visa' allows them no other work than news-vending (Fuchs, 1998b). As soon as they lose their jobs the fact is automatically reported to the aliens' office and means deportation, a regulation making them completely dependent upon their trading company. They are considered to be self-employed although many aspects of their daily working life contradict this position. They don't need a trading licence, are dependent on special publishing houses and are not allowed to sell the products of other publishers.[11] They have to buy the papers from their publisher and sell them on the street at their own risk and have to wear a uniform displaying the name of their publisher, acting therewith as a living advertisement.

Other considerations than their status as self-employed might indicate their work is strictly regulated. At the beginning of the 'newsvendor career' they have to attend a training course, in which the most basic German sales vocabulary and rules of behaviour are taught and a brochure with the most important rules is distributed (Fuchs, 1998b: 51 ff.). Moreover selling hours are fixed and 'well behaving' and 'misbehaving' are sanctioned positively or negatively (for example, by getting better or worse sales pitches) (*ibid.*: 54 f.). The status of self-employment has advantages mainly for the quasi-employers. It saves contributions to the social security system as well as the payment of holiday and Christmas money. The monthly average income of a newsvendor amounts to about €500 to €700 and is subject to income tax. As they are not employed they have to take care of their own insurance and are not entitled to social or unemployment benefits.

Little is known about the exact numbers of newsvendors in Vienna. Publishers are not very co-operative in giving information. In 1995 one representative of Mediaprint estimated that the number of Indian newsvendors working for his company in Vienna amounted to about 200. Taking into account that Mediaprint employs about half of all newsvendors in Austria the total number in Vienna can be assessed at about 400.

The Indian example

It has been mentioned that immigrant entrepreneurship in Vienna is mainly restricted to two sectors: restaurants and food retailing. This can also be proved by the Indian example. The development of a broader Indian ethnic economy began at the end of the 1980s. The first Indian boutique already opened in 1951 and the first restaurant in 1970 (Fuchs, 1998a: 292, 276). While most of the businesses are run by migrants from north India, Malayalees are hardly to be found as entrepreneurs. Reasons for that might be that their qualifications are needed in the 'normal' labour market and the leading role of females in the migration process who are in general to a lesser extent self-employed than men. Indian entrepreneurship in Vienna has to be analytically structured in four sectors:

1   *Wholesale businesses and market stalls.* Wholesale business is exclusively run by migrants from north India. Most of them are specialized in trading in textiles; only one deals with style jewellery. The products are mainly imported from South and East Asian countries but also from Great Britain and Germany. Some of the wholesalers also run retail

shops and market stalls at open markets in Vienna or/and street markets all over Austria. They employ Indian compatriots, who sell the goods. On the other hand there are also market stalls that are owned not by wholesalers but by the retailers themselves. They obtain their goods from different wholesalers in Austria.

2 *Ethnic travel.* Labour migration is an important factor in international air traffic. Travel agencies offer special conditions for migrants and therewith open up a new market and meet special needs of the migrants. There are at least three travel agencies owned by Indian migrants in Vienna. Two of them are run by migrants from Kerala. Aerojet is a branch of a south Indian firm in Frankfurt on Main in Germany. It not only serves as travel agency but also sells food and loans videos in Indian languages. Prompt Reisen Tours & Travel has a branch in Kerala and offers special packaged trips to Kerala. It has two associated companies in Vienna: a supermarket and a restaurant.

3 *Food retailing.* When the first Indian restaurant opened in Vienna in 1970 it was rather difficult to get any ingredients for the Indian cuisine. This situation has changed considerably. At the end of the 1990s the number of Indian food retail shops amounts to about twenty to thirty. Most of them are run by migrants from northern India. The goods sold can be seen as supplement to the Viennese market, with hardly any competition with native Austrian retailers. The shops are mainly addressed to the Indian community. An essential additional service of Indian food retail shops is the hiring out and also selling of video-tapes in different Indian languages. In 1994 an association of Indian shopkeepers (ISA Indian Shopkeeper Association) was founded to establish better co-operation between the shopkeepers, to agree on prices for special products and to minimize ruinous competition. Some agreements were fixed but the association did not even exist one year and has never been officially registered (Fuchs, 1998a: 260).

4 *Restaurants.* Indian restaurants are also mainly a domain of migrants from north India. Of some thirty existing explicitly Indian restaurants only three are run by migrants from Kerala. The first south Indian restaurant (the Kairaly Restaurant)[12] was founded only in 1996. The professional careers of some of the restaurant owners are further empirical evidence of the high flexibility of South and East Asian immigrants on the Vienna labour market. Hardly any of them has a respective professional background, and some of the restaurants have been established by former newsvendors.

Conclusion

There is a lot of empirical evidence on the concentration processes regarding immigrants on the urban labour market, that – in the case of Vienna – is always focused on guest workers from the former Yugoslavia and Turkey, as they still represent the largest immigrant groups. As the present chapter shows, the concentration in only a few sectors of the labour market is even more pronounced in the case of some of the immigrant groups from South and East Asia. But owing to their better educational background – and this holds particularly true for women – they are to some extent more successful and hold better jobs than immigrants from other countries. On the labour market, clearly defined niches from the very top to the very bottom can easily be identified. South and East Asian migrants found their place within the urban economy, both as gainfully employed and as self-employed.

South and East Asian immigrants in Vienna are on the one hand part of a more traditional labour regime, which holds true of the Philippine and Kerala nurses. On the other hand their paths into the Vienna labour market prove very flexible and innovative ways and strategies to earn one's living in an urban economy with very tight and restrictive labour market regulations. The example of the newsvendors clearly shows a new dimension of contemporary working spheres. Without having any other opportunity, they are stuck in that pseudo-self-employment which brings with it more disadvantages than advantages. But as it is the only possibility for them to stay in Austria and to make a living, they seem to have accepted the discriminating rules of that business.

Networking does not only play a decisive role in the migration process as such, but also in the field of labour market integration, which we could prove in our analysis. The Indian nurses are a very good example of this rather general evidence. On an individual basis they would never have made their way to Vienna; individual recruiting of friends or relatives is the name of the game. Being employed with international enterprises, which send them to worldwide destinations, elite migrants from Japan use quite other types of networks. This is one more indicator of the heterogeneity of Asian migration to Europe and Vienna and the variety of networks built and used to get there.

Finally it has to be pointed out that immigration to Austria from countries outside the European Union is almost impossible owing to a whole string of restrictions (which holds true of the whole European Union) and this situation is not likely to change in the future. Thus it is to be expected that people will try to find other informal ways of entering the country and the urban economy.

Notes

1 Statistics on employees are provided by the Arbeitsmarktservice Österreich. They are made topical continuously and record all legally present persons who are allowed to take up a job, also with regard to their nationality. Thus this source is better in covering the ongoing processes than the census or the microcensus. Its weakness lies in the fact that only 'legal cases' are being registered.

2 For complete documentation see Hintermann (1995, 1997).

3 A first wave of Chinese immigration is found already before World War I, when

people from China tried to settle in Vienna, but they faced massive xenophobia and some of them had to leave the country. During the inter-war period, some 600 persons settled in Vienna's fourteenth district, but they left when the German troops invaded Austria. But already in 1947 the Chinese community had forty-seven members and started to run restaurants. Most of them came from Taiwan at that time (see John and Lichtblau, 1990: 61).

4 The immigration quota for the year 2003 amounts to 8,280 people from non-EU countries, 5,700 places being reserved for dependants. In addition there is a quota for seasonal workers (8,000) and a quota for harvesters (7,000) (© der-Standard.at).

5 The numbers for the other countries are as following: 1,794 Indians, 1,869 people from Bangladesh, 229 Chinese, ten Filipinos and 917 people from Sri Lanka. For the time being, only 10 per cent of all asylum seekers are admitted to stay in Austria.

6 Citizenship is granted in Austria after ten years of permanent residence. This time limit can be reduced to four or six years when there are 'special considerable reasons' (for example: birth in Austria, refugee status, 'effective personal and professional integration'). A legal claim the Austrian citizenship exists only after thirty years of residence in Austria.

7 See MacDonald and MacDonald (1969).

8 Translation of interview passages into English in the following by the authors.

9 In 1998 the numbers of students from South and East Asian countries at Vienna universities amount to: China: 200 (6.9 per cent of the resident Chinese), the Philippines: eight (0.3 per cent), Japan: 114 (7.5 per cent), Vietnam: twenty-three (7.2 per cent), Sri Lanka: eight (2.1 per cent), Bangladesh: seventy-eight (7.8 per cent), India: seventy-five (1.7 per cent).

10 Naschmarkt is a market in the centre of Vienna.

11 The most important employer for news vendors in Austria is Mediaprint, which distributes the paper with the highest number of copies in Austria, *Kronen Zeitung*, and many other daily and weekly papers.

12 The owners are one of the first Indian nurses who came to Vienna in 1972 and her husband, a UN employee. The second restaurant (Himalaya) owned by a migrant from Kerala is run by a brother of the just mentioned nurse.

References

Clarke, C., Peach, C. and Vertovec, S. (eds) (1990) *South Asians Overseas: Migration and ethnicity.* Cambridge: Cambridge University Press.

Cross, M. and Waldinger, R. (1997) 'Economic integration and labour market change. A review and reappraisal'. Discussion paper prepared for the Second International Metropolis Conference, Copenhagen, 25–7 September.

Fassmann, H. and Reeger, U. (1997) 'Sources of information on East–West mobility. Compatibility and complementarity of data sources'. Paper presented at the international conference Central and Eastern Europe. New Migration Space, Pultusk, Poland, 11–13 December.

Fassmann, H. and Reeger, U. (2001) 'Immigration to Vienna and Munich: similarities and differences', in H.T. Andersen and R. van Kempen (eds) *Governing European Cities. Social Fragmentation, Social Exclusion and Urban Governance.* Aldershot: Ashgate, 273–95.

Fassmann, H., Münz, R. and Seifert, W. (1999) 'Ausländische Arbeitskräfte in Deutschland und Österreich. Zuwanderung, berufliche Platzierung und Effekte der Aufenthaltsdauer', in H. Fassmann, H. Matuschek and E. Menasse (eds) *Abgrenzen, ausgrenzen, aufnehmen. Empirische Befunde zu Fremdenfeindlichkeit und Integration.* Klagenfurt: Drava Verlag.

Fassmann, H., Hintermann, C., Kohlbacher, J. and Reeger, U. (1999) '*Arbeitsmarkt Mitteleuropa'. Die Rückkehr historischer Migrationsmuster.* ISR-Forschungsbericht, 18. Vienna: ISR.

Friedrichs, J. (1990) *Methoden empirischer Sozialforschung.* 14th edn. Opladen: Westdeutscher Verlag.

Fronek, H. (2000) 'Illegalisierung in Österreich', *SWS-Rundschau*, 40 (1): 89–99.

Fuchs, B. (1997) 'Ethnischer Kapitalismus. Ökonomie der Südasiaten in Wien'. Dissertation, Grund- und Integrativwissenschaftlichen Fakultät, Universität Wien.

Fuchs, B. (1998a) 'Indo-Pakistanische Lebensmittelgeschäfte. Ethnische Strategien in der Ökonomie', *ÖZV*, 52 (101): 433–46.

Fuchs, B. (1998b) *Freundlich lächelnde Litfaßsäulen. Zeitungskolporteure. Typisierung und Realität.* Veröffentlichungen des Instituts für Volkskunde der Universität Wien, 12.

Gurak, D. and Caces, F. (1992) 'Migration, networks and the shaping of migration systems', in M.M. Kritz, L.L. Lim and H. Zlotnik (eds) *International Migration Systems. A Global Approach.* Oxford: Clarendon Press, 151–76.

Haberfellner, R. and Böse, M. (1999) 'Ethnische Ökonomien'. Integration versus Segregation im Kontext der Selbständigkeit von Migranten, in H. Fassmann, H. Matuschek und E. Menasse (eds) *Abgrenzen, ausgrenzen, aufnehmen. Empirische Befunde zu Fremdenfeindlichkeit und Integration.* Klagenfurt: Drava, 75–94.

Hillmann, F. (2003) 'Rotation light? Oder: Wie die ausländische Bevölkerung in den bundesdeutschen Arbeitsmarkt integriert ist', *Sozialer Fortschritt*, 5 (June): 140–51.

Hintermann, C. (1995) *InderInnen in Wien. Eine neue MigrantInnengruppe zwischen Integration, Assimilation und Marginalität.* Diplomarbeit, Grund- und Integrativwissenschaftlichen Fakultät, Universität Wien.

Hintermann, C. (1997) 'InderInnen in Wien. Zur Rekonstruktion der Zuwanderung einer "exotischen" MigrantInnengruppe', in H. Häußermann and I. Oswald (eds) *Zuwanderung und Stadtentwicklung.* Leviathan Sonderband 17. Opladen and Wiesbaden: Westdeutsche Verlagsgesellschaft, 192–212.

Hintermann, C. (2000) 'Die "neue" Zuwanderung nach Österreich. Eine Analyse der Entwicklungen', *SWS-Rundschau*, 40 (1): 5–23.

John, M. and Lichtblau, A. (1990) *Schmelztiegel Wien – einst und jetzt. Zur Geschichte und Gegenwart von Zuwanderung und Minderheiten.* Vienna: Böhlau.

Lamnek, S. (1989) *Qualitative Sozialforschung.* II *Methoden und Techniken.* Munich: Psychologie Verlags Union.

Lebhart, G. and Münz, R. (1999) *Migration und Fremdenfeindlichkeit. Fakten, Meinungen und Einstellungen zu internationaler Migration, ausländischer Bevölkerung und staatlicher Ausländerpolitik in Österreich.* Schriften des Instituts für Demographie, 13. Vienna: Österreichische Akademie der Wissenschaften.

MacDonald, J.S. and MacDonald, L.D. (1969) 'Chain migration, ethnic neighbourhood formation and social networks', *Milbank Memorial Fund Quarterly*, 62 (1): 82–97.

Sassen, S. (2000) 'Dienstleistungsökonomien und die Beschäftigung von MigrantInnen in Städten', in K.M. Schmals (ed.) *Migration und Stadt. Entwicklungen, Defizite, Potentiale.* Opladen: Leske & Budrich, 87–113.

Schnetzer, G. (1994) 'Motive einer Emigration. Zur Konstruktion von Lebens-geschichten Chinesischer Auswanderer in österreichischen China-Restaurants'. Dissertation, Universität Wien.

Stanek, E. (1985) *Verfolgt, verjagt, vertrieben. Flüchtlinge in Österreich.* Vienna: Europa.

Vertovec, S. (2000) 'Transnational networks and skilled labour migration'. Paper presented at the Ladenburger Diskurs Migration, 14–15 February.

Wilpert, C. and Morokvasic, M. (1983) *Bedingungen und Folgen internationaler Migra-tion.* Soziologische Forschungen, Berlin: Technische Universität.

5 Riders on the storm

Vietnamese in Germany's two migration systems

Felicitas Hillmann

Today Asian migrants form the numerically most important non-European immigrant group in Germany. Furthermore, Asian migration shows a continuous increase in numbers since the early 1990s. Nowadays large communities and individuals, out of them an estimated 40,000 persons of Vietnamese origin, live in Germany without regular legal status and add to the officially registered Asian population. Added to this, there are substantial regional differences – while Asian migrants constitute the most important immigrant group in some eastern *Länder* of Germany, there are few in the western *Länder*.

For various reasons, this immigrant population remains largely invisible to the majority of the population. Sometimes echoes of that invisibility create rumours in the media. In the mid-1990s the newspapers confronted us with extreme stories about Thai prostitution and trafficking, Vietnamese mafia-style cigarette smuggling, and, as a result of those criminal activities, murders among the Vietnamese. Some five years later, Indian IT workers were presented in the media as 'good' immigrants, coming in to help the country to fill the gap in this part of the labour market. As part of the Information Age, highly qualified IT workers were highly sought after. In consequence there has been an exhaustive debate in Germany on Indian IT workers: they were thought to be the pioneers in the German debate about highly qualified immigration, the so-called 'green card' debate. In February 2003 less than 3,000 'green cards' were given to Indian IT workers, representing 25 per cent of all permits in this field. In addition to this, the Japanese community of highly qualified workers in Düsseldorf is of certain interest to the media, but it has never attracted public attention and has existed peacefully for many years.

To date, the discourse on the migrant population in Germany has mostly been centred on the Turkish immigrant population. This is due mainly to the overwhelming presence of Turks among the registered foreign population in Germany: 26 per cent of all registered immigrants in Germany in 2002 stem from Turkey. Since the 1990s the position of the Turkish descendant minority within the total of the immigrant population has become even more stable than in the past. But – and this is one of the underlying assumptions of this chapter – this concentration is part of the perception of the migrant population in the scientific literature as a 'Western story'. Scientific attention is paid mostly to the migrants who were part of the (Western) migration system (former 'guest workers' from Turkey, the former Yugoslavia, Italy, Greece and Spain in the western parts of Germany). The contract worker programme of the East seldom enters into scientific knowledge. Asian immigrants often stem from such contract worker programmes.

Asian migrants – and this is the second underlying thesis in this chapter – represent also a more complex migration pattern that is at first sight tricky to investigate. Compared with other immigrant groups, researchers would have to use much more qualitative methodology to gain a valid scientific impression. Quantitative data are sparse and give only limited insight into the current labour market integration of Asian migration. New, more flexible forms of labour market integration characterize the Asians in Germany much more than traditional immigrant groups – as shown in this chapter.[1]

By far the largest national group within Asian immigration are the Vietnamese, who came to Germany as boat people (into the former Federal Republic) or as contract workers (into the former Democratic Republic). The data of the federal statistical office indicate that by December 2002 about 902,000 persons of Asian origin lived in Germany. Vietnamese (87,207 persons) and Chinese (72,094 persons) head the ranking of that immigrant population from South East Asia. In the five eastern *Länder* Vietnamese immigrants represent the most important immigrant group. About 10,000 Vietnamese officially live in Berlin, joined by a substantial share of so-called illegal Vietnamese who are concentrated in a few districts in eastern Berlin. This chapter focuses on the Vietnamese in Berlin as it is this town where the heritage of the two migration systems of post-war Germany (FRG and GDR) can be observed best. This chapter elaborates basically two theses:

1 The division of Germany into an eastern and a western migration system is still of importance for the labour market situation of these immigrant groups and its embeddedness into the local labour markets differs between the two (former) parts of Germany. Most of the literature overlooks the importance of such regional differentiations.

2 Economic marginalization has led to various forms of precarious self-employment and largely makes use of globalized forms of labour organization. In contrast to other immigrant groups, new modes of transnational labour organization are found as well as traditional self-employment. In the case of the Vietnamese the problem of economic integration has been resolved through self-employment, often provoking harsh forms of social disintegration for the second generation.

For this chapter both qualitative and quantitative recent labour market data are presented. Since there is still little empirical evidence on this group up to now the chapter quotes selected interviewed privileged observers and Vietnamese interviewed during the writing of this chapter. The assumptions made here are due to the very limited knowledge of this group, are of a provisional nature and may be restricted to the very special situation of Berlin.

The structure of the chapter is as follows. The following section highlights facts and figures concerning the position of Asian and Vietnamese immigrants in Germany and reflects their immigration history in the two migration systems. The focus then is on the labour market integration of the Vietnamese in reunified Germany. The last main section scrutinizes more specifically the current labour market integration of the Vietnamese immigrants and in particular their activities as entrepreneurs in Berlin.

Asian migration in Germany today

In 2001 Germany hosted about 7.3 million officially registered foreigners, corresponding to 9 per cent of the total population of eighty-two million inhabitants.[2] Germany is by now in quantitative terms the most important immigrant country in Europe, followed by France (3.2 million foreigners), Switzerland (1.3 million foreigners) and Italy (1.2 million foreigners). Migrants from Asia constitute in Germany about 12 per cent of all foreigners (2002, December). They represent the biggest non-European immigrant group in the country.

Within the country the numerically biggest immigrant group are Turks, with about two million persons, followed by immigrants from Italy (610,000 persons) and by immigrants from former Yugoslavia (591,500 persons). Today, immigrants from the states with which the West German government had guest-worker contracts constitute the vast majority of all immigrants in Germany. Four-fifths of all immigrants in Germany stem from Europe, more than half of all immigrants come from EU countries and about a quarter stem from non-EU Europe.[3] Data indicate that one in ten immigrants in Germany stems from Asia (see Table 5.1).

The Federal Statistical Office adopts a very broad definition of what is classified as 'Asian' (see Table 5.2). *De facto* a substantial proportion of the immigration that is counted as 'Asian' falls under 'Arabian states' or is part of the states in transition after the end of Soviet domination. The most relevant group from (South East) Asia are the Vietnamese, followed by Chinese and Thai immigrants. Gender proportions indicate that female migrants dominate in four South East Asian immigrant groups: the Thai (86 per cent female), Filipinas (78 per cent female), Koreans (55 per cent female) as well as Japanese (54.3 per cent female). Proportionally low shares of women are found among Indians (30 per cent female) and Chinese (43.5 per cent). Gender proportions among the Vietnamese turn out to be balanced. The share of Asian migrations has been rising since the middle of the 1990s: in the year 1996 about 150,000 Asians fewer than in 2002 were registered.

Table 5.1 Provenance of the foreign population in Germany, December 2002

Provenance	Total No.	As % of foreign population
	7,335,592	100.0
EU Europe	1,862,066	25.4
Non-EU Europe[a]	3,954,677	54.0
Asia	901,706	12.3
Africa	308,238	4.2
USA and Canada	126,103	1.7
Latin America	97,764	1.3
Australia, Oceania	11,853	0.16
Without identification	55,982	0.76
Stateless	17,203	0.23

Source: Federal Statistical Office, June 2003.

Note
a Turks among this group : 1,912,169 persons.

Table 5.2 Provenance of Asian migrants in Germany, December 2002 (groups with more than 20,000 immigrants)

Country of origin	Total No.	As % of total Asian migration	Gender % (here as % of women in the immigrant group)
Total Asian	901,706	100.0	–
Sending countries with more than 20,000 immigrants in Germany:	–	85.6	–
Iran, Islamic Republic	88,711	9.8	41.7
Vietnam	87,207	9.6	47.0
Iraq	83,299	9.2	32.8
China	72,094	8.0	43.5
Afghanistan	69,016	7.6	44.6
Kazakhstan	53,551	6.0	54.4
Lebanon	47,827	5.3	33.0
Thailand	45,457	5.0	86.4
Sri Lanka	43,634	4.8	44.7
India	41,246	4.5	29.8
Pakistan	34,937	3.8	54.3
Japan	34,689	3.8	41.3
Syria, Arabic Republic	28,679	3.1	41.3
Philippines	23,496	2.6	78.1
Korean Republic	23,292	2.5	55.8

Source: Federal Statistical Office, June 2003.

For our purposes the Vietnamese population is of special interest. Compared with its European neighbours, Germany hosts a very large group of Vietnamese; only the United States with 303,700 Vietnamese immigrants hosts more. Table 5.3. shows that the number of Vietnamese immigrants has varied over the past twenty-five years. After Germany's reunification in 1991 (1990) the number of Vietnamese in Germany nearly doubled from 46,000 to 97,000 in only four years. From 1994 on there has been a continuous decrease in numbers. In only five years more than 10,000 Vietnamese people vanished from statistical representation. There are three possible explanations for this phenomenon. A huge share of the Vietnamese population may have become illegal since their permission to stay was not extended and they decided to stay without any legal permission on German territory. Second, they may have left the country for good by making use of one of the reintegration programmes of the federal government (see below). Or, third, they were naturalized by adopting German citizenship. Interestingly, the sex ratio has varied over the past ten years. The official data show a decrease concerning the total of the community, but an increase concerning the share of women in the total population. One possible explanation for this trend is that Vietnamese women more than Vietnamese men try to legalize their stay in Germany.

The regional distribution of the total of immigrants on Germany territory is generally still sharply biased. The smallest concentrations of immigrants are found in the former East Germany. No urban agglomeration there

Table 5.3 Vietnamese immigrants in Germany in absolute numbers, 1977–2002, by sex

Year	Total	Of whom women	Gender ratio (%)
1977[a]	2,604	n.a.	n.a.
1979[a]	5,751	n.a.	n.a.
1981[a]	17,606	n.a.	n.a.
1983[a]	24,611	n.a.	n.a.
1985[a]	29,551	n.a.	n.a.
1987[a]	27,168	n.a.	n.a.
1989[a]	33,381	n.a.	n.a.
1991	78,139	n.a.	n.a.
1993	95,659	38,325	40.0
1995	96,032	39,540	41.0
1997	87,928	37,866	43.0
1999	85,362	38,105	44.6
2001	85,910	39,559	46.0
2002	87,207	40,989	47.0

Source: Federal Statistical Office, June 2003.

Note
a Data refer to the former Federal Republic.

exceeds an average of about 3 per cent immigrants among the total population – the average is 2.2 per cent for all eastern *Länder* (Mecklenburg-West Pomerania, Saxony, Saxony-Anhalt, Brandenburg, Thuringia). In contrast, the urban agglomerations in the western part of the country show an average of 15 per cent foreigners. Some cities like Frankfurt on Main reach shares of nearly 30 per cent immigrants within the total population. Most big cities like Berlin, Hamburg, Munich and the Ruhr region have long-standing and important immigrant communities. The most important immigrant group in the western part of Germany is the Turks, constituting with more than two million immigrants 28.8 per cent of the total immigrant population. Forty-six per cent of them are women (Hillmann, 2003). This regionally biased pattern also holds true in the case of the Vietnamese immigrants. As shown in Figure 5.1 the highest proportion of Vietnamese among the population is found in the capital city, Berlin. In terms of total numbers Bavaria turns out to be the *Land* with most Vietnamese in Germany. When looking for the predominant immigrant group within the former eastern *Länder*, Saxony-Anhalt names Vietnamese immigrants as its most important group.

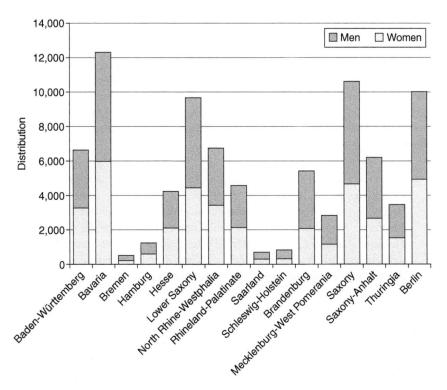

Figure 5.1 Distribution of the Vietnamese population in Germany among the *Länder*, absolute numbers, December 2002 (source: Federal Statistical Office, June 2003).

Two 'German perspectives' on immigration: Vietnamese immigration in the West and in the East

As Table 5.3 shows, before the reunification of the two German states about 33,000 Vietnamese immigrants lived in West Germany. They came as refugees from 1979 on in the context of family reunification. Before that year only around 2,000 Vietnamese people (plus about 4,000 naturalized Vietnamese people) lived in Germany – many of them came for political reasons. Often they made part up of the boat people and came under the Geneva Convention. Many of the boat people were families with small children. The Vietnamese refugees were dispersed by the authorities over the whole territory of the FRG (mostly to North Rhine-Westphalia, Baden-Württemberg, Lower Saxony, Bavaria and Hesse).

They received social benefits and had the possibility to participate in language courses. They had official assistance in their labour market integration as well. They were helped to integrate in all branches of the labour market, with some emphasis on the metal industry. Since they came as political refugees in times of great economic prosperity in Germany and in the middle of the Cold War rhetoric, there was consensus about their desirable integration into western German society. Therefore, Vietnamese immigrants in western Germany were not, as was the case with the majority of immigrants, part of recruitment policies, but they enjoyed a special status that eased integration. In Berlin most Vietnamese immigrants found work in large firms in the field of production and assembling. And another rather forgotten element contributed to their rapid integration into the new environment. One of the responsible managers of the Vietnamese Association of Berlin-Brandenburg (Vereinigung der Vietnamesen in Berlin-Brandenburg) explains the following fact:

> There is a big difference between the Vietnamese that came to the western parts of Berlin and those who went to the eastern parts of Berlin. All such boat people who came at their own risk decided to leave Vietnam. They knew that they could not go back since the government did not guarantee their safety. Those migrants were sure that they would not come back. Everybody who had a little money could come via the sea route. Most of them cut the ties they had with Vietnam.

Of course, the psychological attitude to a migration project makes a difference. Adding to the eased institutional integration, the forced interruption to the country of origin may have had an impact on the speed of integration.

Completely different was the history of the Vietnamese living in the former GDR and joining the reunified Germany in the early 1990s. The eastern German recruitment policies started only in 1973 – when the western German system was already stopping its active recruitment policies.

The workers came to the GDR on the basis of bilateral collective contracts and were chosen by state organizations in Vietnam. They did not come, as did the workers in the western part of the country, because of an individual decision, which then was channelled by an institutional setting – as was the case in the western migration system. In the beginning the political rationale behind the recruitment of the foreign workers in the GDR was not a mere economic project, but thought to 'heighten the economic efficiency in the socialist countries community and to integrate especially those countries of the Third World which were positive about socialist ideas' (Trong, 1998). The integration of immigrants from certain countries into the work process in the GDR was considered to be an important aspect of development aid to the countries of origin. In a second phase the GDR sought to augment the shrinking working population through migration. In quantitative terms the recruitment of workers from East European countries like Bulgaria, Hungary and Poland was most important in the beginning, but since the mid-1980s Vietnam and Mozambique have become the main source countries in the migration policies of the GDR. Enterprises to which the foreign workers (*Vertragsarbeitnehmer*) were assigned had to take care of them. The government and the foreign consulates controlled the implementation of isolation policies as well. There was no possibility to get an individual working permit and forced rotation among the workers was strictly followed. After four or five years the foreign workers were sent back to their country if there was no extension of their working contract. As mentioned before, in the first phase of labour recruitment the qualification of the foreign employees was one key argument for the strict rotation principle. The working contracts could be cancelled because of any violation of the socialist working discipline, in the case of accidents, or because of illness or pregnancy (Krebs, 1999). In a second phase, in the mid-1980s, the supply of consumption goods in the GDR deteriorated. The government tried to heighten the outcome of the production by integrating even more foreign workers into the production process. Between 1985 and 1988 qualifications were no longer at the centre of the immigration policy. A mere unskilled work force was needed to maintain a certain standard of production. One of the supervisors of the Vietnamese working in a state-owned textile conglomerate remembers in June 2003:

> My task was to supervise the incoming Vietnamese. I worked for a state-owned Kombinat [conglomerate] for textile production and it was my job to look after a group of workers coming from abroad. They arrived overnight by plane. I had to supervise 365 Vietnamese workers. I remember them coming along in their flip-flops in the snow. We were told to bring them to a restaurant to make them eat; all of them vomited after the meal. Neither they nor we had any information about the respective cultural habits. Since there was nothing really prepared to host them, one of the dwellings that had recently been finished was

'rented', in order to give them shelter. There was no infrastructure. I had to accompany them to the physician, to supervise their training. Eight persons were divided into apartments with four bedrooms. Every immigrant had the right to receive one chair, one table, one plate, one blanket and a set of cutlery.

In that particular group 90 per cent of the immigrants were women, aged from seventeen to fifty-five. Most of them were between the ages of thirty-two to forty and they had very different educational levels. Some of them were extremely well qualified as textile workers, since they had already worked in that field for many years in Vietnam, others held academic degrees. They were trained for six weeks on the job. Many came direct from the border with South Vietnam, Cambodia and were without an occupation once the war was over. Some had left small children at home.

High unemployment rates in the country of origin at that time made emigration an attractive strategy for many. Once established in the former GDR the Vietnamese workers were paid the same income as their German counterparts: 400 (East German) marks monthly (Spennemann, 1997). And, as our expert, formerly a contract worker in the GDR and now working as manager of a Vietnamese association, puts it:

Most of the contract workers had a better education than the boat people. Often they were engineers or held other degrees. They were emotionally bound to their country of origin and always had the option of going back one day. Why, so many thought, should they learn the difficult German language if they had to go back in five years anyway? They frequently sent money and goods back to their relatives in Vietnam and they were proud of being selected to work abroad.

The social integration of the immigrants into eastern German society was neither favoured by the government nor allowed by the country of origin. Vietnam, like other sending countries, expected the immigrant employees to retain their net income in order to send large amounts of it back to the country of origin. The remitting of 12 per cent of the income earned in the GDR was obligatory for the support of infrastructure measures in Vietnam. It was understood that certain economic problems might be resolved in that way and that the expatriate workers should contribute to the development of their nation (Trong, 1998). Immigrants soon realized that it made little sense to send money back home. Too much was then lost because of inflation. Vietnamese immigrants started to establish sort of a parallel economy in the former GDR (see below), setting the frame for what later became the prerequisites for their self-employment. The immigrants succeeded in sending back home per year the maximum allowed by the GDR authorities,[4] mostly sewing machines, bicycles as well as sugar, clothes for children and soap.

The Vietnamese immigrants in the GDR were hosted exclusively in accommodation provided by the enterprise. The entitled living space of 5 m^2 was often not provided. There was no individual decision about where and with whom the immigrants wanted to live. Outside their accommodation the Vietnamese immigrants were confronted with a xenophobic society that often became violent against them. The Vietnamese embassy on the other hand did not encourage contact between Vietnamese immigrants and the surrounding society. Again, the supervisor cited above remembers:

> At first there were no problems with the host society. On the contrary, there was a certain curiosity on the German side. Problems arose when the Vietnamese immigrants started to selectively buy all the food in the surrounding shops – basically chicken, cabbage, rice, meat. East Berlin, the capital city of the GDR, always had sufficient food resources but in the rest of the country the situation of the food supply was already diffi- cult. Towards the end, it was also the case in Berlin that every purchase of food ate up a lot of time because everybody had to queue up for hours and hours to satisfy their basic needs. At that point it became a problem that the Vietnamese were able to purchase all the goods and nothing was left for the German population.

According to the existing literature, most contract workers were of a young age (90 per cent of them were under the age of thirty) and their stay in the GDR meant being away from home for the first time (or away from their small children). There was neither the possibility of family reunification nor the opportunity to establish a family of their own in the GDR. Female workers who became pregnant during their stay in the GDR were sent back immediately. There were frequent controls in their residential homes and physicians were exempted from confidentiality.

The regional distribution of the Vietnamese immigrants over the terri- tory of the GDR followed the industrial pattern of the socialist state (see Table 5.4). Most Vietnamese immigrants lived in industrial agglomerations like Karl-Marx-Stadt (now Chemnitz), Dresden, Erfurt, East Berlin and Leipzig (Vorsatz, 1995).

Apart from the cultural adaptation the Vietnamese workers undertook (erecting sanctuaries in the residential homes and celebrating their ancestors in a Buddhist manner), there was also a very particular form of economic integration. Many of the Vietnamese immigrants ran their own businesses in the country of origin. While cutting, sewing and preparing clothes, as part of their jobs in the textile industry, some Vietnamese started to transfer some of the material to their homes. Here they designed and fabricated their own collection of clothes and sold it to the local population. As our privi- leged testimony puts it:

> Some Vietnamese developed a parallel economy. They bought some sewing machines and were able to run up fashionable clothes on their

Table 5.4 Flows and stocks of Vietnamese immigrants in the German Democratic
Republic, 1980–89

Year	Inflow	Stock population
1980	1,540	2,482
1981	2,700	5,168
1982	4,420	9,905
1983	150	10,335
1984	330	10,040
1985	0	8,650
1986	0	7,334
1987	20,446	26,001
1988	30,552	53,197
1989	8,688	59,053
1990	48	n.a.

Source: Secretary of the Ministry for Employment and Wages, 1990. Cited in Spennenmann
(1997), also see Krebs (1999).

own. With the money that came out of that business they succeeded in
sending even more goods back to Vietnam. As far as I know, the
common dream to build a house on their own was fulfilled for the
majority of the Vietnamese. They were so clever in the production of
fashionable clothes that they succeeded in imitating stonewashed
trousers!

Even if the GDR authorities tried again and again to suppress such informal
activities, the Vietnamese kept their informal but organized parallel
economy. All experts confirmed that Vietnamese contract workers organized
themselves in a way.

Labour market integration and social insertion of the Vietnamese immigrants in the reunified Germany (after 1990)

When, after the Velvet Revolution, the Berlin wall came down in 1989 and
German reunification took place at the political level in 1990, the whole
setting for the Vietnamese immigrants changed. In the western parts of
Germany former (South) Vietnamese boat people were totally integrated
into the host society. The situation of the (mostly North) Vietnamese contract workers (*Vertragsarbeiter*) in the former eastern *Länder* turned out to be
much more difficult. There were economic, social and, most important, legal
reasons for the growing marginalization of the Vietnamese immigrants. The
years 1990 to 1995 paved the ground for their current situation. The overall
migrant situation during this time was confusing, to say the least. On the
one hand, Vietnamese were encouraged to go back home to Vietnam, but

Vietnam did not take back all of them, while many entered into a bilateral reintegration programme.

On the other hand, Vietnamese in the eastern parts of the country tried to get a foot in the door of the western *Länder*, but they were often not accepted by their co-nationals in the west and faced difficulties with the West German authorities. In addition, Vietnamese from Eastern Europe (Russia, the Czech Republic) came to Germany and made use of the unclear situation, seeking to open up economic opportunities for themselves. Vietnamese immigrants became – in times of transition and change – like riders on the storm.

Under the GDR regime Vietnamese contract workers had not been treated as individuals and they were likely to have avoided any contact with the institutional infrastructure of the host society they were living in. In theory, the contracts they had with their employers were unlimited. A new regulation allowed the termination of employment. As many as 70 per cent of the Vietnamese immigrants lost their jobs after reunification. With the loss of work arose another problem for most Vietnamese immigrants. They were confronted with increasing prices on the housing market. And they were confronted, for the first time, with social realities they had never learned to deal with (Gemende, 1999; Hirschberger, 1997). Very few Vietnamese workers learnt the German language and even fewer of them had even a vague idea about the institutional structure during these times of transition from the planned economy to a Western-style market economy. Since many Vietnamese were not able to survive on the little money they earned in their jobs after 1990, they had to combine different sources of income: all sorts of economic activities and social transfers like unemployment benefits, housing and child benefits.

Vietnamese workers who lost their jobs could decide if they preferred to go back to Vietnam with severance pay if they stayed in Germany until their contract ended. The amount of the severance pay (DM 3,000) was very high compared with what a Vietnamese worker could obtain by their own work. High expectations on the part of the family at home and the need for cash (instead of goods) led to an exodus of Vietnamese. An estimated 19,500 out of 60,000 Vietnamese left the country for good (Hirschberger, 1997). AGEF, an official mediator for reintegration to Vietnam, estimates that nationwide about 12,000 Vietnamese went back to their home country in the decade of the 1990s. There they were inserted into vocational training in about seventy projects in Vietnam. About 4,000 Vietnamese got financial help to set up their own business and one German bank gave loans with favourable conditions. The official side of the programme claims that it was a success and that many problems were resolved. Two-thirds of the former contract workers went back, as one Vietnamese social worker estimates. Though he admits:

> What happened then was that the contract workers who went back first wrote to their colleagues remaining in Germany. 'Don't come back!

There is too much trouble here, too much corruption. You will pay for any step until you succeed in opening up an business. Other Vietnamese will betray you. You are no longer adapted to that mentality.' Of course 3,000 German marks was a lot of money. The normal monthly income in Vietnam during this time was ten German marks. As far as I can judge, most Vietnamese regret that they went back home.

The non-existent social integration into the surrounding society in the eastern *Länder* also led in many cases to an open xenophobic reaction on the German side: many Vietnamese stopped using public transport and had to fear racist attacks when in public or searching for housing. Xenophobic action against Vietnamese-looking people (so-called 'Fidschis') became outrageous during the 1990s. In 1995 half of Vietnamese immigrants in the former East stated that they had had contact with hostile natives and that they had met with discrimination in housing and employment matters. Most immigrants said that this was nothing new for them, but that the intensity of aggression grew (Baumann, 2000).

Legal changes added to this already difficult social and economic insertion and had an enormous impact. Vietnam and Mozambique accepted the end of the bilateral contracts and many immigrants in the former GDR had to accept an unclear legal status (Gruner-Domic, 1999). A substantial number of Vietnamese had political reasons for not going back to their country and sought refugee status in Germany. From 1990 to 1993 about 8,000 to 12,000 former GDR Vietnamese workers (*Vertragsarbeiter*) asked for asylum in Germany. As many as 99 per cent of the applications were rejected. The German authorities tolerated immigrants (*Duldung*) who could not go back to Vietnam (because Vietnam did not accept them) and stopped expatriating them for a determined time. In the early 1990s the legal situation of the Vietnamese contract worker was unclear, as the new foreigners law did not take account of the immigrant groups of the former GDR (Hirschberger, 1997). During this time many Vietnamese became involved in criminal activities, mostly cigarette smuggling and dealing. Some individuals did not understand that this was illegal; others realized that they had to leave the country in any case and that there was little chance to stay. So why not make profits out of illegal deals? As our interviewee tells during the expert interview:

Many Vietnamese immigrants thought that the Germans were stupid. The money was on the streets. You just had to take advantage of the situation. They did not care about the host society. They wanted to make money and had to go back to Vietnam anyway.

All interview partners agreed that up to today cigarette smuggling constitutes one possible survival strategy that is still attractive to some Vietnamese. Often women are involved in this business. And in most cases the cigarette vendors do not hold a residence permit. They do not care about doing illegal things since they have no chance to integrate into the society at all. The peak of the smuggling business was from 1991 to 1994.

After 1993 only those Vietnamese immigrants who could prove they had work which allowed them to maintain themselves and their family were accepted legally. One possibility, often the only possible way because of the closed labour market, was to become self-employed. The German authorities allowed the Vietnamese immigrants to open up their own shop once they could prove that they could support themselves: normally social transfers and a 10 per cent income of their own were sufficient. As a result of that regulation many Vietnamese started up their own businesses, mostly in the food and grocery sectors.

A 1997 study of the Vietnamese immigrant community in Berlin, Brandenburg and Saxony showed that, of 172 interviewees, 145 declared themselves to be self-employed (according to 84 per cent of the total). All others worked in factories; nine of the interviewees were unemployed. The self-employed as represented in this study were concentrated in the textile sector, in the grocery trade, in snack bars and in restaurants. Many of them were in debt but managed to sustain the family and often also succeeded in sending money back to relatives in Vietnam. Most interviewees reported working some twelve to fourteen hours every day plus the time needed to buy the goods. They never go on vacation (Trong, 1998).

In another representative study of the Vietnamese community in Germany, the Vietnamese immigrant community in the mid-1990s shows similar trends of labour market integration. In the early 1990s only one-fifth of interviewees were employed as dependent workers, 7 per cent were in working schemes of the German state. Fifty-five per cent of the interviewed Vietnamese reported being self-employed. On average the income per month was about DM 1,500. This is much lower than the earnings of other self-employed immigrant groups (Bundesministerium für Arbeit und Sozialforschung, 1996).

The tendencies towards marginalization illustrated above became more pronounced in the second half of the 1990s. The next paragraph highlights the situation of the Vietnamese in Berlin with special emphasis on their self-employment. Self-employment turns out to be one element for survival in a difficult local labour market, but at the same moment it seems to be the trigger for growing marginalization and, later on, intergenerational conflicts. The case study of Berlin presented below elaborates this thesis.

Vietnamese immigrants in Berlin

Vietnamese immigrants constitute the most numerous Asian immigrant group in Berlin. According to the official data on the resident foreign population, 56,361 immigrants in total from Asian countries were registered as residents in Berlin in 1999, constituting 17 per cent of the total foreign resident population in Berlin. Out of them, 8,368 were Vietnamese immigrants. The age structure of the Vietnamese community in Berlin is dominated by people between the ages of fifteen and forty-five (63 per cent) and children up to fifteen years old (25 per cent). Only a small share of 1.8 per cent are older than sixty-five. In Berlin the two migration systems came together. The split of the Vietnamese immigrants into a boat people community and into a community of former contract workers can here be observed in a unique way.[6] Even after ten years of coexistence and freedom to move about the city two Vietnamese communities continue to exist. About twenty associations and institutions concentrating on Vietnamese immigrants are at work in Berlin. All experts in the western and eastern parts of the city agreed that there is nearly no exchange between the networks of Vietnamese in West and in East Berlin. Some said that there is a clear-cut division between the Vietnamese in the western part of the city, who mainly stem from South Vietnam, and the Vietnamese in the eastern part of the city, who mainly come from North Vietnam. Mr Nguyen,[7] an experienced social worker in (formerly western) Berlin says:

> Immediately after the wall came down there was a strong interest in that other 'half' of the Vietnamese community. But soon old political cleavages arose. While the former contract workers were never in opposition to the Vietnamese government, the former refugees were often characterized by anti-government sentiment – this led to harsh confrontations in the overall community. Later the involvement of many former contract workers in illicit trading and cigarette smuggling repelled many 'western' Vietnamese.

Two trends emerged clearly from the fieldwork. First, there is extreme concentration of Vietnamese in some parts of the former eastern part of the town, forming more and more a real parallel society. Second, self-employment became the strategy for avoiding unemployment in both parts of the town – with remarkable differences in (formerly) two systems.

First, the distribution of the Vietnamese among the territory of the 3.2 million metropolis is fairly unbalanced to date: in 1996 only 30 per cent of all Vietnamese people lived in the districts of the former West Berlin while 70 per cent lived in the districts of the former East Berlin. Hohenschönhausen is the district with the highest concentration of Vietnamese immigrants. Here many live in run-down dwellings in socialist style, about 10 km away from the core of the city. There is a considerable share of illegal immigrants among the Vietnamese living here, as all experts agree. Vietnamese tend to live where they work, so all activities are restricted to a certain territory. One can find houses lacking windows and with broken doors and foyers where Vietnamese live. A real subculture developed here, fed by always new arrivals of Vietnamese without a permit to stay. Mafiosi-style cigarette smuggling has provoked various cases of murder in the past

(*TAZ*, 27 November 2002), and experts see the danger of a coming generation of Vietnamese drug abusers and pushers. Since there is no hope for illegal Vietnamese to regulate their status (except marriage), problems in this part of the town are expected to grow in the future. While the official statistic counts about 10,000 Vietnamese in Berlin, the estimates by the police department are twice and even triple that number.

Unemployment is a growing problem among the Vietnamese community in all of Berlin. In the year 2000 more than 1,000 Vietnamese people were registered as unemployed (see Table 5.5). Forty-three per cent of the registered unemployed Vietnamese were women. Most of the unemployed Vietnamese persons were between the ages of thirty to forty years. Eighty per cent did not hold a professional degree, the rest held a vocational training degree and, rarely, a university degree. Unemployment (among Vietnamese) was high, especially for those formerly employed in the metal industry, the textile industry, the food sector, in trade, bureau and other services – in branches where the highest rate of self-employment is also found (Landesarbeitsamt Employment Office for Berlin-Brandenburg, 2001). As mentioned above, self-employment constituted for many Vietnamese immigrants in the former eastern parts of Germany the only way out of a closed labour market, and was necessary to obtain a permit to stay. While this trend towards self-employment has been common for many years among the former contract workers, it is also now occurring in the western part of the town. Confronted with problems of unemployment, and unable to find a new job on the Berlin labour market, western Vietnamese immigrants try to establish their own enterprise in the retail food, flower and grocery sectors. Vietnamese in the east already have established Vietnamese whole store centres and may also deliver to the Western market. In the year 2003 we found widespread activities of Vietnamese in most parts of Berlin: groceries and snack bars, but also flower shops, are the main fields of activity. In the south of Prenzlauer Berg, the fashionable student and tourist district, Vietnamese activities concentrate in the Wins *Viertel*. In an area of $0.6\,\text{km}^2$ to $1\,\text{km}^2$ three restaurants, nine grocery stores, four snack bars and four flower shops were run by Vietnamese immigrants. While in the inner districts this kind of business prevails, in the outskirts trade centres have been established. In the mid-1990s

Table 5.5 Registered unemployed resident population, Vietnamese nationality, 1998–2000

Year	1998	1999	2000
Total	974	1,047	1,057
Men	561	587	600
Women	413	460	457

Source: Landesarbeitsamt (Employment Office for Berlin–Brandenburg) Berlin–Brandenburg 2001, Sonderauswertung (special analysis).

a big Vietnamese trade centre was opened up in Rhinstrasse. More than fifty traders with about 100 employees, four translation bureaus and a communal consultancy office were located in the centre. This centre moved its location, but still exists in 2003. According to journalists it is nicknamed 'Little Hanoi' and some politicians see it as a seedbed for further economic activity with Asia (Leinkauf, 2003). Vietnamese small businesses are based on the integration and the exploitation of all potential work forces (like relatives, especially wives and children). Only in rare cases do more than four or five persons work together. Some authors describe the inner structures of this type of small business as comparable to the illegal organization of the cigarette dealers: networking and informal agreements are prominent features of the business (Liepe, 1998). Many of the traders rent market stalls at the weekly markets in the town. In the early 1990s they often bought their goods at the more established Turkish market places. Recent impressions of privileged observers indicate that even if the economic climate in Berlin is tense, only few Vietnamese shut up shop. They complain, but up to now they have not given up. Also data on the registration and cancellation of enterprises by nationality confirm those two trends: Vietnamese are – following the Turkish community – very likely to set up their own enterprises and they have a high turn-over in terms of cancellations (see Figure 5.2). Only in a few cases do the Vietnamese entrepreneurs have a good knowledge of the German language, so their possibilities to enter the formal labour market are very low (Verein für Gegenseitigkeit, 1997).

In both parts of the town problems with the second generation seem to arise. Some descendants of the boat people speak fluent German and have absorbed many German values. These children are much better integrated into German society than those of the other immigrant groups. While their parents had to battle with many social problems and with low status within the native society, the young Vietnamese orientate themselves towards the more individualistically structured German society. The conflict is not so much between the young Vietnamese and the surrounding society as intergenerational (Baumann, 2000). In many cases, so experts explain, they do not speak the Vietnamese language, since their parents themselves tried to speak German to them in the hope of easing integration. In the eastern part of Berlin such intergenerational problems are even more pronounced. The children are often no longer able to communicate with their parents and, owing to their long working hours in their shops, parents have no time to devote to the education of their children. Especially young Vietnamese who came as asylum seekers after reunification did not succeed in integrating. All social centres in eastern Berlin predict that they will see big problems with that group in the years ahead, culminating already now in growing drug abuse. Social centres in western Berlin report similar problems, but the biographical stories they tell during the interviews seem to be far from the dramatic eastern case.

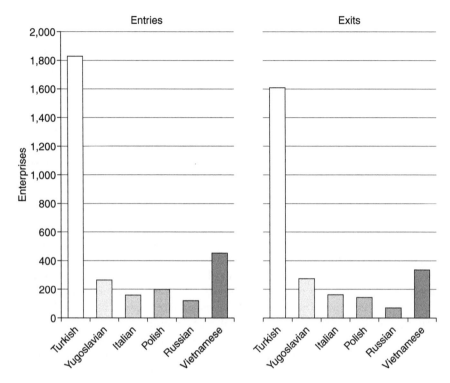

Figure 5.2 Registered enterprises in Berlin, by nationality, 2002.

Conclusion

Even if there is empirical evidence of an enormous expansion of Vietnamese small businesses during the 1990s, especially in Berlin, a systematic and gendered research is still missing. At present most studies look at the performance of the immigrant groups of the guest-worker system. There is further empirical evidence that there is a persistent cleavage of the Vietnamese community in the west and in the east that stems from the days before reunification. Self-employment plays a crucial role for this immigrant group and served first as a forced legal survival strategy for the Vietnamese contract workers in the east.

It becomes clear that the Vietnamese immigrant community was marginalized in its legal status first and was forced into self-employment in order to obtain permission to stay later. Already in GDR days, Vietnamese contract workers established a parallel, in some respects ethnic, economy that flourished and led to a heightening of remittances and allowed a higher living standard in Berlin. Today we find own business centres and wholesale as well as small shop owners especially in the eastern parts of the town.

Even if the Vietnamese people in the western part of Germany who came as refugees in the early 1980s were well integrated throughout the years, the harsh economic crisis in the local labour market has forced them into self-employment starting about two years ago. Compared with the other immigrants groups in Germany, self-employed Vietnamese earn less and have even more difficulty finding a job in the (Berlin) labour market than other immigrant groups. Besides this economic and partly social exclusion of the Vietnamese immigrants there is the phenomenon of spatial exclusion: In parts of Marzahn and Hohenschönhausen substantial subcultures have developed, often of an illegal, sometimes criminal character. Here massive social problems are to be expected for the coming years, or are already part of the story. Given that those areas of the town are distant from the centre, there is very little public attention to the deteriorating social and housing trends.

This chapter also shows how important the perspective of the migrant in the individual migration project is. Germany's hesitation in deciding which legal status the former contract worker should be assigned had a strong impact not only on the immigrants but also on the personal situation. When there is no stimulus to play by the rules of the host society because there is no clear regulation of the permission to stay, there is easily criminal behaviour by the immigrants. As one social worker points out, 'If immediately after the Wall came down or after a year later there had been clear regulations, all that cigarette smuggling would not have become as prominent.' Self-employment turned out to be a sound strategy to cope with harsh labour market conditions and with vacant clear political regulation. The ability to enter self-employment had been experienced in the past and could be adapted to the new environment. The tendency of Vietnamese entrepreneurs to take over grocery stores from Turks and to invest into flower shops in the western parts of Berlin, that means to expand spatially from the east of the town to the west, has been confirmed in a study of Vietnamese ethnic business (Schweizer, 2004). In the western case, the rapid political acceptance of the boat people, symbols of the Cold War, eased integration into the host society. What helped further was the idea of not going back home. Like riders on the storm the Vietnamese immigrants were blown in one direction or the other.

Notes

1 The research results presented in this chapter are part of ongoing research by the author into the migrant integration and local labour market restructuration in the process of globalization. The data on the Vietnamese were collected during the years 2002 and 2003 through seminars, excursions and fieldwork.
2 If not indicated otherwise all data presented in this section stem from the Federal Statistical Office and represent the latest available data on foreigners in Germany.
3 The following countries are classified as non-EU European countries by the Federal Statistical Office: Albania, Andorra, Bosnia and Herzegovina, Bulgaria,

Estonia, Iceland, Yugoslavia, Croatia, Latvia, Liechtenstein, Lithuania, Malta, Macedonia, Moldavia, Monaco, Norway, Poland, Rumania, the Russian Federation, San Marino, Switzerland, Slovakia, the Czech Republic, former Czechoslovakia, former Soviet Union, Turkey, Hungary, Ukraine, City of the Vatican, Belarus, Cyprus.

4 Vietnamese workers were allowed to send abroad up to six packets per year by mail. There was also the possibility to send a consignment of 0.5 tonnes, sized 1 m³, per year.

5 For the natives the entitled living space was $12\,m^2$.

6 Berlin's story of Vietnamese immigration is the miniature of the national story: The very first Vietnamese immigrants came to West Berlin in the early 1970s. They were often students from South Vietnam and were sent to Berlin to study. (In 1974 140 Vietnamese people were registered in West Berlin.) At the end of the 1970s many boat people found shelter in West Berlin. In the eastern part of the city immigrants from North Vietnam were recruited for work programmes (Ausländerbeauftragte des Senats, 1997).

7 All names have been changed.

References

Baumann, M. (2000) *Migration – Religion – Integration: Buddhistische Vietnamesen und hinduistische Tamilen in Deutschland.* Marburg: Diagonal.

Bundesministerium für Arbeit und Sozialforschung (1996) *Situation der ausländischen Arbeitnehmer und ihrer Familienangehörigen in der Bundesrepublik Deutschland.* Forschungsbericht 263: Sigma/FES, Bonn.

Gemende, M. (1999) 'Vietnamesinnen in Dresden', in Marion Gemende and Bildungswerk Weiterdenken (eds) *Migrantinnen in Dresden.* Frankfurt am Main: Sülberg, 39–160.

Gruner-Domic, S. (1999) 'Beschäftigung statt Ausbildung. Ausbildung, Ausländische Arbeiter und Arbeiterinnen in der DDR 1961–1989', in Jan Motte *et al.* (eds) *Fünfzig Jahre Bundesrepublik. Fünfzig Jahre Einwanderung.* Frankfurt am Main: Campus, 215–37.

Hillmann, F. (2000) 'Ausländische Familien und ihre Integration in den formellen und informellen Arbeitsmarkt', in *Expertisen zum sechsten Familienbericht. I Familien ausländischer Herkunft in Deutschland.* Empirische Beiträge zur Familienentwicklung und Akkulturation. Opladen: Leske & Budrich.

Hillmann, F. (2003) 'Rotation light? Oder: Wie die ausländische Bevölkerung in den bundesdeutschen Arbeitsmarkt integriert ist', in *Sozialer Fortschritt*, 5 (June): 140–51.

Hirschberger, M. (1997) 'Zwischen Ausweisung und Duldung', in Tamara Hentschel *et al.* (eds) *Zweimal angekommen und doch nicht zu Hause.* Berlin: 21–43.

Isoplan (ed.) (2000) *Maßnahmen des Bundesministeriums für Arbeit und Sozialordnung zur sozialen und beruflichen Integration von türkischen Arbeitnehmern und ihren Familienangehörigen.* Saarbrücken: Isoplan.

Krebs, A. (1999) 'Daheimgeblieben in der Fremde. Vietnamesische VertragsarbeiterInnen zwischen sozialisitischer Anwerbung und marktwirtschaftlicher Abschiebung'. Master's thesis, Berlin: Fachhochschule für Sozialarbeit und Sozialpädagogik.

Landesarbeitsamt Berlin-Brandenburg (1999) Der Arbeitsmarkt im Landesarbeitsamtbezirk, Referat Statistik, Statistische Mitteilungen, Berlin.

Landesarbeitsamt Berlin-Brandenburg (2000) Bestand an Arbeitslosen nach Staatsangehörigkeiten im Berichtszeitraum April 2000. Sonderauswertung.

Landesarbeitsamt Berlin-Brandenburg (2001) Sonderauswertung.

Leinkauf, M. (2003) 'Guten Morgen, Vietnam', in *Der Tagesspiegel*, 10 August, 5.

Liepe, L. (1997) 'Die vietnamesische Migrantenökonomie', in Tamara Hentschel, M. Hirschberger, L. Liepe and N. Spennemann (eds) *Zweimal angekommen und doch nicht zu Hause.* Berlin: Reistrommel, 44–52.

Liepe, L. (1998) 'Vietnamesische Migrantenökonomie im Ostteil Berlins', in Renate Amman and Barbara von Neumann-Cosel (eds) *Berlin. Eine Stadt im Zeichen der Migration.* Berlin: Verlag für Wissenschaftliche Publikationen, 126–9.

Mitteilungen der Beauftragten der Bundesregierung für Ausländerfragen (1999a) *Daten und Fakten zur Ausländersituation.* Bonn: Bundesregierung.

Mitteilungen der Beauftragten der Bundesregierung für Ausländerfragen (1999b) *Migrationsbericht 1999. Zu- und Abwanderung nach und aus Deutschland.* Berlin and Bonn: BfA.

Mitteilungen der Beauftragten der Bundesregierung für Ausländerfragen (ed.) (1999c) *Migrationsbericht 1999.* Bonn: BfA.

Nonnemann, Thuy (1997) *Vietnamesen in Berlin.* Berlin: Ausländerbeauftragte des Senats, senatsverwaltung für Gesundheit, Soz`iales und Verbraucherschutz.

Sachverständigenkommission der Bundesregierung (2000) Sechster Familienbericht. Familien ausländischer Herkunft in Deutschland. Leistungen – Belastungen – Herausforderungen, 14. Wahlperiode, Drucksache 14.

Schweizer, Julia (2004) 'The Vietnamese Ethnic Economy in Berlin'. Master's thesis, Department of Geography, Humbodlt-Universität Berlin.

Schmalz-Jacobsen, C. and Hansen, G. (1995) *Ethnische Minderheiten in der Bundesrepublik Deutschland.* Munich: Beck.

Spennemann, N. (1997) 'Aufbauhelfer für eine bessere Zukunft. Die vietnamesischen Vertragsarbeiter in der DDR', in Tamara Hentschel, M. Hirschberger, L. Liepe and N. Spennemann (eds) *Zweimal angekommen und doch nicht zu Hause.* Berlin: Reistrommel, 8–20.

TAZ Die (2002) 'Vietnamesen vor Gericht', 27 November, 24.

Trong, Ph. D. (1998) 'Vietnamesische Vertragsarbeitnehmer in der Bundesrepublik Deutschland. Probleme der Integration. Wege und Programme der Reintegration in der SRV'. Dissertation, Humboldt-Universität Berlin.

Verein für Gegenseitigkeit (ed.) (1997) *Stadterneuerung und ethnische Unternehmer.* Berlin: Parabolis.

Vorsatz, R. (1995) 'Die vietnamesische Minderheit', in Cornelia Schmalz-Jacobsen and Georg Hansen (eds) *Ethnische Minderheiten in der Bundesrepublik Deutschland.* Munich: Beck, 535–45.

6 Labour migration into Poland

The case of the Vietnamese community

Krystyna Iglicka

Labour migration in Central and Eastern Europe prior to 1989 and after: some general remarks[1]

During the communist era labour migration in the Central and Eastern European countries (CEE) was regulated by means of a system of intergovernmental agreements and business contracts. The entry of foreign workers was based on the planned economy system and on the provision of international aid to the other countries of the former socialist bloc (IOM/ICMPD, 1999). However, the effectiveness of state-enforced migration control measures was limited to the formal sector of the economy. The crisis of both the economy and the state, which became increasingly visible in the 1980s, led to the mushrooming of informal economic activities. For example, informal trading activities – so-called 'suitcase trading' (Wallace, 1999: 35) – often involved cross-border movements. In the late 1980s, other informal and/or seasonal/temporary activities became more important (Bedzir, 2001).

The political and economic reforms implemented since the symbolic fall of the Iron Curtain amounted to a complete reorientation of economic relations, relinking the former Comecon countries with the West. The incorporation of the previously 'detached' economic space formed (the Council for Mutual Economic Co-operation) by the Central and East European countries (CEEC) and the Soviet Union into the world economy as well as the liberalization of migration controls led to increased mobility of labour, capital and goods. At the same time, the globalization of market economies also brought about structural changes in the Central and Eastern European labour markets and led to the emergence of dual labour markets in Central and Eastern Europe (Iglicka, 2000; Piore, 1983).

Thus, large-scale labour migration to the Central and Eastern European countries has to be considered a rather recent phenomenon that gathered momentum only in the 1990s. At the same time, labour migration is highly differentiated according to duration, the skills and the origins of migrants. In general, migrants occupy positions in the labour market, which are distinct from those of natives. Migrants are most often found in the private sector, whereas the employment of compatriots dominates in the state sector. In addition, firms employing migrant workers tend to be small, whereas natives most often work in large firms (Wallace, 1999: 38). It appears that migrants are found in a labour market segment whose main characteristic is flexibility, which translates into insecurity, whereas native workers seem to prefer secure though often poorly paid jobs in the state-dominated sector. This 'parallel labour market' for migrants, however, is likely to gradually disappear, or at least be significantly altered, as employment in the private sector becomes more accepted or even a necessity for a larger proportion of the native work force. This is especially true in the Polish case. During 2001–04 the unemployment rate exceeded 15 per cent and topped 18 per cent in 2003. Some fractions of the local labour force have started to accept insecure and risky employment conditions.

Data drawn from various sources suggest that the structure of employment of migrants is marked by the polarization of different and sometimes overlapping categories, namely skilled/unskilled, short-term/long-term work/stay, legal/illegal work/stay (Nesperova, 1997: 292). The result is segmentation of labour markets into distinct segments. At least two different segments can be observed:

1   The labour market for highly skilled migrants forms part of the globalized labour markets most often organized as internal markets of large multinational companies. Highly skilled migrants ('expats') usually come from the United States or the European Union. This type of labour market is special in that it involves the migration of highly skilled young persons who either set up their own businesses (mostly in the tourist industry) or are employed by their compatriots in these businesses. Migrants of this type usually stay for a limited period of time and regard their stay and job as a sort of curriculum upgrading ('life stage migration'). A lot of young Americans and Britons fall into this category (see Wallace *et al.*, 1998: 16; Rudolph and Hillmann, 1998).

2   The second segment is composed of low-skill labour. This is again split into several distinct segments, for example long-term/permanent versus seasonal and temporary workers; workers employed in the formal economy versus workers engaged in the informal sector; legal versus illegal employment, etc. Workers in this category most often find themselves on the lower echelons of the labour market, as is the case, for example with Ukrainian and Romanian workers in the Czech Republic, Hungary and Poland. It is reasonable to locate self-employment within this segment of the labour market, as it is often a strategy for circumventing barriers to formal employment. Migrants engaged in low skilled/manual work most often come from neighbouring countries but also from further abroad, for example from Vietnam.

The segmentation of labour markets is also, to some extent, linked with an ethnicization process, particularly in respect of certain types of activities and certain countries of origin (Iglicka, 2000, 2001). This is most obvious in the case of Romanian and Ukrainian casual, seasonal and/or construction workers (especially in Hungary, the Czech Republic and Poland) as well as in the case of Vietnamese and Chinese traders and their co-national employees (especially in Hungary, Bulgaria, Poland, Romania and Slovakia). Often – as is the case in Poland – such migrants tend to be employed in the informal sector. This ethnicization of certain parts of the labour market has to do with legislative measures to limit labour migration and the important part played by migrants of certain countries as migration channels for their compatriots, usually via clientilistic networks (see Wallace *et al.*, 1998: 38).

The relatively large size of the shadow economy in CEECs as well as inadequate or inefficient labour regulations in this region makes it relatively easy for foreign workers to be employed illegally. According to the Polish Ministry of Labour and Social Policy's estimates there are about 100,000–150,000 foreigners working illegally each year in Poland, accounting for between 12 per cent and 18 per cent of Poland's estimated 'informal work force'[2] (*Informal Labour Market*, 1995). The majority of them came from Vietnam and from the Ukraine and other parts of the former Soviet Union. The men work mainly in the construction industry, in forestry and fruit farming. Women are employed in agriculture but also work as seamstresses, housekeepers and babysitters. Clandestine employment is probably substantial in scale throughout Central and Eastern Europe. Apart from employment in the industries mentioned, a sizeable number of illegally employed migrants work in small business, often established by their compatriots. At the same time a number of migrants may have formal employment and additionally engage in informal activities (Wallace, 1999).

In Poland the largely temporary or pendular movement of migrants from the successor states of the former Soviet Union had a considerable positive effect on some sectors of the economy while it also led to increased competition in labour markets. Citizens of the former Soviet Union (the bulk being Ukrainians) massively engaged in trading activities, thereby fostering local economic growth. Foreign demand for textile and leather products was, for instance, one of the main factors behind the boom in small, private textile and shoe businesses. Trading[3] was significant also in monetary terms, thus contributing to the alleviation of the chronically negative balance of payment (Iglicka and Sword, 1999). It is probable that the same effects of increased economic interaction could be observed in other countries where similar conditions prevailed.

Vietnamese in Poland

As far as the history of the presence of Vietnamese in Poland is concerned, from the beginning of the 1960s until the beginning of the 1990s the immigration flows to Poland were statistically insignificant. Nevertheless, the movement of students from Vietnam, who arrived in Poland under a government-sponsored programme of 'socialist co-operation' or academic exchange, was quite conspicuous. In general, the Vietnamese newcomers were very well received in Polish society. In the 1960s and 1970s in the Polish mass media much attention was focused on the 'American war in Vietnam'. There was newspaper coverage at the time and front-page articles in which Polish society was portrayed side by side with the history of Vietnam as a nation fighting for its freedom. Generally Vietnamese were perceived as victims. Furthermore, the Vietnamese students in Poland were seen as quiet, calm, diligent and honest. After graduation the majority of them returned to their home country, where they gained a higher place in the social hierarchy, due to having a European diploma. Moreover, the years spent in Poland created some strongly built-up network connections.

After 1989, irrespective of still being a communist country, Vietnam started to pursue a more liberal migration policy that caused new inflows to Poland. These included not only students, most of them from privileged families, but also people in search of new employment and place to live. Notwithstanding the fact that since the early 1990s Poland and Vietnam found themselves located in divergent political structures and trends, the general image and stereotypes of the Vietnamese remained similar to those which had developed during the 1970s. This may explain the rapidly growing number of arrivals from Vietnam in the 1990s. At the dawn of a new century the Vietnamese in Poland form one of the biggest Asian groups in the Central and Eastern European region (Table 6.1).

The Vietnamese living in Poland come from different provinces in the north of Vietnam: Ha Noi, Thanh Hoa and Nghe Tinh (to the south of Hanoi), Son La (to the west of the capital), Lang Son (to the east), Hai Duong, Nam Dinh (very close to Hanoi in the south-east), Ha Tay (next to Ha Noi in the south) and Ha Bac (Grzymala-Kazlowska, 2002).

As the data from Table 6.1 prove it is very difficult to estimate a proper number of Vietnamese in Poland. Different estimations of the Vietnamese community vary seriously. At the beginning of the 1990s many Vietnamese migrants came to Poland as 'tourists' but remained here and started to work either illegally or legally. Since late 1993 Vietnamese citizens applying for work-permit visas have increased notably and this has contributed to a rise

Table 6.1 Chinese and Vietnamese migrants in CEECs (resident population), 2000

Country	Bulgaria	Czech Republic	Hungary	Poland	Romania	Slovakia
Chinese	1,891	3,551	8,861	n.a.	5,231	n.a.
Vietnamese	n.a.	23,556	2,447	20,000– 100,000	137	1,258

Source: Kraler and Iglicka (2002: 41).

in the number of Vietnamese entering on a legal basis. Until 1996 Vietnamese were the second largest group (after Ukrainians) as far as the number of those granted visas with work permit is concerned (see Table 6.2). Between 1997 and 1999 the Vietnamese took first place. Since 2000 they have been in fourth place (after Ukrainians, Russians and Belorussians).

Strategies used by some of the clandestine migrants in an attempt to legalize their stay included applying for a permanent residence permit and through marriage with a person legally residing in Poland. This is especially visible in the case of Vietnamese women married to Polish men; their number grew from fifteen in the year 1995 to 310 in 1998, a twentyfold increase. Similarly, the number of Vietnamese men married to local women in Poland increased almost sixfold between 1995 and 1998. This is in contrast to mixed marriages between Poles and other nationalities, which remained relatively stable or decreased (Table 6.3). However, since 1999 a declining trend has been observed in the number of marriages between Poles and citizens of Vietnam. In 2001, 107 such marriages were contracted, which is forty-one fewer than in the previous year. Nevertheless, it is not very likely that in the immediate future the number of such marriages will be comparable with the numbers of the peak years: 1997 and 1998 (see Table 6.3).

As far as applications for a residence permit are concerned, in 1993 the Vietnamese constituted only 4 per cent of the total. At the end of 1995, however, they were the third largest group of immigrants receiving residence permits. Three years later, they already held second place, right after Ukrainians (see Table 6.4). Since 1 January 1998 the former category 'permanent residence permit' has been replaced by two categories: 'permission to settle' and 'fixed-term residence permit'. At present (mid-2003) Vietnamese form one of the biggest elements as far as immigrants receiving

Table 6.2 Visas with work permit granted 1994–98, by most numerous nationalities (%)

Country	1994	1995	1996	1997	1998
Ukraine	13.0	14.0	15.8	15.2	13.1
Vietnam	11.0	13.0	14.6	17.8	15.1
Russia	8.5	7.5	7.6	6.5	5.6
USA	7.0	7.0	6.0	5.3	4.1
China	7.8	7.0	7.8	6.5	6.7
Great Britain	7.0	7.0	6.0	5.0	5.0
Belarus	5.0	3.5	3.0	3.3	4.0
Germany	4.6	5.0	6.0	6.0	5.5
Other	36.1	36.0	33.2	34.4	40.9
Total	100.0	100.0	100.0	100.0	100.0

Source: Poland – statistical data on migration 1994–98, Office for Migration and Refugees, Warsaw, 1999.

Table 6.3 Mixed marriages by most numerous nationalities, 1990–98 (selected years).

Country	1990	1995	1996	1997	1998
Foreign wife					
Ukraine	–	331	340	456	537
Russia	–	119	151	127	142
Belarus	–	95	104	122	124
Lithuania	–	41	40	33	41
Armenia	–	27	28	42	53
Latvia	–	6	10	9	10
Kazakhstan	–	13	11	10	23
USSR	255	–	–	–	–
Vietnam	–	15	42	110	310
Others[a]	656	273	251	257	301
Total	911	920	977	1,166	1,541
Foreign husband					
Ukraine	–	89	108	106	119
Russia	–	51	38	38	46
Belarus	–	18	21	26	35
Lithuania		8	15	15	15
Armenia	–	44	64	75	140
Latvia	–	–	–	–	–
Kazakhstan	–	–	–	–	–
USSR	210	–	–	–	–
Vietnam	–	45	79	152	251
Others[a]	3,119	2,065	1,852	1,794	1,822
Total	3,329	2,320	2,177	2,206	2,428

Source: *Roczniki Statystyczne, GUS* (Statistical Yearbooks, Central Statistical Office), Warsaw, various years.

Note
a Including Western European countries.

either type of permit are concerned. They are in second place as far as permission for settlement and in fourth place as far as foreigners granted fixed-term residence permits. The Vietnamese in Poland are mainly located in the large urban areas, but particularly in the Warsaw conurbation. In big cities with more than 500,000 inhabitants, 60 per cent, and in medium-size cities 71 per cent, of Poles have met Vietnamese in person at least once (Grzymala-Kazlowska, 2002).

Labour market performance

As was mentioned earlier, access to the various economic sectors for foreigners is segmented according to ethnicity. As far as the movement from the

Table 6.4 Foreigners granted the permanent residence permit (PRP) in Poland according to the most numerous nationalities, 1993–98 (%)

Country	1993	1994	1995	1996	1997	1998
Ukraine	15	21	19	22	23	24
Russia	11	12	11	10	8	7
Belarus	7	6	7	7	8	7
Germany	5	5	6	5	4	4
Vietnam	4	4	7	9	8	10
Kazakhstan[a]	1	2	8	8	15	10
Lithuania	3	3	2	3	2	2
Armenia	1	2	2	2	2	5
Total (No.)	1,964	2,457	3,051	2,844	3,973	1,567

Source: Poland, Statistical Data on Migration 1994–98, Office for Migration and Refugees, Warsaw, 1999.

Note
a Ethnic Poles and their family members migrating to Poland on the basis of repatriation resolution issued by Polish government in summer 1996.

'east' is concerned Eastern Europeans prevail as unskilled manual workers and skilled manual workers. Thus even those with work permits so far mainly occupy the secondary sector of the Polish labour market.

Unlike the Eastern Europeans other nationalities from the East, such as the Vietnamese and Chinese, find a job not only in the secondary sector but in the primary markets as well. In many cases this is because they hold executive positions in sectors in which they invest, for example restaurants and trading companies (see Table 6.5). Asian nationals are most often employed by food or by trade companies (see Figure 6.1). The Vietnamese also play a significant role in Polish street markets, especially in Warsaw. In the main

Table 6.5 Work permits granted individually by occupation, Poland, 2000

Origin	Total	Manager	Owner	Expert/ consultant	Teacher	Skilled worker	Unskilled worker
Former Soviet Union[a]	4,479	244	851	1,214	568	894	158
Ukraine	2,927	122	624	951	405	582	117
Asia (Vietnam, China)	1,747	113	846	104	74	472	83
Western Europe	4,560	1,493	1,244	1,087	631	143	30
Other	7,016	1,707	1,361	1,900	844	866	273
Total	20,729	3,679	4,926	5,256	2,522	2,957	661

Source: Polish Ministry of Labour and Social Policy, Warsaw.

Note
a Former Soviet Union denotes Russia, Ukraine, Byelorussia, Armenia.

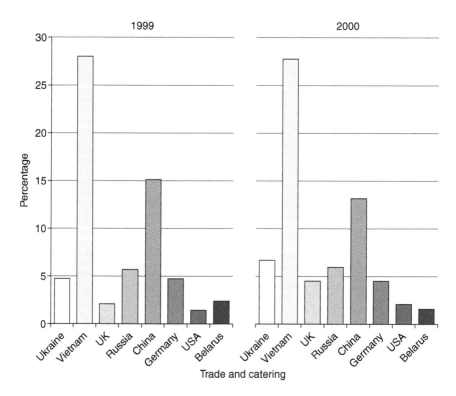

Figure 6.1 Work permits granted individually, by branch of economic activity: main countries of origin.

Warsaw market there are more than 1,200 Vietnamese stalls. Though males probably predominate, according to some estimates one in three or one in four open market traders is constituted by women. Furthermore, according to leaders of the Vietnamese community there are between 300 and 400 Vietnamese fast-food restaurants and thirty to forty big restaurants in the Polish capital targeting both Polish and Vietnamese clientele.

Another rather special case of Asian immigrants in Poland is that of South Koreans. Since 1996, South Koreans are visible among those Asians who hold executive positions. In contrast to Chinese and Vietnamese, however, they are employees of large transnational companies. This is connected with the huge investments made by large South Korean concerns such as Daewoo in Poland. The Vietnamese and (to a lesser extent) Chinese are hired most often in the private firms based on foreign capital. This is in contrast to Ukrainians who are mainly employed by Polish state and private companies (Table 6.6 and Table 6.7).

The 2000 statistics allow us for the first time to take a look at the educational attainment of foreigners granted employment consent. A striking

Table 6.6 Work permits granted individually by ownership of enterprise (eight top countries of origin), 1995 and 2000 (%)

Country	Total	Ownership				
		State	Private Polish	Private foreign	Private mixed	Other
1995						
Total (No.)	10,441	1,000	3,456	3,846	1,554	585
Ukraine	17.0	36.6	34.8	1.3	4.7	20.5
Vietnam	10.3	0.6	3.9	22.8	2.6	3.7
UK	8.5	6.2	9.3	7.4	8.7	13.0
Russia	6.8	11.3	7.6	2.8	11.1	8.2
China	6.5	0.1	1.6	14.4	4.6	0.3
USA	5.5	4.9	3.3	5.8	9.7	6.2
Germany	5.5	7.9	1.4	6.8	10.4	6.5
Belarus	6.4	6.3	14.4	0.8	1.2	10.5
2000						
Total (No.)	12,771	1,190	3,960	4,968	1,980	613
Ukraine	20.2	32.5	40.6	3.4	5.4	29.1
Vietnam	13.8	0.4	4.8	24.2	3.1	2.4
UK	7.7	5.8	2.8	7.1	9.6	6.8
Germany	8.5	9.7	8.3	7.8	12.1	8.4
Russia	6.4	12.3	7.1	3.1	8.4	5.7
China	5.8	0.7	4.8	12.1	2.2	0.4
USA	5.9	7.1	7.9	5.8	8.2	12.9
Belarus	4.5	8.3	10.1	1.3	3.1	6.4

Source: Ministry of Labour and Social Policy, Warsaw, various years.

majority of those persons (59 per cent) have a tertiary school diploma. Only 7 per cent have attended only a basic vocational school and 6 per cent have never attended a secondary school.

Workers who were the citizens of the United States represented the highest educational background. The higher educational attainment category was also characteristic of Ukrainians (54 per cent), Germans (67 per cent), Russians (55 per cent) and Chinese (49 per cent). Secondary education was most typical for Vietnamese (51 per cent), Turkish (44 per cent) and Indian (48 per cent) workers. The majority of Bulgarian immigrant workers had only elementary education (57 per cent) (Table 6.7).

Adaptation and integration

The Vietnamese in Poland form a group that is well organized socially, and unlike other ethnic groups now in the process of formation in Poland, for example Armenians and Ukrainians, has not been involved much in crime or

Table 6.7 Work permits granted to foreign individuals, by educational attainment, Poland, 2000

Country	Educational attainment				
	Total	Post-secondary	Secondary	Vocational	Elementary
Total	17,082	10,031	5,569	1,202	1,000
Ukraine	2,927	1,588	1,070	205	64
Germany	1,336	904	300	129	3
Vietnam	1,230	361	631	223	15
UK	1,218	980	232	3	3
France	1,217	942	248	15	12
Belarus	796	445	270	64	17
Bulgaria	773	33	183	114	443
Russia	756	420	300	16	20
USA	634	555	75	2	2
Turkey	604	143	268	50	143
China	517	252	180	62	23
Italy	402	216	166	15	5
India	370	133	179	23	35

Source: Central Statistical Office, Warsaw, 2000.

abuse. Data on crime committed by foreigners issued yearly by the police do not include Vietnamese.

Results of anthropological studies on this ethnic group show that there is much to indicate that they feel 'at home'. The majority of Vietnamese send their children to Polish schools and learn the language quickly. Data issued by the Ministry of Education show a growing presence of Vietnamese students enrolled in both state and private universities, especially in Warsaw. Those who work carry on lively economic activity and declare a feeling of 'well-being' within the cultural sphere. They generally stress the tolerance of Polish society and the possibility of building up an 'economic existence' in Poland (Halik and Nowicka, 2002). It is especially true in the case of those Vietnamese who are in marital relations with Poles. The leaders of the community are willing to meet Polish journalists, politicians and researchers, and present the advantages of a Vietnamese presence in Poland (such as for example the enrichment of Polish culture or creation of new jobs) (Grzymala-Kazlowska, 2002).

Notwithstanding these facts, however, the Vietnamese still form a fairly close-knit society, which 'externally' adapts itself to the dominant culture. Internally, however, the Vietnamese still maintain their individuality in the ethnic and cultural sense and as far as language use is concerned. According to the results of anthropological studies there are more than fifteen Vietnamese organizations. In both cultural and labour-market performance level (as data show) it is a self-contained ethnic group with a high level of mutual co-operation and self-organization. It is an example of a well organized, thriving ethnic community that has achieved economic success in an almost completely new environment.

Conclusion

Analysing the official data on Vietnamese society in Poland and examining the results of research on their cultural awareness, three scenarios of future development of the Vietnamese community are probable (Halik, 2000). First, in case the majority of Vietnamese immigrants are staying illegally in Poland and are young, poorly educated men with the main aim of earning money, it is likely they will eventually decide to return home or move on to Western Europe. This scenario will mean that in the foreseeable future the Vietnamese presence in Poland will decrease or the numbers arriving will remain stable. Second, the Vietnamese legally residing in Poland will probably consider their stay merely as a 'stop-over', especially those Vietnamese who have permanent residence and with network connections consisting of fellow countrymen residing in Germany or France. Thus in a 'friendly environment' and in 'an atmosphere of tolerance', they can save up money and wait for Polish citizenship in order to subsequently enter another Western European country more easily. This again means stabilization or a slow decrease in numbers. At present, it is impossible however to estimate the possible flow of this delayed transit migration; respondents in the above-mentioned anthropological study (Halik and Nowicka, 2002) did not want to talk about it openly. Third, along with the further thriving of the Polish economy there is also the possibility of an increase in the number of Vietnamese permanently resident, which will gradually lead to their integration into Polish society.

Notes

1 Fragments of this section were published in Kraler and Iglicka (1999).
2 The Central Statistical Office estimates the size of the informal sector to be around 850,000, or 5 per cent of the total Polish work force (SOPEMI Country Report Poland, 1999).
3 Goods are usually bought by traders in Poland and then resold in the Ukraine and other countries.

References

Bedzir, V. (2001) 'Migration from Ukraine to Central and Eastern Europe', in C. Wallace and D. Stola (eds) *Patterns of Migration in Central Europe*. Basingstoke: Palgrave, 277–92.
Central Statistical Office (2000) *Statistical Yearbook*. Warsaw: Central Statistical Office (GUS).
Grzymala-Kazlowska, A. (2002) 'The Formation of Ethnic Representations. The

Vietnamese in Poland', Sussex Migration Working Paper 8. Brighton: Sussex Centre for Migration Research, Sussex University.

Halik, T. (2000) 'Vietnamese in Poland: images from the past, present and future', in F.E.I. Hamilton and K. Iglicka (eds) *From Homogeneity to Multiculturalism. Minorities Old and New in Poland.* School of Slavonic and East European Studies (SSEES) Occasional Papers 45. London: UCL Press, 225–40.

Halik, T. and Nowicka, E. (2002) *Wietnamczycy w Polsce. Integracja czy Izolacja?* (Vietnamese in Poland. Integration or Isolation?). Warsaw: Warsaw University Press.

Iglicka, K. (2000) 'Ethnic division on emerging foreign markets in Poland', *Europe–Asia Studies*, 52 (7): 1237–55.

Iglicka, K. (2001) 'Migration movements from and into Poland in the light of East–West European migration', *International Migration*, 39 (1): 3–33.

Iglicka, K. and Sword, K. (eds) (1999) *The Challenge of East–West Migration for Poland.* London: Macmillan; New York: St Martin's Press.

Informal Labour Market (1995) Warsaw: Central Statistical Office.

IOM/ICMPD International Organization for Migration and International Centre for Migration Policy Development (1999) *Migration in Central and Eastern Europe. 1999 Review.* Geneva: ICMPD.

Kraler, A. and Iglicka, K. (1999) 'Labour migration in Central and Eastern European countries (CEECs)', in F. Laczko, I. Stacher and A. Klekowski von Koppenfels (eds) *New Challenges for Migration Policy in Central and Eastern Europe.* Geneva: International Organization for Migration and International Centre for Migration Policy Development.

Ministry of Labour and Social Policy (various years) Warsaw.

Nesporova, A. (1997) 'Education systems and labour mobility: the particularities of Eastern Europe', in G. Biffl (ed.) *Migration, Free Trade and Regional Integration in Central and Eastern Europe.* OECD/WIFO Schriftenreihe Europa des Bundeskanzleramts. Vienna: Verlag Österreich, 291–4.

Piore, M. (1983) 'Labour market segmentation: to what paradigm does it belong?' *American Economic Review*, 73: 249–53.

Poland (1999) statistical data on migration 1994–98. Warsaw: Office of Migration and Refugees.

Roczniki Statystyczne (Statistical Yearbooks) (various years) Warsaw: Central Statistical Office.

Rudolph, H. and Hillmann, F. (1998) 'The invisible hand needs visible heads: managers, experts and professionals from Western countries in Poland', in Khalid Koser and Helma Lutz (eds) *The New Migration in Europe. Social Constructions and Social Realities.* London and New York: Macmillan, 60–84.

SOPEMI (1999) *Country Report. Poland.* Paris: OECD.

Wallace, C. (1999) *Economic Hardship, Migration, and Survival Strategies in East Central Europe*, Vienna: Institut für Höhere Studien (Institute of Advanced Studies).

Wallace, C., Bedzir, V. and Chmouliar, O. (1998) *Some Characteristics of Labour Migration and the Central European Buffer Zone.* Vienna: Institut für Höhere Studien (Institute of Advanced Studies).

Part II

'Coming in'

Specific incorporation trajectories of
Asian migrants

7 South Asian women and employment in Britain

Diversity and social change

Fauzia Ahmad, Stephen Lissenburgh and Tariq Modood

Background

This study builds on the Policy Studies Institute's Fourth National Survey of Ethnic Minorities[1] in the United Kingdom (conducted in 1994) and other studies that point towards great diversity and polarity in the employment profiles of South Asian women in Britain, especially the markedly different economic activity rates of Pakistani and Bangladeshi women on the one hand and Indian and African Asian women on the other (Modood *et al.*, 1997). A number of explanations have been posited; for instance, Owens's analysis of 1991 census data emphasizes the economic and socio-demographic factors (1994), whilst a series of PSI studies have found that the lower economic activity rates of Bangladeshi and Pakistani women are partly explained by cultural factors (Brown, 1984; Jones, 1993; Modood *et al.*, 1997). These would suggest that Muslim communities espouse gender roles that firmly locate women within the domestic sphere. However, the issues are far more complex than the above studies have shown. Religion cannot be the only factor, since employment patterns in Muslim countries also exhibit great diversity and, as Bruegel (1989) points out, Muslim women from East Africa and India have higher activity rates than other Muslim women (though Modood *et al.*, 1997, found that it was not as high as non-Muslims from the same groups). Similarly, Brah and Shaw (1992), in a valuable qualitative study on South Asian Muslim women and the labour market, found that demands of housework and child care, language and qualifications, family and community pressures and discrimination in the labour market were also relevant factors. In examining differentials in economic activity in Bristol, West and Pilgrim (1995) identified similar patterns and variables affecting economic activity in their research, and also proposed that local economies and male unemployment may have a bearing on levels of labour market participation.

Holdsworth and Dale (1997), in a quantitative analysis using the 1 per cent Household Sample of Anonymized Records (SARs) from the 1991 census to explain variations between ethnic groups in economic activity rates, suggested that it was not child care as such that took Pakistani and

Bangladeshi women out of the labour market but marriage. Findings from their later work (Dale *et al.*, 1999, 2000), based in Oldham, Manchester, looking more specifically into the labour prospects and educational aspirations of young Pakistani and Bangladeshi women, are largely consistent with those produced by cross-tabular analyses found in the Fourth National PSI Survey (Modood *et al.*, 1997) and our London-based study, described here and elsewhere (Ahmad *et al.*, 2003; Lissenburgh *et al.*, forthcoming). Holdsworth and Dale were keen to avoid reductionism in explaining the significance of economic, demographic and cultural factors such as the finding that Pakistani and Bangladeshi women most likely to withdraw from work as soon as they married while Indian women continued to work until the birth of their children, concluding that 'simplistic, monocausal explanations of ethnic differences in women's employment based either on cultural differences or economic necessity are inadequate' (1997: 40).

When looking at figures for those working and the type of work engaged in, the Fourth PSI Survey has shown, on the whole, that a certain degree of social mobility is evident from previous surveys, where minorities, except African Asians and Caribbean women, were overrepresented in manual work.[2] The survey found that both men and women from all the minority groups were considerably less likely than white men to be in the top occupational categories as employers and managers of large organizations. Where they were present as employers, it was largely as a consequence of self-employment, and this was more significant among South Asian groups and the Chinese. Overall, Pakistani and Bangladeshi men were among the most disadvantaged educationally and economically, which, in turn, impacted upon the economic positions of women from these groups in various ways.

Turning to the mode of labour market participation among South Asian women, the Fourth PSI Survey noted a wide range of activities both between and within ethnic groups. Bangladeshi women were most likely to be looking after the home or their families at just over 80 per cent, while this figure represented 70 per cent for Pakistani women, 26 per cent of African Asian and white women, and 36 per cent of Indian women. These figures were naturally reflective of family structures, with most Asian women being married, with the Pakistanis and Bangladeshis more likely to have larger families. The survey noted that gender divisions within the labour market were more profound than ethnic differences apart from those in the higher-level non-manual jobs. Here there were about half the number of women than men in the higher-level, professional, managerial and employer categories, though across ethnic groups the variation was similar to that of men, with about 10 per cent of South Asian women in the top occupational group. There were lower proportions of women in manual work in general, with the exception of jobs such as cleaning, and machine sewing, leaving over half the women from each ethnic group working in intermediate or junior-level non-manual work such as clerical and secretarial work or sales-related jobs. Just as for men, participation in higher levels of professional employment was far more likely to be a feature of self-employment, and this was particularly so for Pakistani and Indian women, but once this factor was accounted for they made up only 6 per cent of this job category, compared with around 10 per cent for African Asians, 15 per cent for white women and 25 per cent for Chinese women. While the proportion of South Asian women in part-time work was around 25 per cent for all occupational categories, 36 per cent of white women worked part-time in non-manual jobs and 55 per cent worked part-time in manual jobs, compared with Caribbean women, who represented around half these totals for both manual and non-manual part-time working. The impact of qualifications showed similar patterns to men, with larger proportions of those with higher-level qualifications less likely to be in top jobs when compared with white women, with the exception of Chinese women with whom they shared similar occupational patterns. The survey data suggest that the higher economic inactivity rates from less qualified Pakistani and Bangladeshi women could be attributed to a preference for domestic roles, whilst for qualified South Asian women in general, poor job prospects were likely to deter them from seeking employment that might be less suitable – a finding supported by Afshar in her work on Pakistani women in Bradford (Afshar, 1989). Discrimination is likely to be a contributory factor in discrepancies at all levels, but in the top occupational category this may also be compounded by 'social–educational' networks associated with private education and prestige universities such as Oxford and Cambridge.

In terms of higher education, a number of sources such as the Universities and Colleges Admissions Service (UCAS) and the Higher Education Statistics Agency (HESA) began to collate data in the early 1990s on the ethnicity of applicants and noted a strong drive for qualifications among ethnic minority groups as a whole. However, their distribution across the university sector has not been consistent, with some evidence of the older universities exhibiting bias against minorities (Modood and Shiner, 1994). This has led others to suggest that ethnic minority women, in particular, suffered from a 'double disadvantage' in terms of university admission (Taylor, 1993). Although the Fourth PSI Survey showed that South Asians as a group had the highest rates of participation in post-compulsory education for the sixteen to twenty-four age group, there still remained a striking degree of polarization between British South Asians with higher qualifications and those with few or none. Indian and African Asian men were the most likely to possess degrees, whilst Pakistani and Bangladeshi men were the least likely. Some other and more recent studies suggest that the participation rates for Pakistani and Bangladeshi women in higher education are on the increase (Gardner and Shukur, 1994; Modood *et al.*, 1997; Modood and Acland, 1998, 2003; Dale *et al.*, 1999). In comparison with Indian and East African Asian women, although Pakistani and Bangladeshi women were among those possessing the least number of qualifications *overall*, Modood *et al.* (1997) found that Pakistani and Bangladeshi women were beginning to be well represented at degree-level study. Qualitative and local studies also pointed to the high aspirations of young South Asian Muslim women and their participation in higher education and professional employment (Basit, 1997; Ahmad, 2001). So, by the end of the decade, despite continuing evidence of bias against South Asians by the old universities, there was a noted increase in higher education admissions among South Asians in general (Modood, 2003).

The purpose of this study then, was to explore this acknowledged diversity and social change in South Asian women's education participation rates and economic profiles and to understand the processes and dynamics involved, especially as understood by South Asian women. It similarly aimed to explore the variables involved that act as barriers in preventing some women from achieving their aspirations. By focusing on the complexity of the lived experiences of Asian women, the study sought to move beyond statistical correlation's and attempted to highlight culturally and religiously sensitive issues affecting South Asian women relevant to sociology as well as policy.

The methodology built upon the analysis of the Fourth Survey in order to achieve sophisticated models of South Asian women's employment and educational profiles and to explore attitudes to employment and study and how these impact on women's lives. An initial phase of quantitative analysis was undertaken where data from the Fourth Survey was revisited, the implications of which were used to inform the qualitative phase of the research in which detailed interviews with a sample of South Asian women from the Fourth Survey were conducted (see Lissenburgh *et al.*, forthcoming). Various themes from these interviews were, in turn, used to suggest areas for further, improved quantitative modelling. By taking the unique step of utilizing qualitative data to inform a further level of quantitative modelling, our study sought to employ a far more rigorous approach to explaining the diversity in South Asian women's levels of employment participation.

Explaining the employment gap

Using a sample extracted from the Fourth Survey of 1,306 South Asian women, the initial phase of quantitative analysis involved building econometric models in order to isolate particular variables that were instrumental in determining South Asian women's employment profiles. The analysis involved the construction of models which explained why women were either working or not. These models were then used to quantify the relative importance of factors and to 'decompose' the employment gap between Indian and East African Asian women on the one hand and Pakistani and Bangladeshi women on the other. These models identified a number of correlations relating to employment choices for South Asian women that further statistical modelling sought to modify on the basis of ethnicity. These results were compared with work by Holdsworth and Dale (1997).

Variables such as age at migration and whether the respondent was born in the United Kingdom were not found to be significant in our study, though they have been important in other studies such as Holdsworth and Dale's (1997). However, educational qualifications and fluency in English were found to play a very important role in explaining the extreme dichotomies existent between the Indians and East African Asians on the one hand and Pakistanis and Bangladeshis on the other, accounting for 20 per cent and 12 per cent of the employment gap respectively. The results of this exercise are shown in Table 7.1. Whilst household and personal characteristics partly explained these differences (about 18 per cent of the gap), in our preliminary analysis, *the most significant variable was found to be religion*. This accounted for just over a quarter (28 per cent) of the employment gap between the two groups. Whilst being Muslim reduces the likelihood of employment for both groups, the fact that 97 per cent of Pakistanis and Bangladeshis are Muslim compared with 15 per cent of Indians and African Asians explains why this single factor makes such a large contribution to the employment gap. While this seems to confirm the importance of cultural factors, an additional factor may be that Pakistani and Bangladeshi women, as Muslims, experience higher levels of hiring discrimination than other South Asian women (see later and Ahmad *et al.*, 2003).

By making use of the wider range of variables available in the Fourth Survey as compared with other data sources, we have been able to show that once variables such as English language fluency are brought into the analysis, factors such as whether the respondent was born in the United Kingdom do not have as much importance as suggested by earlier studies (for example, Holdsworth and Dale, 1997). These findings naturally have a number of implications for the qualitative phase of the research and can be found in Lissenburgh *et al.* (forthcoming). This chapter though, will focus on the qualitative aspects to the study, which was composed of seventy in-depth semi-structured interviews with South Asian women based in London and conducted between 1999 and 2000 and aimed to explore some of the reasons for the diversity found in South Asian women's employment in Britain and to highlight areas of social change.

Table 7.1 Decomposing the employment gap between Indian/African Asian and Pakistani/Bangladeshi women

Factor	Contribution to the employment gap (%)
Qualifications	20
Language ability (English)	12
Household characteristics	16
Other personal characteristics	2
Religion	28
Unexplained	22

The characteristics of women who participated in the qualitative study reported here differed substantially from those of Fourth Survey respondents in a number of ways. As well as being geographically concentrated in London (the home of about a third of the South Asian women interviewed for the Fourth Survey), the qualitative sample were considerably more likely than the Fourth Survey sample to be employed at the time of interview (64 per cent against 34 per cent) and to be graduates (37 per cent against 12 per cent). The qualitative sample were also younger and less likely to have children than the Fourth Survey sample. Therefore, single women and women without children were targeted, as their labour market participation is not affected by childcare issues. We also wanted to ensure that the majority of young women sampled were either born in the United Kingdom or had the majority of their school and socialization in Britain. So our sample was deliberately biased towards young women and resulted in a disproportionate number of graduates. Over half our respondents were in full-time or part-time employment (57 per cent), working for an employer, and, of these, 37 per cent were graduates. A further 7 per cent were self-employed, and all these were graduates who also possessed additional postgraduate qualifications, or were studying for further qualifications. Students studying for first degrees represented 14 per cent of our sample. There was only one woman who was unemployed and seeking employment at the time of the interview after having recently completed a PhD. Women who were not seeking employment comprised 20 per cent of our total sample and were all born outside the United Kingdom in rural or semi-rural areas of the Indian subcontinent.

In terms of religion, just over 50 per cent of our sample came from Muslim backgrounds, just under 40 per cent came from Hindu backgrounds, with 10 per cent coming from Sikh religious backgrounds. For those employed, 23 per cent of the sample came from Muslim backgrounds (3 per cent worked part-time) and of these, 15 per cent were graduates (with 1 per cent working part-time); 27 per cent were from Hindu backgrounds (with 3 per cent working part-time) and of these, 14 per cent were graduates (with 1 per cent working part-time) and 7 per cent of our sample came from Sikh backgrounds, all of whom were graduates working full-time. The only unemployed woman in our sample, as mentioned above, came from a Hindu background. Of those who were self-employed, 4 per cent came from Hindu backgrounds, all of whom were graduates; just over 1 per cent each came from Muslim and Sikh backgrounds, with the latter being graduates. Students represented just over 11 per cent of the Muslim sample and just under 3 per cent of the Hindu; none came from Sikh backgrounds, whilst just over 1 per cent of the Sikh sample were classed as not seeking employment but possessed a first degree. Just under 3 per cent of the Hindu sample were not seeking employment, compared with just under 16 per cent of the Muslim sample, of which just over 1 per cent were graduates. All respondents lived in areas with a concentrated Asian population at the time of interview. Further breakdowns of the sample can be found in Ahmad *et al.* (2003).

Our interviews covered a range of issues, from family histories, detailed personal employment and educational histories, experiences, views and attitudes to work and study, domestic roles and family life, views on marriage, and identity.[3] Whilst the qualitative phase of the research was not large enough to enable a meaningful statistical comparison, our interview data shed light on various relevant aspects, such as the barriers and constraints that prevent some Pakistani and Bangladeshi women from improving their qualifications. Some of these will be elaborated upon below. In order to respect confidentiality, all respondents cited here have had their names changed.

Education and motivations: why choose to study and work?

In order to fully appreciate why some women are economically and educationally active and others are not, and whether attitudes towards education and employment were changing, our sample contained a large proportion of students and graduates (ten and thirty respectively). Key questions then were: what are the advantages and disadvantages of additional study or formal employment? Have attitudes changed and, if so, why? How do these relate to notions of 'traditional' domestic and familial obligations?

The vast majority of women in our interview data, including those not seeking employment themselves, identified the prospect of financial and personal independence and increased social status as significant incentives to continue studying beyond school. Those with experience of higher education spoke in overwhelmingly positive terms about their experiences. Higher education was perceived to bestow benefits such as an increased knowledge base, increased confidence, self-esteem and self-awareness, as well as preparation for future employment. Improved marriage prospects, encouragement and support from the family and positive role models were among other positive factors that were mentioned.

Expectations and the support of parents

The supportive role played by parents in encouraging their daughters to pursue higher education and a career was significant in our interviews. Most respondents came from working-class backgrounds but expressed 'middle-class' aspirations when speaking about higher education and the pursuit of professional careers, as has also been found by other research (for example, Basit, 1997).

> I think we're all influenced by parents, the fundamental influence was that we were going to get further education, we were going to continue to get qualifications and continue in education. That was always the

> overriding thing ... the fundamental nature of the encouragement was
> that they made it such an innate expectation to complete this course
> that you never even question it! Skilful on their part.
>
> (Humaira, Muslim Pakistani, solicitor, thirty-three years of age)

A large number of women also talked about the particular role played by
their *fathers as primary motivators*. Again, this feature is not restricted to only
'professional' families; that women from *all* religious and socio-economic
backgrounds talked about the positive role their fathers played marks a profound shift away from stereotypes about 'overbearing fathers' and signifies a
call for a re-evaluation of static definitions, especially in relation to Muslim
fathers. Other research would also suggest that the roles Muslim fathers play
in motivating their daughters into higher education should not go unnoticed (Ahmad, 2001).

> My Dad certainly is very ambitious for me. He's more ambitious than I
> am! And he is a major force because as far as he's concerned, I can do
> everything – you know, I can train ... and become a lawyer ... he was
> an influence in my politics and my beliefs as well.
>
> (Nadira, Muslim, Bangladeshi, senior research executive, thirty-two
> years of age)

This emphasis on high educational aspirations was, however, described as
'too much pressure' by some women. Ethnic minority teachers were also
viewed as positive role models and were accorded particular respect if they
were perceived as women who had achieved success and professionalism
without compromising their religion or culture. Other positive role models
included female relatives working and studying within the family, such as
older sisters who had 'paved the way', acting as 'pioneers', mothers and
female peers in local and social networks.

'I want you to become a doctor'

The 'stereotype of children becoming doctors' (Humaira, Muslim Pakistani,
solicitor, 33 years of age) was discussed by many women regardless of cultural, religious or educational background. Women spoke of their parents'
overwhelming preference for their children to study what were regarded as
the 'traditional' high-prestige, high-status vocational degrees such as medicine, law, accounting and dentistry. The following sentiment was echoed by
several respondents in our sample and supports other research findings
(Basit, 1997; Dale *et al.*, 1999).

> Its always been highly considered being a doctor – you know, 'My son's
> a doctor' kinda thing – but that's sort of been the first profession a
> father would boast if one of their children became one. I think there's

less hype about dentistry or pharmacists or anything like that, but if one child became a doctor it would be out in the whole of the community, that such and such son is a doctor, kinda thing!

(Vaishaly, Hindu, Indian, senior lecturer, thirty-one years of age).

The professions mentioned here are those that Asian communities recognize and are familiar with; for parents who were not able to offer their daughters careers advice, the high-profile professions may have been the main careers to remain uppermost in parents' minds. This was found regardless of parental education levels or class and is supported by other research (for example, Afshar, 1989; Brah and Shaw, 1992; Brah, 1993; Basit, 1997 and Ahmad, 2001).

'Standing on my own two feet'

Another feature to emerge from our interviews that is sometimes associated with South Asian families is the differential level and nature of encouragement that can exist between men and women (cf. Taher, 2000). One of the possible reasons why men in some families may be more actively encouraged, supported and even pressured into higher education could lie in their perceived future 'breadwinner role', and again, the manifestation of this expectation varied from family to family. However, very few women actually gave concrete examples from within their own families. There was considerable diversity and complexity in the nature and degree of this 'gendered encouragement' which is elaborated upon elsewhere (Ahmad *et al.*, 2003).

However, the importance placed on not 'depending on a man' was voiced by both younger and older women. Economic independence, or at least economic activity, was further viewed as a source of increased autonomy and social status within the household. It gave women (both working full-time and those working outside formal economies) an identity that was not circumscribed solely by their domestic roles and afforded them the opportunity to forge a set of relationships that were distinct from family networks. This prospect of 'financial independence' was viewed as advantageous by women from all religious and socio-economic backgrounds in our sample and was most commonly expressed by our respondents in terms of 'being able to stand on my own two feet'. By encouraging daughters to seek a certain degree of financial independence through education and a career, parents were able to 'rest assured' that their daughters were at least financially secure and independent, especially in the event of unforeseen circumstances, such as divorce, widowhood or continued single status.

Improved marriage prospects

For some women and their families, a positive consequence to higher education and economic activity was improved marriage prospects and greater choices in issues of marriage. These choices could range from freedom to marry the partner of one's choice, articulated as 'love marriages', to the prospect of stipulating requirements and conditions in marriages that were 'arranged'. The most frequently cited expectation was one where women envisaged marrying a partner of equal, if not higher, economic status. The following is representative of this view:

> If she was educated … she can sort of be a bit fussy and choosy as to who she wants to marry and doesn't have to say 'yes' to the first available person … once you're educated and that, you become a bit more knowledgeable as to who you want to marry and who you don't.
> (Aswa, Muslim, Bangladeshi, community safety worker, twenty-seven years of age)

There was some concern expressed, though, that higher education may result in women becoming 'too educated' and 'too old' for marriage. For example:

> I am twenty-three and my mother already thinks I should be married and have a child. I think I am too young.
> (Sangeeta, Hindu, Indian-Pakistani, research adviser, twenty-three years of age)

Younger women in professional employment seemed, however, able to reach a satisfactory balance between themselves and their parents on the question of marriage. Contrary to Bhopal (1997, 1998), our data do not suggest that employment and higher education are associated with the 'rejection' of arranged marriages or that marriage results in withdrawal from the employment market. Our data suggest that women are negotiating alternative approaches to marriage that remain sensitive to familial values and obligations. (For further discussion see Ahmad, forthcoming.)

Attitudes to employment

Virtually all women across religious, ethnic, employment and age divides agreed that employment and educational opportunities should be available to all women *if* they want to pursue these avenues. This 'if' and the importance of personal choice are the crucial point of juncture between those women who spoke about choosing to stay at home and those who chose to work.

> I think it is a very good, very positive thing. I think women have to gain control of their lives rather than depending on men. I think women needing financial security is very important.
> (Mumtaz, Muslim, Pakistani doctor, forty-one years of age)

Respondents not seeking employment also voiced positive opinions about paid employment outside the domestic realm, though many prefaced their comments by stating the primacy of the domestic role.

> Well, I think women have every right to work if they choose to work . . . If they are not married, and they have no children, then they can work as they choose. There's no hindrance there, it's to do with choice, really.
> (Hamida, Muslim, African Asian, housewife, thirty-six years of age)

Our interviews also reveal a marked *gendering of obligations* expressed by women from diverse educational, ethnic and religious backgrounds, though this may well become increasingly dependent upon individual experience (Ahmad *et al.*, 2003).

Employment preferences: good jobs/bad jobs

Employment preferences largely followed study choices, described earlier, with a definite preference for high-prestige occupations. Teaching and community work were also highly regarded by most in our sample. Muslim women further identified careers on the basis of social acceptability on the grounds of religion and 'suitability'. For example:

> My dad would never allow me or be happy with me to work behind the bar because it involves working with alcohol and for the same reasons I wouldn't work there.
> (Khalida, Muslim, Bangladeshi, civil servant, twenty-three years of age)

Some women, as we have already indicated, suggested that a gendered division of employment *should* exist, based on concern with 'female honour'. Examples of compromising professions or 'bad jobs' were those that were considered 'masculine' such as manual work (working on a construction site, lorry driver, dustman), or jobs where certain uniforms would contravene accepted norms of respectability, for example, jobs with short dresses, or jobs where few or no clothes were required such as lifeguard, modelling or stripper. Attitudes towards men and careers were expressed along similar lines in terms of the high-prestige professions, but also included other lucrative careers such as engineering and computing, again stressing responsibility to future financial obligations within the family.

Tensions in employment and study

Community pressure and gossip: 'success in a Prada bag'

Jobs that involved international travel and living away from home were a source of tension for some women and their families. An example was cited of one successful architect in her late twenties who designed catwalks for fashion shows around the world. Despite her success, her parents' concern revolved around her single status whilst travelling around the world unchaperoned, which was also a source of local gossip. The respondent expressed some frustration with this attitude:

> I mean, she comes to weddings and looks fabulous and she's got a Prada bag and she's got really expensive shoes, but our people don't recognize things like that; *they don't see success in a Prada bag* [laughs], they think, 'Oh God, she's travelling around the world.'
> (Nadira, Muslim, Bangladeshi, senior research executive, thirty-two years of age).

This quotation highlights the complexities associated with the status of 'professional', where the acquisition of certain 'success symbols' such as designer bags, as in the example cited above, may not be accepted as a universal sign of success.

Another cause for gossip was the question of working wives and mothers in general. For some women with young children (such as Bangladeshi women) who sought to continue their working lives, there was a cost in terms of personal credibility and loss of 'face' in their communities, even if they chose to rely on the assistance of close relatives for child care. Gossip has therefore emerged as a constraining feature for many women in our sample, acting to organize, control or interfere with the aspirations and everyday lives of South Asian women sometimes in extreme and overt ways. This was especially so in closely knit communities such as the Bangladeshis in Tower Hamlets.

Juggling study, work and domestic obligations

Many women spoke of expectations and obligations of domestic and caring responsibilities, most of which are discussed in some detail under 'Barriers to employment'. Women from all backgrounds agreed that institutions did not do enough to support women with families. For some the pressure to continue working in order to support and contribute to the family finances was a source of tension, especially if child care was an issue.

In many cases women were happy to continue working from choice but expressed some resentment over the lack of support they received, both institutionally and privately, when attempting to manage the multiple responsibilities of work and home. For some women, this resulted in a considerable amount of personal sacrifice:

> If you are willing to go to the top, there are a lot of sacrifices that you will probably have to make with regard to – you might not be able to see your family, your children, and things like that. You might be at

meetings, your house might be probably suffering, going to the top, so it depends on what your priorities are.

(Shailu, Hindu, African Asian, careers adviser, forty years of age)

Women who were self-employed

A small number of our respondents – four – were self-employed and were graduates. None of the women in our sample worked in a 'family business'. Obviously, it is difficult to draw any substantive conclusions from their very specific work experiences, but some general themes to emerge as reasons for self-employment were issues of time management, especially in order to study further, as two were reading for postgraduate degrees; financial needs; and the ability to pursue an area of work that was personally fulfilling and remain in control of their working hours. It is interesting to note that the women here had *chosen* self-employment as a response to *being employed*. Studies such as those conducted by Dhaliwal (2000) or Hardill and Raghuram (1998) dealing specifically with Asian women's entrepreneurial activity shed more light on these issues whilst more detail on women's working experiences, including those who were self-employed, can be found in Ahmad (forthcoming).

Barriers/alternatives to employment and study

Caring responsibilities

These consisted of caring either for children still at school and/or elderly relatives such as a parents or in-laws. If we examine these responsibilities in detail, we are able to offer some explanations for these women's absence from the labour market, though we need to bear in mind that *personal choice* is another important pervading factor and is apparent in many responses. The point is also made by Hakim (1995). An opinion that was often expressed was the belief that the role of career to children and or elderly relatives was more important than the financial gains that could be accrued from formal employment.

The burden of housework

Housework emerged as a significant activity for most women regardless of background. There was much variety in the levels of housework done by men in families, but men were generally reported to do the least amount of housework. Many working mothers in our sample spoke of the limited amount of free time that was available to them, but many women not seeking employment were also constrained by the lack of spare time they were able to allocate for themselves. Why should this be the case? The Fourth Survey noted that many Bangladeshi and Pakistani families lacked various domestic electrical appliances and goods such as dishwashers and food processors. The present study did not explore this aspect any further but was able to offer some insights into why women from low-income families often spoke of having no free time or time to look for employment or study. Many housewives in our sample emphasized the preparation of fresh food on a daily basis – and given the nature of South Asian cooking, preparation times can be fairly lengthy. The relative absence of Asian convenience pre-prepared foods may also be relevant, though for some older women they may not be viewed as an attractive alternative.

However, many younger women working in professional careers had managed to negotiate the sharing of domestic chores with their partners. Students living at home were able to successfully negotiate a reduction in household chores especially when study deadlines and exams were close by.

Early marriage and early pregnancy

Our qualitative data suggest that early marriage and the early birth of children may be key factors in delaying or preventing labour market entry and pursuing further qualifications. Although early marrying ages appear to be more common among those from the older age groups, early marriages were not uncommon for some younger women, notably Bangladeshis, who had received most of their secondary schooling in the United Kingdom. Women in our sample who were in their early forties also had young children in infant or primary schools and their domestic and childcare responsibilities were therefore prioritized.

> Well, I got married, didn't I, and I had my children and then I came to this country. When I was in Bangladesh, my mum was still alive and, do you understand, in Bangladesh, when girls get older, their education is stopped.
>
> (Qudsia, translated, Muslim, Bangladeshi, housewife, twenty-six years of age)

For communities with a noted low level of female economic participation, such as Bangladeshis and Pakistanis, childcare and domestic obligations were likely to begin early in adult life. There is some evidence from the interview data to suggest that younger UK-schooled women were able to negotiate entry into the labour market once their children were of school age. Further research would be needed to ascertain if working motherhood was becoming a significant trend.

Lack of qualifications and fluency in English and other personal constraints

Fluency in English has obviously had an impact on entry into paid employment and was cited by women as a major obstacle to their entry into formal work arenas. Some women also cited it as a source of difficulty when needing to speak to their children's teachers. Although most expressed ambitions to learn or improve upon their English, various circumstances such as domestic responsibilities and caring for children or relatives made attendance at classes difficult. Women here appear to be caught in a cycle of barriers where time to pursue necessary qualifications is taken up by domestic responsibilities and obligations that revolve around childcare. Others, however, did not feel inhibited in their everyday activities by their lack of fluency in English.

A small number of women believed that education had 'corrupting' and 'Westernizing' influences. This set of opinions, if not challenged, may account for why significant numbers of young women are withdrawn from school at the earliest opportunity, and why some parents seek early marriages for their daughters. Religious concerns were cited by a minority of Muslim women, but also acted to constrain the ambitions of other women in areas that were deemed 'unsuitable', such as a Hindu Brahmin woman in our sample who was prevented from pursuing a nursing career by her family. A few women also spoke of constraints from their husbands. Concern over the potential loss of benefits, ill health and lack of work experience were also cited as contributory factors in preventing women from seeking employment.

Domesticity as an alternative to paid employment

Not all women viewed economic participation as necessary to *their* lives. Many housewives expressed a positive sense of pride in their domestic accomplishments and associated any paid employment on their part as an insult to their husband's role as 'breadwinner'. In addition, the status of a 'housewife' was positively perceived and generally associated with the 'gentrified' classes of the subcontinent. Non-employment was also articulated by some as a lack of 'need'.

> Well, we would have [learnt English], had we felt the need. If we felt it was necessary to study further, we would have tried to study further here. I have maintained the level of knowledge that is necessary in my lifestyle.
> (Firdoz, translated, Muslim Pakistani, housewife, forty years of age)

Employers – should they be doing more?

Throughout our interviews, a number of issues were raised on the role employers, colleges and universities could play in attracting more South Asian women into the labour market and on experiences of discrimination in the employment market. *Racial and sexual discrimination* was cited by a number of respondents despite the existence of Equal Opportunities policies in the workplace. These were not always found to be satisfactory in dealing with certain experiences and were found to be difficult to implement. Furthermore, the nature of discrimination described in our interviews was often diffuse and was not limited to 'white organizations' but cited by a small minority who worked for Asian organizations. Whilst some challenged discrimination head-on, others channelled their energies elsewhere.

Many women noted how, despite being well qualified, and in some cases over-qualified, they were still not being called to job interviews in comparison with their white peers, or were being overlooked for promotion (see also Runneymede Trust, 1997). However, experience of discrimination, whether direct or indirect, had not 'discouraged' women in our sample from pursuing their goals as they have done to women elsewhere (Afshar, 1989).

Religious discrimination was noted by a number of Muslim respondents. Most examples cited revolved around issues of alternative uniforms and the wearing of the *hijab* (headscarf). This would appear to support the contention raised in the earlier quantitative analysis and can be most explicitly seen in the following example:

> I remember going to this one job after I had started wearing the Hijab. I remember going to the interview and before I was even interviewed there, the person made a comment about how I was dressed ... when they looked at me, the person who was there said, 'Oh, we're not actually looking for someone with this appearance' sort of thing.
> (Neelam, Muslim, Bangladeshi, English teacher, EFL, twenty-five years of age)

The question of dress was one that also affected non-Muslim respondents or their mothers, who were forced to abandon their Punjabi suits and saris for Western dresses and uniforms. In cases of overt and extreme religious discrimination, as experienced by some of our Muslim respondents, the situation was further compounded by the absence of specific legislation outlawing religious discrimination, leaving victims open to further hostility. Other issues featured the right to observe religious rituals such as prayers, breaking the fast in Ramadan and taking religious holidays.

Working identities

The issue of 'fitting in' with British society surfaced in discussions on whether employers should offer training and recruitment in ethnic minority languages, gendered segregation and whether employers should do more to learn about South Asian cultures and religions. There were two distinct responses to this question. Those who felt that this would signal a positive way forward for multicultural Britain; and those who felt this was a step backward and was unnecessary. The relaxation of uniforms and dress codes that respected cultural and religious sensitivities was cited as one way employers could signal that they were 'open to difference'. However, some others in our sample, in recognizing the discreet nature that racism and stereotypes sometimes manifested themselves, advocated an approach that required *them* to modify cultural differences in order to 'fit in' with their environments. One respondent was able to articulate this particularly clearly:

> I think the more we try to fit in with them, the more we try to be like them – that is the only way of getting somewhere. Like in my role, if I, for example, did not socialize with my managers, or I decided to go in the next day covering my head, as they know it could be part of my religion, then yes, it would be regarded as oh she looks different, she does not want to fit in; she does not fit in.
>
> (Priya, Hindu, African Asian, financial analyst, twenty-nine years of age)

Clearly, 'fitting in' would cause serious personal dilemmas for some women who choose to express their religious identities through clothing practices such as the *hijab* and other religiously or culturally circumscribed practices that challenge Western work practices, such as the office party (Carter *et al.*, 1999). Given the findings of the Fourth Survey, that Pakistanis and Bangladeshis, especially women, were more likely to be culturally conservative and to assert their right to 'difference' (Modood *et al.*, 1997: chapter 9), this issue of identity and the willingness of some minorities to 'fit in' and adapt accordingly may well be a crucial feature behind the socio-economic 'success' Indian and East African Asian Hindus and Sikhs have enjoyed relative to other South Asian groups.

Referring back to our original quantitative analysis, then, it would appear that not only do religious factors play a role in determining whether certain women enter the labour market, but also for those that do, their relative 'success' may be dependent upon their working environment's ability to accommodate them.

Local initiatives: responding to social change

The London Borough of Tower Hamlets (LBTH) in the East End of London is an ethnically diverse area and home to a large number of Bangladeshi migrants, including a notable number of our respondents. It has been described by some researchers as an 'encapsulated and insular community' largely in reference to the first generation of Bangladeshi migrants (Eade *et al.*, 1996), and has been subject to long-term under-investment, social deprivation and exclusion, racial tension and high unemployment whilst ironically being located 'next door' to the City of London's lucrative financial centre. Within Tower Hamlets though, we find a particularly good example of a local initiative that has sprung up in response to the growing numbers of local graduates – approximately 600 per annum (as quoted in Tower Hamlets Graduate Forum *Annual Review*, 2000) who were either unemployed following graduation, or in jobs where they were 'under-employed'.

The Tower Hamlets Graduate Forum was launched in 1997 and with funding from the LBTH, European Social Fund, various regeneration budgets, and more general support from major international companies such as ABN Amro, ING Barings, Ernst & Young and the local London Guild-hall University, seeks to assist graduates into employment by developing projects and schemes with key employers in both the local borough and the City of London. Concentrating mainly in four key areas – IT, finance, law and teaching – it aims to prepare and empower students and graduates for the various recruitment processes of employers. It does this by providing advice, training and support and through strengthening links between the graduate community of Tower Hamlets and the business community via networking opportunities. In this way the Graduate Forum acts to raise the profile of its graduates – many of whom are Bangladeshi in origin – to employers who may still perceive Bangladeshis from this part of London through popular stereotypes and otherwise fail to consider recruitment from this highly educated and motivated potential work force. So apart from acting as a general 'careers advice centre' it also promotes the skills of the local population, and encourages college and school students into higher education. Apart from maintaining links with the business community and other employers, the Graduate Forum also works at a community level with local employers and organizations. At the time of writing, the Graduate Forum was itself engaging in research to assess specific issues relating to the Tower Hamlets graduate community and employment.

At the other end of the educational spectrum, other community-based initiatives geared to migrant women with limited English or qualifications are also sources of local engagement with community needs. However, their success appears to be dependent on the 'acceptability' of the perceived social environment women may find themselves in and whether further learning would compromise domestic obligations. Some evidence would indicate the informal community initiatives are more successful in attracting and main-taining potential students than formal learning environments. One of our Bangladeshi-speaking respondents worked part-time as a teacher of English to Bangladeshi women in a local community centre in Tower Hamlets and spoke of the client group it served and the scheme she was involved in:

> The ladies were really keen to learn English; they wanted to learn like how to phone up a doctor's surgery and make an appointment, so we used to concentrate on things like that . . . the whole point was to help them learn the language, in a way, help my community, because a lot of these women, they couldn't get out of the house, and because they couldn't go to a college and this was just a ceneer, and it was all females and it was a female teacher, so a lot of their mothers-in-law didn't have much of a problem. And it was very local. A lot of them tried to go to

college, but it was just one thing or another that was stopping them. But the Chestnut Centre was ideal for these ladies.

(Neelam, Muslim, Bangladeshi, English teacher, EFL,
twenty-five years of age)

Refining the quantitative analysis

Following insights from the qualitative research, further analyses were carried out of the Fourth Survey data. These analyses involved computing separate employment entry models for Indian/African Asian and Pakistani/Bangladeshi women who were: (1) aged twenty-three to thirty-five and (2) had educational qualifications obtained in the United Kingdom.

These additional analyses were not always consistent with the main thrust of the qualitative findings. Women aged twenty-three to thirty-five did not have employment rates substantially different from those of the sample as a whole. In addition, the employment gap between Indians and African Asians on the one hand and Pakistanis and Bangladeshis on the other was as large for younger women as for the sample as a whole. It was also the case that educational differences between the two ethnic groupings accounted for a larger proportion of the employment gap among younger women than for the sample as a whole. This is somewhat different from the qualitative research, which suggested that younger Pakistani and Bangladeshi women, especially those with qualifications, were showing levels of employment participation that were not markedly different from those shown by Indian and African Asian women with similar characteristics.

In findings that *are* consistent with the qualitative research, women with educational qualifications obtained in the United Kingdom had higher employment rates than the sample as a whole and the employment gap between educated Indian/African Asian women and their Pakistani/ Bangladeshi equivalents was somewhat smaller than for all women. Surprisingly, however, religious differences explained a greater proportion of the ethnic employment gap for this sub-group than for the sample as a whole. In contrast, the qualitative research found that Muslim women with relatively high levels of education were finding new ways of engaging with the labour market that were minimizing religious differences between South Asian women in terms of employment participation and career advancement.

The qualitative findings suggest that the new generation of Pakistani and Bangladeshi are increasingly pursuing higher education and careers. This seems to be supported by a trend analysis of UCAS university entry data (Modood, 2003) and the Manchester-based study conducted by Dale *et al*. With the Fourth Survey data having been collected in 1994 and there being too few graduates in their twenties to carry out separate analyses of this group, it has proved difficult to uncover this nascent development. This is perhaps a task for further surveys in the series though some preliminary confirmation of this should be derivable from the 2001 census. Yet the qualitative work shows that religion is potentially a shaping factor in employment choices and progression and it may still be the case that qualification levels and religion will continue to be among the most important variables in the analysis of South Asian women's employment profiles.

Conclusion: social changes

The data discussed above are indicative of a period where significant shifts and social changes are taking place as far as South Asian women and education and employment are concerned. Although our sample was biased in favour of younger working women, our study has shown that there is much diversity of opinion around a whole host of issues, including marriage choices and concerns, lifestyle choices and identities. Contrary to studies that present a simplified dichotomy between 'educated' and 'uneducated' women where the former are said to be actively choosing to reject their cultural and religious backgrounds, our work suggests a more complex picture and calls for a more sensitive interpretation. This is most explicitly revealed in the apparent contradiction between the quantitative analysis and findings from the interviews. Paradoxical and ambivalent attitudes among some women were much more common than some approaches to modernization and gender allow. By trawling in such a disproportionate number of Pakistani and Bangladeshi university students and graduates in employment we may have identified an emerging generation of women who were not accounted for in the Fourth Survey.

Not only was there evident commitment to family and domesticity across a wide range of women but sometimes a commitment to customs such as arranged marriages – suitably adapted – and to norms of 'modest dress' amongst graduates and career women. This suggests that cultures that till recently might have been portrayed as opposed to the education of and employment of women seem to be producing growing cohorts of highly motivated young women. This is particularly evident in some localities such as London and Manchester and highlights the significance of local studies that can account for regional developments.

For instance, in Tower Hamlets, we can see evidence of this marked social change. In what we call the 'Tower Hamlets phenomenon' we see a 'drive for qualifications' matched by a number of new employment initiatives that have sprung up in that area, such as the Tower Hamlets Graduate Forum and the City Circle – a network of Muslim professionals largely (though not exclusively) drawn from London's multinational financial City firms who annually 'sponsor' younger, new graduates through summer work placements. More recently there has been another business initiative called 'People into Management'. All these local services are clearly meeting a recruitment demand for lucrative positions in the heart of London's financial base, which ironically, is located within minutes of its much poorer neighbour.

Women from all ethnic and religious groups were, however, modifying social and cultural norms in the light of new opportunities and social contexts. Social change was an undisputed phenomenon, but it manifested itself in differing ways and was dependent upon specific psychic, religious, cultural and social contexts. Our data would suggest that younger South Asian women are confidently expressing their identities through at least two facilitative frameworks: *gender mediated through ethnicity* and *gender mediated through religion*. These frameworks are dynamic and represent on-going processes of redefinition and renegotiation of boundaries and identities.

Notes

1 The National Survey cited here is the fourth in a series of major reports conducted by the PSI examining the experiences of ethnic minorities in Britain in a variety of areas such as employment, housing and neighbourhoods, education, language and qualifications, health, family and household patterns. The Fourth Survey has for the first time also included chapters on cultural identities and racial harassment. The findings act to provide a comparative platform to assess the experiences of ethnic minorities against those of the white majority and also acts to complement data from other sources such as the census and Labour Force surveys.
2 See the Fourth PSI Survey (1997) for definitions of what constituted manual and non-manual work (p. 100).
3 Further details of the methodology can be found in Ahmad *et al.* (2003).

References

Afshar (1989) 'Gender roles and the moral economy of kin among Pakistani women in West Yorkshire', *New Community*, 15: 211–25.

Ahmad, F. (2001) 'Modern traditions? British Muslim women and academic achievement', *Gender and Education*, 13 (2): 137–52.

Ahmad, F. (forthcoming) *British South Asian Women and Marital Choices.*

Ahmad, F. (forthcoming) 'To work or not to work? Educational and career experiences and choices of British South Asian women'.

Ahmad, F., Modood, T. and Lissenburgh, S. (2003) *South Asian Women and Employment in Britain. The Interaction of Gender and Ethnicity.* London: Policy Studies Institute.

Basit, T.N. (1997) *Eastern Values. Western Milieu. Identities and Aspirations of Adolescent British Muslim Girls.* Aldershot: Ashgate.

Bhopal, K. (1997) *Gender, 'Race' and Patriarchy. A Study of South Asian Women.* Aldershot: Ashgate.

Bhopal, K. (1998) 'How gender and ethnicity intersect: the significance of education, employment and marital status', *Sociological Research Online*, 3 (3): 1–16.

Brah, A. and Shaw, S. (1992). *Working Choices: South Asian Women and the Labour Market.* Department of Employment Research Paper 91. London: DoE.

Brah, A. (1993) 'Race and culture in the gendering of labour markets: South Asian young Muslim women and the labour market', *New Community*, 29: 441–58.

Brown, C. (1984). *Black and White Britain.* London: Policy Studies Institute.

Bruegel, I. (1989) 'Sex and race in the labour market', *Feminist Review*, 32: 49–68.

Carter, J., Fenton, S. and Modood, T. (1999) *Ethnicity and Employment in Higher Education*. London: Policy Studies Institute.

Dale, A., Fieldhouse, E., Shaheen, N. and Kalra, V. (1999) *Routes into Education and Employment for Young Pakistani and Bangladeshi women in the UK*. Occasional Paper 19. Manchester: CCSR, University of Manchester (also in *Ethnic and Racial Studies*, 25 (6) 2002: 942–68).

Dale, A., Fieldhouse, E., Shaheen, N. and Kalra, V. (2000) *The Labour Market Prospects for Pakistani and Bangladeshi Women*. Occasional Paper 18. Manchester: CCSR, University of Manchester (also in *Work. Employment and Society*, 16 (1) 2002: 5–26).

Dhaliwal, S. (2000) 'Asian female entrepreneurs and women in business: an exploratory study, *Enterprise and Innovation Management Studies*, 1 (2): 207–16.

Eade, J., Vamplew, T. and Peach, C. (1996) 'The Bangladeshis: the encapsulated community', in C. Peach (ed.) *Ethnicity in the 1991 Census*. II *The Ethnic Minority Populations of Great Britain*. London, HMSO.

Ermisch, J. and Wright, R. (1992) 'Differential returns to human capital in full-time and part-time employment', in N. Folbre, B. Bergmann, B. Agarwal and M. Floro (eds) *Issues in Contemporary Economics*. Basingstoke: Macmillan, 195–212.

Gardner, K. and Shukur, A. (1994) 'I'm Bengali, I'm Asian and I'm living here: the changing identity of British Bengalis', in R. Ballard (ed.) *Desh Pardesh. The South Asian Presence in Britain*. London: Hurst.

Graduate Forum Annual Review (2000) Tower Hamlets Graduate Forum, London.

Hakim, C. (1995) 'Labour market and employment stability: is there a continuing sex differential in labour market behaviour?' Working Paper 1. London: Department of Sociology, London School of Economics and Political Science.

Hardill, I. and Raghuram, P. (1998) 'Diasporic connections: case studies of Asian women in business', *Area*, 30 (3): 255–61.

Holdsworth, C. and Dale, A. (1995) *Ethnic Homogeneity and Family Formation. Evidence from the 1991 Household SAR*. Occasional Paper 7. Manchester: CCSR, University of Manchester.

Holdsworth, C. and Dale, A. (1997) 'Ethnic differences in women's employment', *Work, Employment and Society*, 11: 435–57.

Jones, T. (1993) *Britain's Ethnic Minorities*. London: Policy Studies Institute.

Joshi, H. (1991) 'Sex and motherhood as handicaps in the labour market', in D. Groves and M. Maclean (eds) *Women's Issues in Social Policy*. London: Routledge.

Lissenburgh, S., Modood, T. and Ahmad, F. (forthcoming) 'South Asian women and employment: quantitative analysis of the fourth National Survey of Ethnic Minorities', PSI Research Discussion Series.

Macran, S., Joshi, H. and Dex, S. (1996) 'Employment after childbearing: a survival analysis', *Work, Employment and Society*, 10: 273–96.

Modood, T. (2003) 'Ethnic differences in educational performance', in D. Mason (ed.) *Explaining Ethnic Differences*. Bristol: Policy Press.

Modood, T. and Acland, T. (1998) *Race and Higher Education*. London: Policy Studies Institute.

Modood, T. and Shiner, M. (1994) *Ethnic Minorities and Higher Education*. London: Policy Studies Institute.

Modood, T. *et al.* (1997) *Ethnic Minorities in Britain. Diversity and Disadvantage*. London: Policy Studies Institute.

Owen, D. (1994) *Ethnic Minority Women and the Labour Market*. Manchester: Equal Opportunities Commission.

Runneymede Trust (1997) *Islamophobia: a Challenge for us All*. London: Runneymede Trust.

Taher, A. (2000) 'Stuff of dreams', *The Guardian*, 7 November.

Taylor, P. (1993) 'Minority ethnic groups and gender in access to higher education', *New Community*, 19: 425–40.

West, J. and Pilgrim, S. (1995) 'South Asian women in employment: the impact of migration, ethnic origin and the local economy', *New Community*, 21: 357–78.

8 The Milky Way to labour market insertion

The Sikh 'community' in Lombardy

Maria Josè Compiani and Fabio Quassoli

This chapter focuses on the socio-economic and the legal/institutional characteristics of international migrants' incorporation into the Italian labour market. Our interpretation of the statistical evidence available with respect to immigrant's labour insertion combines the role played by the local characteristics of labour force demand, the normative constraints generated by immigration policies – and more broadly by the institutional framework which organizes economic relationships – and the role performed by migrants' network. The assumption upon which the analysis we present is concerned with the complexity of the ways in which migrants have accessed the Italian labour market so far. In general, it is linked with the whole set of changes that have produced the shift of Western economies from a Fordist to a post-Fordist model (Mingione, 1997). Reference is also made to a series of expounding factors that overcome the limitations of former approaches. These include, first, interpretations that solely consider the needs of the economic systems of the countries where migrant flows are spontaneously directed, or where they are attracted/kept away by means of incentives/disincentives policies (Moulier-Boutang, 1998; Piore, 1983). Second, it concerns alternative views boasting the capacities of migrants' social networks – that are often 'ethnically' typified – to define the conditions where the newcomers' economic insertion occurs (Massey and Espinosa, 1997; Portes, 1995).

In order to understand what is happening in different territorial contexts and with respect to different migration flows, it is necessary to jointly examine at least three frameworks that contribute to define the conditions of migrants' labour insertion in Italy (Ambrosini, 1999). First, there are those transformations occurring during economic organization processes, such as deindustrialization/tertiarization, the decentralization of production, and the widespread use of different types of outsourcing and delocalization. They indicate labour needs on the demand side as well as the migrants' modalities of functional incorporation into the production system as far as dominant patterns and economic actors are concerned. The second factor is the institutional framework, which includes labour laws and regulations, welfare policies and immigration policies, but also the characteristics of local economic systems. They can influence the presence, the relevance and the outcome of different types of immigrant insertion into the labour market and the importance of informal arrangements and they can be more or less functional with respect to the needs of the economic system (Morris, 2002; Sassen, 1996). Third, there is a wide range of networks (made up of persons as well as institutions) which encompass 'ethnic' and autochthonous networks and different types of institutions and special services for migrants. Together, these elements can play a crucial role in creating the social capital that is necessary to establish some reliable economic relationships between labour demand and supply (Barbieri, 1997; Granovetter, 1985, 1995; Mutti, 1998).[1]

On the basis of such an interpretative framework, we analyse here a rather peculiar case of immigrant labour market insertion: the Sikh community which, in the last decade, has been growing in the provinces of Cremona, Brescia and Mantova. Our analysis is based on two quantitative and qualitative investigations that were carried out in 1997 (Compiani, 1998) and 2000 and it focuses on the socio-economic aspects of Sikh settlement in Lombardy.

In the following paragraphs, we will first sketch an outline of migrants' labour insertion models in Italy, specifying the three dimensions introduced above (the local socio-economic-systems, institutional framework and social networks) in more detail. The second part of the chapter will focus on Indian immigration in the province of Cremona, with a twofold objective. On the one hand, it will show how the 'productive vocation' of the Cremona province, together with the role played by social and economic actors and the institutional/regulatory framework, has had a crucial influence on the methods through which Indian citizens residing in that area have entered the work force and to generate the local socio-economic arrangements which accompany immigrants' settlement. On the other hand, it will reveal how Sikhs represent a deviant case compared with the salient characteristics of migrants' descriptive labour insertion models discussed in the first part.

Immigrants' insertion in the Italian labour market: an outline

As anticipated in the previous paragraph, in order to understand the patterns of immigrants' insertion into the Italian labour market and more broadly in the Italian society, three areas should be considered. The first relates to the significant characteristics of territorial models of socio-economic organization and the types of economic activities that prevail in each area. As far as immigrants are concerned, we can rely upon a typology proposed by Ambrosini (1999: 156), who has identified three general models of labour insertion: a 'metropolitan model', an 'industrial model' and a 'seasonal/rural model'.[2]

The first model, defined as 'metropolitan', is typical of large metropolitan areas (like Rome or Milan); it is characterized by widespread jobs in low-level personal services (housekeeping, health care, etc.) and to companies (janitorial services, fast delivery services, etc.). These jobs are mainly done by women from the Philippines and South America. Let us consider, for example, house workers, who represent the most important component of this labour market segment. At the end of 1999, out of a total of 114,182 foreign house workers – which represents by itself a high percentage of foreign workers living in Italy – about one third came from the Philippines alone, 10.4 per cent from Peru and 8.6 per cent from Sri Lanka (Table 8.1).[3]

Moreover, almost 80 per cent of these jobs are done by women (Table 8.2). The territorial distribution of these data shows the importance of large metropolitan areas: as a matter of fact, a high concentration of house workers is found in the north, especially in Lombardy (24 per cent of the total) and in the centre, particularly in the Lazio region (29 per cent of the total). In Lombardy and Lazio alone, foreign housekeepers account for over 50 per cent of the people employed in this sector; in all the other regions, and mainly in those without any large urban areas, their number represents a minority figure.

In spite of their quantitative significance, their symbolic importance (social representations of immigration) and the key role they have played in migration policies since 1990, the figure concerning domestic servants basically remained stable for all the second half of the 1990s and the percentage of new insertions was undoubtedly lower than in other sectors.

In the second model, which is typical of highly industrialized areas (some

Table 8.1 Documented foreign housekeepers, by country of provenance, at 31 December 1999

Country	A.V. No.	%
Philippines	36,606	32.1
Peru	11,847	10.4
Sri Lanka	9,791	8.6
Romania	5,591	4.9
Poland	4,533	4.0
Albania	4,530	4.0
Morocco	4,292	3.8
Ethiopia	3,204	2.8
Dominican Republic	2,985	2.6
Ecuador	2,887	2.5
Somalia	2,771	2.4
Cape Verde	2,216	1.9
Brazil	1,424	1.2
Nigeria	1,309	1.1
Mauritius	1,235	1.1
El Salvador	1,196	1.0
Total	114,182	100.0

Source: Caritas (2002: 298; elaborations by Caritas on data from INPS).

for the tourist industry, foreign workers are employed in low-level jobs that do not involve direct contact with clients, like floor waiters, laundry workers, porters, commis chefs, dishwashers, etc. (Caritas, 2002: 275). On the contrary, the use of foreign workers in various tasks of the agricultural sector (fruit and vegetable picking, driving transport vehicles, processing industry and animal husbandry) appears to be much more important. Since the late 1980s, this sector has been one of the biggest employers of foreign work force in Italy. It is characterized by high exploitation levels and informal arrangements, especially in southern farming areas, whereas in the central and northern regions regular seasonal work contracts are more common.

The second area we should mention is the political-institutional framework, whose characteristics are easier to recall as far as their general principles are concerned. By 'institutional framework' we mean different political, institutional and administrative spheres. Taken as a whole, these form the structure of the constraints and opportunities that are intertwined with paths of social insertion and the economic and job-related strategies used by immigrants in our country. Even in this case, we can identify three relevant spheres of action.

The first sphere solely concerns foreign citizens from outside the European Union and it is represented by the totality of the laws, regulations, rules and administrative practices establishing the legal position of foreign citizens in Italy, as well as the collection of rights and duties they are subject to (Sciortino, 2000). Some especially interesting aspects of our analysis concern the complex legal-administrative systems that are directly connected with the foreign citizen's contact with the labour market, whether as an employee or as an independent contractor (Quassoli, 1999). A fundamental aspect, which may help to explain both the continuing reproduction of illegality and problems of legal insertion of immigrants into the job market, lies in the fundamental contradiction between the characteristics of the job market and residence laws. On the one hand, we have a job market which is noteworthy for the unstable nature of its contractual agreements, the widespread nature of atypical forms of labour, the proliferation of informal working relationships and the extremely weak position of the employee (Mingione and Pugliese, 2002). On the other hand, we have residence regulations modelled on the formal and substantial integration path typical of Fordist systems which was widespread in continental Europe in the 1950s and 1960s, despite the fact that the labour market has changed dramatically. The Fordist system is characterized by employees with stable employment, a legal employment contract and the opportunity to benefit from a series of welfare resources.

The second sphere is, of course, the system of administrative rules which regulate financial transactions and labour relationships. This area is, for instance, directly relevant to the question of the informal economy under at least two aspects. On the one hand, there are the limits and the obligations imposed by compliance with the administrative procedures regulating every economic activity. On the other hand, there is a tax system that affects salaries and corporate income and that gives entrepreneurs and employees an incentive, where and when possible, to favour relationships of an informal nature which evade both registration procedures and administrative controls (Portes, 1994).

The third sphere, which is closely connected with the two described above, albeit in some respects independent from them, is represented by welfare policies, both generally and with regard to specific resources aimed at the resident foreign population. The opportunity to access the resources made available through welfare policies may work in a number of different ways with respect to immigrants' economic strategies. Welfare resources have a direct influence on the strategies employed in the search for employment: They produce a rise in the minimum acceptable conditions to access the job market and they favour mixed strategies between a formal and an informal presence into the labour market (Mingione and Pugliese, 2002; Pugliese, 2000). Besides, they are intricately intertwined with the strategies of insertion into the labour market, be it formal or informal, thus creating the 'informal labour market', where many foreign immigrants have gradually inserted themselves since the early 1990s (Mingione, 2001; Reyneri, 1998a). Finally, some entire occupational segments can be directly dependent on welfare policies' characteristics. The expansion of the sector of personal services (home care, babysitting, etc.), which is often characterized by informal work arrangements, is directly connected, on the one hand, with widespread cultural models whereby women keep an active role also after marriage and, on the other hand, to its role as a substitute for those social policies that the family (namely its female component) has traditionally been required to carry out (Saraceno, 1998).

The third area relates to the social embeddedness of immigrants' economic strategies and the central importance of social networks in explaining individual choices such as the decision to emigrate, the organization of the journey, the collection of information relating to opportunities – work-related or otherwise – in various local contexts, and so on. The immigrant worker, like his Italian counterpart, is not an economically rational protagonist, who evaluates the costs connected with the acquisition of information on job opportunities, assesses the available choices and works out strategies for entering the labour market. On the contrary, the perception of the socio-economic characteristics of local contexts, of the statutory framework which governs economic activity and employment relationships and of the employment paths taken by fellow immigrants is filtered through the cognitive and normative mediation offered by the particular social network which is the individual's point of reference (Sassen, 1995). The role of these networks is even more important in a situation like the Italian one, which is worthy of notice for both the low level of efficiency with which economic activities are regulated by government authorities and the extensive use of informal arrangements. As a result, immigrants' access into the informal sector represents the persistence of an existing state of affairs rather than an innovation in the Italian employment market (Reyneri, 1998b; Mingione and Quassoli, 1999). At the same time, it is made possible by the existence and operation of social networks made up of compatriots (ethnic networks) or assistance organizations and agencies (private social associations), which mediate and facilitate the insertion of immigrants into the work environment and help to 'lower the transaction costs connected with the search for, and selection of, a "reliable" workforce' (Ambrosini, 1999: 95).

A case study in Asian migration: the Sikh community in Cremona province[4]

The members of the Indian community, which has established itself in the province of Cremona, come almost exclusively from the north of India: from the states of Haryana, Himachal Pradesh, Delhi and, in particular, from the Indian part of the Punjab. Over the last thirty years, Sikh emigration, which began in the last century (Sarmiento, 1991; Wilson and Samuel, 1996; McLeod, 1989b), has leaned strongly towards northern Europe and this has happened for three reasons.

First, we should take into consideration the fragmentation of ownership of agricultural land in the Punjab, due to Jat[5] traditions of inheritance, which provide that when a parent dies the land is divided equally among all sons. The result of this practice is a progressive reduction in the size of agricultural properties to such an extent that succeeding generations are unable to maintain a decent living standard (McLeod, 1989a). It must also be added that the Jat code of conduct discourages these individuals from working for others in their village, and pushes them to seek their fortune elsewhere.

Second, we have the economic and social transformations which occurred in a number of Indian states following the so-called Green Revolution of the 1960s (Spagni, 1984; Dhami, 1988). This encouraged the development of a capitalistic type of agricultural economy and favoured large agricultural interests (Gallucci and Venturi, 1984), thus carrying negative consequences for small and small to medium landowners. It was from this category of farmers, who had been impoverished by the mechanization and the capitalistic transformation of agriculture, that a part of the Sikh immigrants in the province of Cremona originally came.

Third, between the mid-1980s and the end of the 1990s, there was an involution of the political situation in the Punjab (Wallace and Chopra, 1988) with an intense militarization of the territory by the Indian army commencing at the beginning of the 1980s,[6] which was justified by the need to combat Sikh terrorist organizations. The situation rapidly deteriorated, to the point where a civil war began in the true sense, with devastating effects on the population and the economy of the region (Restelli, 1990; Dhami,

1988). The reduction in employment opportunities, which were already limited when compared with the availability of skilled labour in the Punjab, was one of the reasons why some people sought work abroad.[7] Finally, as far as the Cremona case is concerned, among the main causes of migratory flows from India to Italy, we should mention both the growing demand for skilled workers on the part of local dairy firms and the attractiveness of Western consumer models, fuelled by the success stories of the first emigrants, which raise hopes of much higher incomes than normal Indian standards, generated by emigration to foreign countries.

Therefore, in general terms, Sikh immigrants in the province of Cremona have not left a situation of total poverty (*garib*) behind them, but have chosen to remove themselves from prospects characterized by a progressive reduction in job opportunities and growing professional dissatisfaction.

> I left India not because there was no money, but because I could not find the job I wanted to do. We could not work in the family business because it wasn't big enough. If I had worked there, I would have earned the same amount my brother now earns working there alone. Since in India there are low-cost workers, here I earn more.
>
> (R., 1997)

In the accounts we have collected, the emphasis is not on serious economic problems at home, but on the desire to earn more than would be possible in the Punjab or in Delhi, and on the fact that it is impossible for them to attain a higher social position than that of their father, or, due to limited industrialization in the Punjab, to find a position worthy of their educational level.

Migrant projects and chains

While the emigration paths which the Sikhs follow with notable determination[8] may be reduced to the formula 'earn money to make investments', there emerge two underlying patterns. On the one hand, we find the classic 'birds of passage' (Piore, 1983), whose original project is based on the accumulation of money and forced savings, with the purpose of returning to India as soon as possible. Their life is private and dedicated entirely to work, and their children return to India to study. In these cases, family ties play a fundamental role with respect to migratory projects. The decision to emigrate is often taken by the family or within a community group, and it is the family that acts to collect the funds required for the move. Migration is, therefore, a kind of collective investment, of which the main beneficiary is the family, which derives financial resources and social prestige from the investment of its human resources abroad (Barrier and Dusembery, 1989: 5).[9] The relationship between the member who migrates and the family involves moral obligations based on reciprocity.

As has emerged from other studies of Sikh communities abroad (Barrier and Pashaura, 1996, 2001), the family often has real social control over its members, and frequently over generations, and the family members live according to parameters, values and customs imported from India and jealously maintained (Dusembery, 1989). It follows that social conduct and restrictions, and the important moments in the life of a Sikh, are seldom based on factors acquired in the country in which he/she lives, but remain anchored in his/her home country, which is on occasion an imaginary and idealized place, and sometimes one to which one must return to live crucial events in one's life.[10] In actual fact, this first model currently typifies a small section of immigrants and is associated by several respondents with the past and the early period of the migratory experience in Italy.

The majority of Punjabis, however, show a different path: a settlement migratory project that foresees a long-term stay in the country of immigration. The causes of this attitude are to be found in the progressive settling of work and professional perspectives and in the presence of family members. In particular, children become the key factor around whom the decision on whether or not to go back to India revolves. Consequently, they are a crucial variable in the decisions of Indian immigrant families concerning ways to invest their savings.

> Then, when somebody comes and stays here for four or five years, their life changes. They are not well here, but they are not well in India either. They are stuck in between. If they leave here, they will not be well there; if they leave there, they will not be well here. This is because it is difficult to settle down quickly and well. Settling down in my condition is very hard, life is very hard. It is not easy to leave all your friends, not knowing anybody, and especially the language.
>
> (G., 1997)

> Now we are planning to stay because our girls are studying here and then we will choose according to their wishes. If they had started their studies in India they would have finished them there, but since they have started them here they have to finish them. Otherwise they will not study well.
>
> (S., 2000)

These statements show how the tendency to save money too is becoming gradually aimed at the host country instead of the country of origin. In most cases, they do not send all their savings to their families in India: in fact, their plan is to invest in Italy for their future and that of their children. In this typology of immigrants it is possible to recognize the 'businessmen', those who have invested part of their savings in some business operations in the immigration country. Some of them have opened shops in Italy selling ethnic goods (mainly foodstuffs), restaurants, phone agencies, etc.

I will surely start something, small but mine. I don't know, if I integrate all right I might stay for a long time and have my son study here. If I don't integrate well, I will have my son study in India, I'll send him to college. Many people send their children to study in India. Here, you don't have assurances: it's not like other countries. We don't have a right job: we don't have a great future; a *bergamino* will always be such.

(G., 1997)

The need to go up on the Sikh social ladder, which could be achieved through a successful migratory experience, is very manifest and shows through in several interviews. The aim of migration is to have success, which, most important, needs to be tangible and measurable by compatriots, by relatives who have remained in India and – once definitely established in the host country – by the same relatives in Italy. This aim can be ascribed to the dynamics of the system regulating the attribution of *izzat* (status): 'Migration was thus a family strategy, an honourable alternative to the reduction of status at home ... The objective was enhanced status or prestige (*izzat*)...' (Dusembery, 1989: 5). One of the criteria of status attribution resides in the immigrants' economic success, which ends up involving all the extended family.

Strategies for economic and social integration

If we move on to organizational set-ups in Italy, we may see a number of interesting aspects characterizing the Sikh social network in Cremona. We use the term 'social networks' to identify four areas of relationship (Piselli, 1990): 'Workplace', 'origins' (parents, relatives and friends connected with the family, 'chosen' (friends, sentimental relationships and acquired family – that is, strong acquired ties) and 'community' (neighbours and the residents of the neighbourhood).

Our interviews show that in the case of the local Sikh community only one of these networks is really strong and complex. This is the 'origins network', which here means belonging to a clan from the caste point of view.[11] Caste hierarchies play a fundamental role in the life of Indian immigrants in the province of Cremona. While the statements that have been gathered confirm the importance of marital endogamy within a caste, its significance is played down in other social contexts, like work, but also in social relationships among compatriots both in Italy and in India. Actually, some studies carried out on the immigrant Sikh population in the province of Reggio Emilia (Bertolani, 2002) highlight the crucial importance of caste relationships both when selecting potential migrants (who usually also belong to the same *got*[12]) and in interpersonal relationships within the immigrated population itself (Marenco, 1976).[13]

Inter-caste relationships are questions of primary importance during the early phases of expatriation. Within the network of relationships among members of the community, in fact, genuine migratory chains are created which allow the community to integrate itself with ease in the host country. When an Indian arrives in Italy, he either has direct contact with an acquaintance, or has a series of addresses of compatriots who live in Italy. These people are expected to offer him assistance in the early stages and there is an excellent chance that he will get it. It is common practice to play host to compatriots, even for long periods of time, to assist them with bureaucratic procedures and to look for a job for them. This type of solidarity, although it implies no material obligation to repay the favour, creates a form of debt between the individual providing the assistance and the person who receives it. It extends to the entire family, and may be cashed in by the creditor at the appropriate moment.

This type of network operates in a number of areas, such as labour insertion. Some members of the community call relatives and friends (directly from India or from other areas of Italy) when employment opportunities arise. Once the mechanism of the migratory chain has established itself, it is, of course, self-perpetuating, since it creates (or reinforces) the social networks which sustain it (Portes, 1995; Reyneri, 1979).

Women's social insertion is a different matter that deserves particular attention (Galloni, 2000). Women's situation presents some problems linked with relationships with Italian fellow citizens and with the public administration. They live in their husband's workplace, usually some isolated farmhouse poorly connected with towns and they do not have access to any services. In most cases, they do not work outside the house; therefore they are not encouraged to learn Italian (they often start learning it when their children go to school) and they do not have occasion to come into contact with public services. Their husbands are usually concerned with this aspect; they take care of interacting with the institutions (police headquarters, hospitals, schools, municipality where they live), thus filtering and mediating their wives' entire social relationships. Few women also work outside the house and acquire an autonomous ability to move within the territory. In the majority of cases, this difficulty is caused by reasons of an objective (being isolated) and a subjective (looking after the children and the house) nature; these factors do not leave Indian women any room for autonomy of leisure time.

Work conditions

Before going into the details of employment activities of Cremonese Sikhs, however, we should describe the employment paths they have followed, and above all those of individuals who have been in Italy for many years and who have had a varied work experience. Thanks to research carried out in the province of Cremona (Fondazione Cariplo-I.S.Mu. and Provincia di Cremona, 2000), we may have an idea of the date of arrival of members of the community in Italy and, later, in Cremona. As much as 60 per cent of the Indian resident population came to Italy before 1995, while, at that time, only 28 per cent lived in the provinces. The boom in arrivals to the area took place between 1995 and 1997, when 35 per cent of current residents arrived. One may assume from the available data that, given the large number of Indians who have been resident in Italy for many years but who have only recently arrived in Cremona, the employment path followed by Sikhs in the Italian labour market – as is the case with other communities – developed across a large number of work experiences (Ires Piemonte, 1991). In the first stages of their immigrant experience (which often occurred between the end of the 1980s and the beginning of the 1990s), Sikhs settled principally in central Italy (Lazio and Tuscany), where they found work in the greenhouses and farms of Rome, Latina and Viterbo. The prevailing employment method was undocumented labour.

The employment paths, however, were very different: from unskilled workers in the flower business to tractor drivers, from metalworkers to bakers, from workers in cheese factories to workers in travelling circuses. It is in this area that many lived through the change from being an illegal to being a legal worker. Most obtained their work permits in the 1990 amnesty and then moved north in search of more remunerative, and hopefully legal, employment. The move north took place in most cases thanks to relatives or friends who offered lodging. This illustrates the strength and the cohesiveness of the social networks.[14] One of the goals of this internal migration to the north was Cremona. Cremona is a town mostly involved in agricultural activities situated about 100 km south of Milan, in the alluvial Padana plain. The economy of the province of Cremona – with the exception of the two largest municipalities – is mostly agricultural, and it is in certain agricultural activities that the Indian population has found significant opportunities for entering the work force.

After World War II the countryside around Cremona, due to the mechanization of agricultural work, started becoming less and less populated. At the same time, jobs were scarcer, while the demand for skilled work force started to grow. In zootechnics, this process carried even greater consequences. Milking and milk storage processes evolved enormously, but there was not a corresponding class of skilled workers who could take over from those who were. Working as a *bergamino*[15] does not nowadays have characteristics such as to encourage local youths to start this profession, which – although it is well paid – has a number of characteristics that make the job demanding and insalubrious. The term *bergamino* refers to cattle workers in the Lombardy plain. Working as a milker, or *bergamino*, is extremely demanding and tiring, although these days it is well remunerated.[16] Work is done in double shifts: from one or two o'clock in the afternoon until six in the evening or later, and a night shift from one or two o'clock until six or seven in the morning.

The employment contract concerning *bergamini* provides for a minimum wage of 1,600,000 lire a month, paid on the basis of a working day of six hours and forty minutes, with a weekly rest day. These rules, however, are rarely followed. In actual fact, a worker works an average of two to three hours' overtime a day, and does not always receive his day off, but this happens by agreement between the employee and his employer. Immigrants who do this work want to earn as much as possible; they do not refuse overtime, and have an interest in missing their days off.

Indians began to work in this sector at the beginning of the 1990s. Until a few years previously, this type of job had been well covered by the local work force, but as the older workers reached their retirement age it became increasingly difficult to find young workers willing to replace them. A few farmers recall that there were other immigrants – Moroccans, Thais and Ivorians – but only for a short time and only until the Indians who now monopolize this sector came on to the scene. There is a great demand for skilled Indian workers, because they guarantee both commitment and great flexibility.

The work of the *bergamini*, in addition to the significant financial advantages, also offers workers the opportunity of tied accommodation. This old custom is not an official part of the employment contract, but is a tradition in the Lombardy countryside; milkers must live with the herd so as to be ready for any eventuality; therefore, when they are hired they are given lodging (which is often in a dubious hygienic state).[17] Accommodation makes this work extremely practical, if we remember that housing is one of the main problems facing immigrants, due to the practical difficulties in finding it and to high rents.

Another attraction, as we have mentioned, is the wages, which are considerably higher than the average earnings of an immigrant in northern Italy.

Notwithstanding the belief (which is particularly common among employers and our colleagues) that Indians are especially good at this kind of work (veneration for cows is mistakenly linked with a professional skill which, in truth, most Indians acquired in Italy), not all Sikhs are satisfied with the work, which is considered by many to be unhealthy and out of line with their initial expectations. The majority dream of independent activities. Many of the interviews show this desire for independence and a leaning towards entrepreneurship, which the Sikhs have demonstrated in other countries. Some would be prepared to modify their immigration plans and invest their savings in Italy rather than in India, provided that they had the opportunity for a significant professional growth.

Documented and undocumented work

There are two main sources through which Sikhs find work: a highly developed internal information network and trade union associations. Trade unions have played a historically important role in agriculture in Lombardy. Such a role is mainly ascribable to the farmers' struggles that occurred in the twentieth century. Therefore, trade unions, through their offices are spread over the territory, act as consultants to protect workers and sometimes they help joining demand and supply. Many Indians, ever since their arrival, came into contact with trade unions (especially CGIL CISL and UIL); they developed these relationships to their advantage, both in terms of protection of workers' rights[18] and in the use of some consolidated networks for finding jobs for some workers. In the former case, we are referring to the network of friends, relatives and families which ensures that anyone who receives any news of job vacancies passes the information on to a relative or friend. The mechanisms of this social network also facilitate another extremely delicate relationship, the one with the employer. The likelihood that a worker is serious and trustworthy increases if he is introduced by a friend, who takes upon himself part of the responsibility for the reliability of the new worker. This confirms the fact that the agricultural sector in Cremona, notwithstanding the modernization, which has taken place over the past few years, is still somewhat based on traditional economic relationships, which are not determined by market rules.

We may identify various cases in the area of employment relationships between entrepreneurs and *bergamini*, which range from complete legality to total illegality. Legal employment is not the exception to the rule, despite the fact that various elements of illegality permeate some relationships, one of the most widespread of which is the payment of overtime 'outside the pay packet', that is, as undeclared income. This is a very common practice, which suits both the employer and the employee.

There are also cases where the worker is forced into a situation which results in exploitation. This often occurs even with documented workers. Cases where a legally hired worker is not paid for overtime worked, or is not regularly paid for rest days or holidays not taken are far from rare. Fear of being replaced by a more flexible substitute discourages workers from reporting these occurrences to the union.

The general picture of employment integration is, therefore, extremely varied and complex, and is characterized by an indissoluble mix of the formal and the informal. Sikhs are very good at combining legal and illegal work, not unlike many Italians who do two jobs for the same reasons. The best combination seems to be a regular job with overtime and holidays paid under the counter; this ensures payment of contributions and a continuing work permit (which gives access to the national health service[19] and other services provided by the state). At the same time, more money can be earned, so that sufficient saving to satisfy individual immigration projects can be accumulated more quickly. This system creates collaboration between employer and employee which is very similar to that which exists among Italians. An undocumented worker does not attempt to report his situation to the authorities, or else he would not be able to find other work in the same area. The same person who finds him the job, usually a compatriot, guarantees that an individual offered to an employee will not report his undocumented status. There is a delicate equilibrium in these cases, which involves methods of social control in two areas: in the demand for employment, and within the community itself. For this reason, one might say that the basis of the relationships is 'extra-economic'.

The employment situation described here leaves us in no doubt as to the beneficial results which the Sikhs bring to the Cremonese farming industry. They have helped solve a labour shortage, which had become chronic over the past years. One may speak here in terms of a substitution effect, and, in certain cases, of a complementary relationship with the local work force. It is also possible, despite the cases of undocumented work, to exclude the hypothesis that they might have introduced new forms of undocumented labour. They have, in fact, inserted themselves into a sector which was already well developed, and have continued to exploit possibilities which had previously been opened up by Italians, and have, for the most part, sought the same assurances and tried as far as possible to exploit a pre-existing favourable economic situation.

Conclusion

Sikh immigration in Lombardy shows some original traits within the general framework of immigrants' economic and social insertion in Italy. One of the most distinctive features is the ethnic monopoly of a specific sector of labour insertion. In contrast to what is widely believed at a local level, this does not result from a sort of 'vocation' of a religious nature; rather, it is part of a precise migratory project, whose strength characterizes this immigrant population and relates it to other expressions of the Sikh diaspora in the world.

Although the job of *bergamino* is not held in very high regard by Sikh people, who, just like local people, realize what low social prestige the job holds, this type of work carries a number of advantages that facilitate the realization of an ambitious migratory project. The possibility to have free accommodation, where bills are often paid by the employer, combined with wages above the average, are all elements whereby – through cautious control of expenses – some profitable types of investment can be made in a relatively short time. All this makes the job of *bergamino* a good starting point for strategies of economic insertion aimed at the pursuit of material success and social prestige.

Furthermore, the case of Sikhs does not conform to any of the labour insertion models illustrated in the first part of this chapter. On the one hand, the importance of the factors that have been shown to be at the base of migrants' socio-economic insertion in the Italian society – like the local production system, the institutional framework, social/institutional networks – is corroborated by what has occurred in the Cremona context. On the other hand, however, the interaction among these factors seems to have produced a kind of virtuous circle that has encouraged some unusual forms of socio-economic insertion – especially when compared with the conditions of legal and social precariousness and to the severe economic exploitation which can be noticed in several other contexts – and the development of complex migratory projects in relatively short periods.[20]

Notes

1 Actually, there are two additional key factors for any analysis of international migrations, which cannot be dealt with in detail here. One of them is the socio-economic context where the decision to emigrate is made; such context is often politically, economically and/or culturally dependent from the immigration country (as in the case of ex-colonies) and it undergoes some deep political, economic, social and cultural transformations as a result of the mass emigration of a part of the economically active population (Sayad, 1999). The second factor is the more general immigration context, whose boundaries can be defined by the existence of some form of diaspora and which help form the overall reference framework for all those who have chosen a country as the destination of their own migration project (Appaduray, 1996).

2 The main patterns of migrant incorporation in the Italian labour market can be reconstructed by two main data sources. The first source is the INPS (the National Social Security Agency) archives regarding subordinate employment in non-agricultural firms. The second source is the Ministry of Labour registration of job entry. Quantitative information about size and characteristics of the informal sector can be obtained from the inspections carried out by the local agencies of the Ministry of Labour and official data collected by ISTAT (the Italian National Statistical Institute).

3 Generally speaking, statistics on the number of migrants and their occupations are only partially available and they have a very problematic nature. Besides, the way immigrants obtain residence permits gives rise to distortions in official statistics. The number of regular housekeepers, for instance, is greatly overestimated as for many years it represented the only way of entering Italy legally. On the other hand, the actual number of housekeeping jobs may also be subject to underestimation, since many migrants, as well as many Italians, are taken on as housekeepers on an informal basis.

4 Here, the term Sikh is used to identify all the people coming from Punjab who live in the territory and are the object of our research (Osservatorio sull'immigrazione, 2002). Although obviously not all the Indians coming from Punjab and living in the Cremona province belong to the Sikh religion (in fact, many of them are Hindu), it has been chosen to define this population by the Sikh religious and cultural identity, since this category seemed to be able to sum up the historical and cultural characteristics of this population. This choice is shared by much international literature, which generally includes migration from Punjab within the framework of the 'Sikh diaspora'. With 2,172 immigrants of Indian origin, Cremona ranks at the top positions concerning the presence of migrants from India, who form the largest national group and account for 20.2 per cent of the total foreign resident population. They live in as many as ninety-six out of 115 communes of the Cremona province and they predominantly live in rural areas (93 per cent of them live in small towns).

5 Jat is a caste of Sikh landowners.

6 1984, the year in which Indira Gandhi was assassinated, was one of the most difficult periods in the history of relations between the Sikhs and the Union of India.

7 It should not be forgotten that internal migration of low-cost labour from other states (Bihar and Uttar Pradesh) further reduced wages and the number of jobs available (Wallace and Chopra, 1988: 44).

8 In any event, the destination of their savings may change: from the original idea of investing in their home country, they sometimes pass to investments in activities in Italy, but all the interviewees indicated that the purpose of emigrating was to make investments and to improve their current economic situation.

9 From this point of view, the results which emerge from our research are perfectly in line with the studies carried out in other countries where there has been immigration, and the members of the Sikh community of Cremona maintain intensive and continuous links with family members who remain in India (Barrier and Dusembery, 1989; Sarmiento, 1991; Wilson and Samuel, 1996).

10 Marriage is one of these events, and it is almost completely controlled by the relatives in India (parents, grandparents and siblings), who, in addition to finding a suitable partner, organize every aspect of the ceremony, which absolutely has to take place in India.

11 Although the Sikh doctrine denies the continued existence of castes (Delahoutre, 1994) there is among Sikhs a hierarchical system where the Jat (and not Brahamans) are at the top of the social pyramid. Moreover, Sikh caste groups show more indulgent endogamic ties than among Hindus (some inter-caste marriages are tolerated); they do not practise separation based on food (in *langars*, gurudwara kitchens, members of any religion and nationality are granted access without any form of discrimination). For an analysis of castes in Hinduism, please see Dumont, 1991.

12 The *ghot*, 'lineage', is the word used to describe some blood relation belonging to the same caste. Marriage must take place within castes according to endogamic principles but without the *ghot* (Angelo, 1997).

13 Within this framework, marriage comes to represent a useful tool to strengthen internal relationships within the group: the event coincides with the time where the families involved bring all the weight of their caste heritage into play.

14 The experience of Indians who have recently arrived in Italy and Cremona is, of course, different. These individuals, who are often young and single, and looking to improve their financial situation, they have often entered the local labour market immediately, initially through illegal work, and then, taking advantage of the various amnesties, by regular jobs.

15 This term comes from a particular type of cow called precisely *bergamina*.

16 The local population still vividly remember the struggle of farm workers in the Cremonese countryside to increase the miserable wages of agricultural workers (Ongaro, 1985)

17 There is another reason why employers are prepared to offer lodging: cottages are considered as rural housing for tax purposes, and cannot be rented out through regular rent contracts.

18 Our interviews with trade unionists show several cases of disputes initiated by Indians against their employers who had not complied with their work contract.

19 The national health service guarantees the protection of health for all Italian citizens and for foreign citizens who regularly live on the national territory. Registration is compulsory and free of charge when the foreigner holds a residence permit; in that case, registration also covers dependants. According to the same principle, the children of foreign citizens are guaranteed access to education.

20 For other cases of 'successful integration', see Palidda (2000).

References

Ambrosini, M. (1999) *Utili invasori*. Milan: Franco Angeli.

Angelo, M. (1997) *The Sikh Diaspora. Tradition and Change in an Immigrant Community*, New York and London: Garland.

Appaduray, A. (1996) Modernità *at Large. Cultural Dimensions of Globalization*, Minneapolis MN and London: University of Minnesota Press.

Barbieri, P. (1997) 'Non c'è rete senza nodi: il ruolo del capitale sociale nel mercato del lavoro', *Stato e Mercato*, April, 67–110.

Barrier, G.N. and Dusembery, A.V. (eds) (1989) *The Sikh Diaspora. Migration and Experience beyond Punjab*. Delhi: Chanakya.

Barrier, N. and Pashaura, S. (eds) (1996) *The Transmission of Sikh Heritage in the Diaspora*. New Delhi: Manohar.

Barrier, N. and Pashaura, S. (2001) *Sikh Identity: Continuity and Change*. Delhi: Manohar.

Bertolani, B. (2002) *Immigrazione. Dossier Statistico 2002*. Rome: Anterem.

Bertolani, B. (2003) 'Indiani punjabi in provincia di Reggio Emilia. Etnicità e parentela come nodi di inserimento in contesto migratorio'. Ph.D. dissertation, Università degli studi di Parma.

Caritas (2002) *'Lavoratori e cittadini'*, *Dossier Statistico Immigrazione*. XII *Rapporto Caritas – Migrantes sull'immigrazione*. Rome: Caritas Italiana.

Compiani, M.J. (1998) 'L'immigrazione sikh in Italia. La comunità dei bergamini', M.A. thesis in Political Science, Università degli Studi di Pavia.

Delahoutre, M. (1994) *I Sikh*. Rome: Interlogos.

Dhami, M.S. (1988) 'Caste, class and politics in the rural Punjab: a study of two villages in Sangrur district', in P. Wallace and S. Chopra (eds) *Political Dynamics and Crisis in Punjab*. Amritsar: Guru Nanak Dev University.

Dumont, L. (1991) *Homo Hierarchicus*. Milan: Adelphi.

Dusembery, A.V. (1989) 'A century of Sikhs beyond Punjab', in G.N. Barrier and A.V. Dusembery (eds) *The Sikh Diaspora. Migration and Experience beyond Punjab*. Delhi: Chanakya.

Fondazione Cariplo-I.S. Mu. and Provincia di Cremona (2000) *L'immigrazione straniera in provincia di Cremona*. Milan: I.S.Mu/Franco Angeli.

Galloni, F. (2000) 'Minori sikh a Cremona: Inserimento sociale e scolastico', M.A. thesis in Psychology, Università degli Studi di Padova.

Gallucci, F. and Venturi, P. (1984) 'L'economia indiana: un tentativo di bilancio', in E. Collotti Pischel (ed.) *India oggi. Lo sviluppo come speranza e come dramma*. Milan: Franco Angeli.

Granovetter, M. (1985) 'Economic action and social structure: the problem of embeddedness', *American Journal of Sociology*, 91: 481–510.

Granovetter, M. (1995) 'The economic sociology of firms and entrepreneurs', in A. Portes (ed.) *The Economic Sociology of Immigration*. New York: Russell Sage Foundation.

Ires Piemonte (1991) *Uguali ma diversi. Il mondo culturale, le reti di rapporti, i lavori degli immigrati non europei a Torino*. Turin: Rosemberg & Sellier.

Marenco, K.E. (1976) *The Transformation of Sikh Society*. Delhi: Heritage Publisher.

Massey, D.S. and Espinosa, K.E. (1997) 'What's driving Mexico–US migration? A theoretical, empirical and policy analysis', *American Journal of Sociology*, 102: 939–99.

McLeod, W.H. (1989a) *The Sikhs. History, Religion, and Society.* New York: Columbia University Press.

McLeod, W.H. (1989b) 'The first forty years of Sikh migration', in N.G. Berrier and A.V. Dusenbery (eds) *The Sikh Diaspora. Migration and Experience beyond Punjab.* Delhi: Chanakya.

Mingione, E. (1997) *Sociologia della vita economica.* Rome: Nuova Italia Scientifica.

Mingione, E. (2001) 'Labour market segmentation and informal work', in H.D. Gibson (ed.) *Economic Transformation, Democratization and Integration into the European Union* Basingstoke and New York: Palgrave.

Mingione, E. and Pugliese, E. (2002) *Il lavoro.* Rome: Carocci.

Mingione, E. and Quassoli, F. (1999) 'The insertion of immigrants in the underground economy in Italy', in R. King, G. Lazaridis and C. Tsardanidis (eds) *Eldorado or Fortress? Migration in Southern Europe.* London: Macmillan.

Morris, L. (2002) *Managing Migration. Civic Stratification and Migrants' Rights.* London: Routledge.

Mottura, G. and Pinto, P. (1996) *Immigrazione e cambiamento sociale. Strategie sindacali e lavoro straniero in Italia.* Rome: Ediesse.

Moulier-Boutang, Y. (1998) *De l'esclavage au salariat. Économie historique du salariat bridé.* Paris: Presses Universitaires de France.

Mutti, A. (1998) *Capitale sociale e sviluppo. La fiducia come risorsa.* Bologna: il Mulino.

Ongaro, E. (1985) *Camera del lavoro e lotte nelle campagne cremonesi.* Milan: Franco Angeli.

Osservatorio sull'Immigrazione, Provincia di Cremona (2002) 'Turbanti che non turbano. Indagine sociologica sugli immigrati indiani nel cremonese', Unpublished research report.

Palidda, S. (ed.) (2000) *Socialità e inserimento degli immigrati a Milano.* Milan: Franco Angeli.

Piore, M. (1983) *Birds of Passage.* Cambridge: Cambridge University Press.

Piselli, F. (ed.) (1990) *Reti. Le analisi dei network nelle scienze sociali.* Rome: Donzelli.

Portes, A. (1994) 'The informal economy and its paradoxes', in N.J. Smelser and R. Swedberg (eds) *The Handbook of Economic Sociology.* Princeton NJ: Princeton University Press; New York: Russell Sage Foundation.

Portes, A. (ed.) (1995) *The Economic Sociology of Immigration.* New York: Russell Sage Foundation.

Pugliese, E. (ed.) (2000) *Rapporto immigrazione. Lavoro, sindacati, società.* Rome: Ediesse.

Quassoli, F. (1999) 'Migrants in the Italian underground economy', *International Journal of Urban and Regional Research*, 23 (2): 212–31.

Restelli, M. (1990) *I Sikh fra storia e attualità politica.* Treviso: Pagus.

Reyneri, E. (1979) *La catena migratoria.* Bologna: il Mulino.

Reyneri, E. (1998a) 'Immigrazione ed economia sommersa', *Stato e mercato*, 53: 287–313.

Reyneri, E. (1998b) 'The role of the underground economy in irregular migration to Italy: cause or effect?' *Journal of Ethnic and Migration Studies*, 24 (2): 313–31.

Saraceno, C. (1998) *Mutamenti della famiglia e politiche sociali in Italia.* Bologna: il Mulino.

Sarmiento, J. (1991) 'The Asian experience in international migration', *International Migration*, 29 (2): 195–204.

Sassen, S. (1995) 'Immigration and local labor markets', in A. Portes (ed.) *The Economic Sociology of Immigration.* New York: Russell Sage Foundation.

Sassen, S. (1996) *Migranten, Siedler, Flüchtlinge. Von der Massenauswanderung zur Festung Europa.* Frankfurt am Main: Fischer.

Sayad, A. (1999) *La Double Absence.* Paris: Seuil.

Sciortino, G. (2000) *L'ambizione della frontiera.* Milan: Franco Angeli.

Spagni, P. (1984) 'Sviluppo e miseria dell'agricoltura indiana', in E. Collotti Pischel (ed.) *India oggi. Lo sviluppo come speranza e come dramma.* Milan: Franco Angeli.

Wallace, P. and Chopra, S. (eds) (1988) *Political Dynamics and Crises in Punjab.* Amritsar: Guru Nanak Dev University.

Wilson, W.R. and Samuel, T.J. (1996) 'India-borne immigrants in Australia and Canada: a comparison of characteristics', *International Migration*, 34 (1): 117–42.

Zanfrini, L. (1999) *La discriminazione nel mercato del lavoro.* Milan: Franco Angeli.

Zincone, G. (ed.) (2001) *Secondo rapporto sull'integrazione in Italia.* Bologna: il Mulino.

9 Filipinas in Spain

Learning to do domestic labour

Natalia Ribas-Mateos and Laura Oso

In the past twenty years, foreign female immigrants have started to settle in Spain, mainly as workers in the expanding service economy of the larger metropolitan areas such as Madrid and Barcelona. López de Lera (2004) indicates an increase in the foreign population of 265 per cent for the period 1996–2001. At the beginning of 2002 there were two million foreigners registered in the local census (Padrones Municipales de Habitantes-INE), four times more than the 1996 census. These data also show that most of the entrepreneurs are African men, mainly Moroccans and Senegalese, and the few remaining women entrepreneurs are mainly from Asia, especially Chinese (41 per cent).

The situation of the Filipino community in Spain is characterized by an overwhelming representation of women employed in the domestic help sector. Furthermore, Filipina women are often characterized by a downward professional trajectory when considering the educational standards and skills they previously obtained in their country of origin. This explains why the main aim of this chapter is to present from another perspective, through an underutilized approach, the entrepreneurial conditions that have placed Filipino women at a disadvantage in terms of labour and social mobility.[1] The case of Filipinas in Spain was chosen because it seems to shape a particularly clear empirical example of the existing obstacles for the labour integration of migrants in ethnically and gender-specific segmented markets. Paradoxically, in contrast with other Asian communities, Filipino female migrants are the main exponents of non-entrepreneurial strategies, which are generally associated with Asian populations, e.g. the alleged entrepreneurial ethics of the Chinese (Beltrán, 1998; Teixeira, 1998). The absence of such an entrepreneurial ethic and the present mechanisms of professional and social stagnation by remaining in the domestic service sector have led the Filipino community to remain as dependent workers. For this chapter we will consider Filipino female migrants in Spain as an illustrative example of female Asian entrepreneurship in Europe by underpinning the few opportunities they have on the occupational ladder, peculiarized by tight exit barriers leading to blocked social mobility. Therefore, this case is taken as an anti-entrepreneurial example in contrast to other chapters of this book, in a way because studies of Filipinas have already focused on their status as dependent workers. Nevertheless, we decided not only to take this fact into account but also to address future changes in migration dynamics. This change would especially consider the processes of upward social mobility through entrepreneurial activities.

What kind of framework helps us to answer our question on entrepreneurship conditions? In the sociology of migration there are numerous theories and wide empirical research which help us to discern migrant entrepreneurship as a means of achieving social mobility. For our purposes, it is helpful to distinguish here four groups of theoretical thought: first, middleman theories are to be found as the main precedent for the study of social mobility in migration studies, through the focus on the small business dynamics as a mode of incorporation (Bonacich, 1973). Second, ethnic enclave theories see internal solidarity as the main element which facilitates incorporation in the labour market. Conditions in the country of origin (economic and political conditions) and in the country of immigration (culture, social class, skills obtained) are the key factors structuring the incorporation of migrants into the receiving context. Instead of viewing the ethnic enclave as a site for exploitation, the ethnic enclave concept allows us to see the ethnic business as an entrepreneurial engine, as a ladder for social mobility and as a potential means of becoming an entrepreneur. It is along these lines that Wilson and Portes considered the ethnic business as a space for training rather than for exploitation (Wilson and Portes, 1980). Third, structural theories have emphasized contextual external factors, especially focused on the split labour market theory. They argue that the socio-economic exclusion of migrants, especially in the labour market, makes migrants search for other types of resources. Again, as with the middleman theories, the main explanatory weight is placed firmly on the limited opportunities in the economic structure instead of in the cultural and group traits (Bonacich, 1973).

Fourth, the mixed-embeddedness approach has opened up recent debates on the interplay of economy and ethnicity. Approaches based on mix-embeddedness try to combine different factors based on structural factors, social factors as well as personal characteristics, by interrelating urban economic conditions and political-institutional factors (Rath, 2000). Niches would develop in the interaction between the group and its surrounding society, in which the embeddedness in social networks is of crucial importance. Therefore, the type of economy, the type of the welfare state we are referring to, as well as the group characteristics, would all be key contextual factors in understanding the possible means for social mobility using the ethnic business strategy.

Moreover, in spite of all the above-mentioned four sets of arguments, the literature which has been used to explain the phenomenon of immigrant entrepreneurship, especially in the United States, the Netherlands and the United Kingdom, has not really dealt with the gender dimension in understanding ethnic/immigrant business. Only a few studies take a gender perspective (among them Morokvasic, 1991; Anthias, 1992; Hillmann, 1999; Oso, 2004).

In our understanding, the case of Filipino women represents a good example to apply the 'mixed embeddedness' approach (Waldinger, 1995; Rath, 2000). Thus, we seek to identify the strategies of Filipino women in relation to the structural, urban and politico-institutional environment, while integrating the gender dimension. The main issue of this chapter is to analyse the opportunities and strategies of Filipino women in relation to the structural and politico-institutional factors and urban economic conditions (in Madrid and Barcelona). We will then explain their stagnated mobility related to those obstacles, which prevent them from escaping the domestic service sector. We give special weight to the importance of social factors (community cohesion, personal characteristics) and women's strategies in shaping migratory and mobility strategies. Furthermore, Filipino women seem to be closely linked to their ethnic social networks which subject them to traditional family and community obligations.

Filipinos in Spain: a southern European perspective

New recent migrations in Europe pose a challenge in drawing up the various models of migrant incorporation, and especially in creating a suitable model in which we can consider this new, yet diversified, immigration phenomenon in southern Europe (King *et al.*, 1997; Ribas-Mateos, 2001). Such studies have focused particularly on the characteristics of heterogeneity, which can be defined by the multiplicity of nationalities and types of migrants (from both rural and urban origins), the feminization of flows and the important gender asymmetry depending on different origins.[2] In spite of stringent border controls in all four countries, their important informal economies play host to a large number of irregular immigrants, with a high representation of this labour force in certain sectors: services, agriculture and construction (see Solé *et al.*, 1998, for the Spanish case). As in the case of Italy, the Spanish economy is crosscut by pronounced structural divisions, and it is split by a technologically advanced primary sector which is highly unionized and state-regulated, and an extensive underground economy. Among the new changes in this division is the fact that in recent years 'irregular' work has become more organized; rather than representing a marginal sector it has become a central component of Spain's industrial strategy within the activity of small enterprises. Simultaneously, Spain has become a new destination for foreign migrants. Consequently, the socio-economic transformation of southern Europe with regard to post-Fordist structures (tertiarization, flexibilization and informalization of the market, affecting above all women and youngsters) has occurred hand in hand with the role of the country as a receiver of immigrants (Ribas and Díaz, 2001; Mingione, 1999).

An analysis of the data dealing with foreigners holding a residence permit in Spain in 2001 shows that the foreigners are predominantly of European origin (412,522), followed by Africans (304,149) and Americans (298,798). Asian immigrants form the smallest group, but show highly dynamic growth; in just seven years the total number of Asian residents almost doubled, rising from 35,742 in 1994 to 91,552 in 2001. Whilst immigration from the African continent is predominantly male (69 per cent males), immigration from Latin America is overwhelmingly female (58 per cent females). In the case of Europeans and Asians the distribution according to gender is more evenly balanced, although men form the more numerous group (54 per cent and 60 per cent of men respectively). Regardless of the continent and country of origin, the majority of foreign female workers are employed in domestic help (62 per cent), whilst male employment is more widely spread among varying economic sectors (*Estadística sobre permisos de trabajo a extranjeros*, 1999).

Among the group of Asian residents, the Chinese form the most numerous group (36.143 persons), followed by the Filipinos (14,716 persons). Males are predominant among those from China (56 per cent), whilst women constitute the largest part of Philippine migrants (60 per cent). Both those from China and those from the Philippines reside mainly in Catalonia, where the city of Barcelona weighs heavily as the preferred host community (33 per cent and 37 per cent respectively) and in Madrid (29 per cent and 41 per cent respectively), with, as we can see, a fair-sized Filipino community residing in the Spanish capital (*Anuario Estadístico de Extanjería*, 2001). The Filipino community shows a considerable attraction to the borough of Madrid, where a total of 81 per cent of its members are concentrated, and in particular to the district known as Almendra Central, where 78 per cent of all the Filipinos in the town are registered (Lora-Tamoyo, 1999). The district of Ciutat Vella, covering the historic centre of the city of Barcelona, represents the largest increase in the number of foreigners in the city. Since 1986 their number has multiplied by seven (Fundació CIDOB, 2001). The specific pattern of migrant concentration in Ciutat Vella is especially seen in the case of the Pakistanis (71 per cent). They are followed by the Filipino community, with 65.5 per cent of the total number of residents living in this district. Filipinos are one of the best graphical expressions of those concentration tendencies, with 53 per cent of the total number of residents living in the Raval district (Barcelona City Council, 2001). Whilst the streets of Madrid and Barcelona are witness to flourishing ethnic businesses (and mostly the traditional neighbourhood of Lavapiés–Almendra Central in Madrid and the Raval–Ciutat Vella in Barcelona), Filipino migrants remain firmly entrenched in the domestic service sector.

When considering self-employment within the various immigrant groups, migrants of Asian and North American origin turn out to be most active. They represent the largest group of self-employed workers in possession of a work permit (19 per cent and 19 per cent respectively). This leads us to conclude that the Asian community represents a considerable percentage of self-employed workers, in comparison with the total number of immigrants in Spain. The Philippines stand out from other Asian countries in the sense that the vast majority of workers in possession of work permits are employees, demonstrating a far lower level of ethnic entrepreneurial activity and self-employment (0.7 per cent) than among migrants to Spain from the other principal Asian countries, such as China (26 per cent). The total proportion of foreign male workers with a permit for self-employed work in Spain is far higher at 12 per cent than that of women, namely 5 per cent (*Estadística sobre permisos de Trabajo a Extranjeros*, 1999).

The domestic service sector absorbs practically all female immigrant workers in Spain. Self-employment is rare in the case of Filipinas and at even lower levels than for their male counterparts. This high demand for domestic workers has been promoted by specific migratory policies, especially since the introduction of the quota policy in 1993. This policy officially acknowledged the fact that there are job vacancies which are not covered by the native work force and for which there is a demand for foreign workers, providing certain economic sectors with greater labour flexibility. Since the introduction of this policy, the annual quotas have been filled mainly by foreign domestic service workers. By selecting three time periods we can follow the logics of the recruitment of foreign workers: in 1993, 84 per cent of work permits were issued for foreign workers in this sector, compared with only 61 per cent in 1995 and 52 per cent in 1999 (Ministry of Employment and Social Services). Data on the distribution of foreign workers according to gender and marital status, which is not normally published in official statistics, can, however, be sketched at least for the year 1998 (Izquierdo, 2001). The largest group of married Asian women came from China (61 per cent). In contrast, Filipino immigration appeared to follow the pattern of feminized labour-oriented trends independent from those of their male counterparts, due to the considerable proportion of single women (53 per cent).

To sum up, an analysis of the statistical data presented here allows us to define a series of trends in Asian immigration in Spain related to gender, marital status and occupation. First, male-dominated immigration, mainly Chinese, where the high number of married men and women could indicate family reunion or migration of the couple. These migrants include a fairly large number of self-employed workers.[3] Second, it is composed of highly feminized migration flows, including a large share of single women. This immigration is basically work-motivated and is composed mainly of domestic workers, with a very low number of self-employed workers. This general profile of Asian immigration brings with it a series of questions worth asking. Why has self-employment developed among Chinese immigrants in Spain, whilst Filipinos mainly work as salaried employees? Is it the gender difference in these two migratory trends that determines this issue? Are job opportunities more restricted for Filipino women than for the predominantly male Chinese community? We now will continue considering different explanatory factors: mainly the ethnic background and the labour market conditions.

First explanatory factor: the 'ethnic' background

Background or conditions in the country of origin

The fieldwork carried out a decade ago in the country of origin (Ribas, 1993) allows us to analyse the differences in the social capital component of migration and its contribution to the migration project. Most of the current bibliography on international migration refers to the conditions of the reception contexts in relation to immigration policies and the demand for labour (Table 9.2). This can be explicitly seen in the case of Spain, where the feminization of flows responds, as we have already pointed out, to the inter-action of the following factors: the job vacancies officially established and job specialization within a quota system. Nevertheless we can see (Ribas, 2000) how gender discrimination in immigration policies also corresponds to emigration policies in the countries of origin, as in the case of the Philip-pines, where the government has become a mediator in international migra-tion. Apart from recruitment agencies operating in their home countries (as in the case of the Filipino women, and those from Sri Lanka and India, see Anderson, 1996), many domestic workers have entered the country through family networks (called 'direct hire' by them) and have emigrated for family reasons (Table 9.1). As shown in many studies of these women from less developed countries, the effect of this work is the opposite. It means that other women have to replace them in their places of origin, taking over domestic chores and childcare responsibilities. The ethnic background is supposedly the guide for the business immigrant's behaviour, connected with the pre-migration experience, as it is also important for understanding the connection between the self-employment business and the economic

Table 9.1 Social factors

The woman within the context of origin
Filipino women's traditional leadership role and community cohesion
Women's work ethic in the country of origin
Higher female literacy rate than men
The idea of sacrifice by women, the figure of the woman and elder sister bearing the responsibility for the whole of the immediate family.

Social networks
The absence of business networks between the country of origin and destination
The influence of family networks in the Philippines and their importance for women migratory projects.

Women strategies
The strong family element contained within female geographical and social mobility strategies for Filipinas in Spain, which prevents them from leaving domestic service, due to the fact that this job sector is better suited to their strategies, strongly focused on saving and sending money back to the Philippines.

Table 9.2 Structural factors

Structural factors
Segmented labour demand for domestic help in Madrid and Barcelona due to:
Access by Spanish women to skilled, non-domestic work, combined with the lack of state support for reproductive services and an equal distribution of domestic tasks between partners
The established image of the Filipino immigrant as a status symbol among upper-middle class urban families and the channelling of migrant flows towards the service niche

Politico-institutional factors
Migration policies:
The implementation of quota policies has helped the legalization of women within the domestic service market and has channelled them towards this sector of the job market
The determination of the receiver context as a 'gateway', influenced by migratory policies, which create future obstacles to mobility within the labour market

Lack of social integration policies:
No training schemes for migrant women
No help for micro-credits
The difficulty involved in validating studies which would allow female Filipino migrants to look for jobs more in keeping with their qualifications

The general context of a weak welfare state:
The concept of citizenship
The limitations of universal social services
Lack of family policies

Urban economic conditions: Barcelona–Madrid
Growth of urban service economy, especially affecting migrant women
Migrants' spatial concentration in the centre of the cities
Zones of immigrants' business

---

development of the areas of origin. We might differentiate various patterns of pre-migration.

A *internationalized migration pattern*

The Philippines are considered to be one of the world's leading countries in numbers of emigration. The tradition of emigration and the penetration of migration in all aspects of society (from government to all types of employment agencies, travel agencies, banks, business administrators, street hawkers, etc.) have produced a so-called 'migration mentality' in the population. A range of possible migratory destinations is open depending on the socio-economic situation of the family. From the perspective of a person who lives in the countryside, the cheapest option would be to migrate to Manila (although it might not be such a profitable move in material terms). The second option for emigration would be Hong Kong or Singapore, the third option Saudi Arabia (especially in the overseas contract work mode), the fourth place Europe and the fifth North America. On many occasions the choice of the latter country is based on the highest possible level of salary. (According to official data Spain offers medium-range salaries, Canada very high ones and Singapore and Kuwait low levels.) Often the standards of working conditions are ignored by the coming migrants, showing a low degree of awareness regarding the crucial aspects of the migrant worker's life in the country of destination (such as working hours, contact with the employers). This female internationalized pattern through contract work, especially for Asian emigration from the Philippines, is seen through the profile of migrant women aged between twenty-one and thirty, 18 per cent of whom are married (Ribas, 1993).

Gender divisions from the sending context

Female literacy rates are very high, as is the number of women with a university degree. The Philippines has the highest educational levels in South East Asia and shows even less contrast than Spain in the gender distribution of literacy rates. Women also show higher educational attainment than men, but enjoy fewer opportunities than their male counterparts when finding a position on the labour market, particularly one that matches their educational and training profile. The gender-based division of work in the country of origin shows an over-representation of women in subcontracted industries and in informal industries such as the textile sector. The dual presence of women in productive and reproductive tasks intensifies the pressures on women and adds to their vulnerability on the labour market. Partly the answer to the economic crisis is expressed in the feminization of the rural exodus (as also occurs with emigration flows to Catalonia, which, in turn, is often preceded by the Ilocano emigration to Metro Manila) and in the feminization of labour exports to foreign countries. Even in the case of rural to urban migration, women often choose a migration project driven by a family strategy.

Women's work ethic

Fieldwork in the Philippines showed a general perception among the families interviewed of migration as easier for women than for men. The reasons for this include the fact that job opportunities in foreign countries are mainly limited to domestic help, which is still regarded as typical women's work that meets the qualities attributed to the Filipino woman, such as perseverance, courage and a work ethic. (At the same time as they are also tagged as 'those who sacrifice themselves for the family'.) The level of education is an important criterion for migrant selection by recruitment agencies, with women having normally higher educational standards than men.

Generally speaking, the number of children may also have a positive effect on the migration project; the more children, the greater the need, and this becomes another variable of the pressure to emigrate.

The reason for migration is the difference in salaries and the fact that in the immigration country this sort of job is not perceived negatively. Moreover, the legitimization of domestic work outside one's place of origin can be seen for most types of female migration; it does not imply a loss of status in the eyes of their community, for it is well looked upon (even when being over-qualified for the job). However, it is not held in high esteem if done in the Philippines. This is because the women who do domestic work in the Philippines, besides being poor, are not educated. Curiously enough, many of the women who migrate employ domestic workers in their Filipino homes, but not under the same working conditions as in foreign countries. As in the case of Morocco, these workers are mostly family members – young girls who have to leave home due to a poor family economy, or in order to be able to study in the town; most come from remote places.

Women's entrepreneurial activities

The woman, often considered in the receiving context as the main agent of social integration, is seen in the country of origin as an agent of modernization and development. Labour market conditions, social and family organizations are keys for analysing gender in these contexts. The low-income women obtain from their businesses (*carinderia* and *sari-sari* – small shops – crafts, peddlers and shops in local markets), together with low wages and bad working conditions and the discrimination of Filipino women in the labour market, are some of the main obstacles to economic integration. The typical small business of Filipino women, both in rural and in urban areas, is the *sari-sari* store, the small kiosks scattered all over the country.

Work in the rural areas includes the whole life cycle, both children and the elderly. In the city a climate of job dissatisfaction is an added pressure on the 'push' migratory forces. The recipe, which combines 'socio-economic frustration–emigration', is seen through the high cost of goods and very low salaries, poor working conditions and discrimination against women in the job market. The interviews show how it is usually the more educated and socially aware women who refer to discrimination: 'in the bank they don't want you if you are married, they want you single' (interview in Pangasinan, Philippines).

The family and the community strategy

Even though women emerge as leaders of the migratory chains in Spain and seem to be more active in creating integration strategies, their migratory project is not always an autonomous expression. In the migration system, the family chain offers the context of support for the individual in an unknown place, and the house of the employer comes to symbolize the security offered by its living quarters. Many interviewees express the additional help in migration as the need for a *bahala na* attitude ('there is always hope'). Migration from the Philippines shows the importance of the *bahala na* attitude even in the worst situations. 'In the foreign country God will take care of my situation.' Despite the fact that the majority of female migrants are single, we chose to examine the migratory strategies of married Filipino women with children because it is in this context that the family migration issue can be better understood. Migratory strategies are usually similar for all women, in the sense that they involve a family-related goal: from the role of the eldest daughter as a family supporter as well as other forms of support within the extended family. The 'immediate' economic reasons for the interviewees in the Philippines were the following: nutrition and education of the children in the family and survival. Secondary reasons include the possibility of a higher standard of living and the desire to travel. In a society without a welfare state, the woman is frequently responsible for carrying out the migratory project. She has to support the family, make economic pressures more bearable and shape a better economic future, either in or beyond the Philippines. However, for women the children are the most important factor. It is therefore understandable that the role of the mother and maternal sacrifice is the driving force behind migration. Nevertheless, other objectives are not abandoned, such as helping other members of the family, or migration as an 'autonomous act'.

In general, the migratory project is threefold. On a short-term basis, survival – the feeding of the family – is important, 'We are supposed, as Filipino women, to stay at home, but we have to go abroad in search of a good income, due to unemployment, because of the children' (Filipino woman interviewed in Manila). On a medium-term basis, children's education and, on a long-term basis, improving social status and social position are important. The long-term objective is understood through a conception of the future, a future of family stability, achievable through sacrifice as a mother and through a work ethic based on saving for a future life (that of her children, her own family, and even including the rest of the extended family). There is strong hope of return to the homeland, but this is not usually conceived as an organized project. Generally speaking, the idea of return depends on the economic situation of the Philippines and on the obstacles to occupational mobility in receiving labour markets.

As aforementioned, one of the most alarming results of the fieldwork conducted in the Philippines has been to verify the sacrifice involved in long-term separation in order to achieve migration objectives: the migration of married women with young children and the separation of the couple, even for periods as long as ten or twelve years. Migration and the temporary break-up of the family are accentuated especially in the case of married women with children. The children who stay in the Philippines are often left in the care of the father, the grandparents or other members of the family while the woman works in Catalonia. It is understood that the separation of the couple due to migration is part of a couple's commitment to a family project. On the other hand, there are cases that suggest an unofficial separation of the couple, taking into account the fact that in the Philippines there is no divorce. In many cases, the married woman lives in Catalonia and has small children in the Philippines. One of the forms of community reproduction in the immigration country is organization on ethnic grounds, especially through the Filipino traditional value of *bayanihan*, a strong sense of interaction inside the community and the sense of a common struggle. However, in the Philippines ethnic diversity can also be found, and, in spite of the great variety of languages, the Tagalog and the English language appear to be most dominant. Tagalogs are also those who enjoy an easier path to decision making, which can also be reflected through their dominant role when looking at Filipino associationism in Catalonia. Examining the various forms of association in each of the three countries has shown a noteworthy type of social reproduction in the country of destination. From the Filipino idea of *pakikipagkapwa* (human preoccupation and interaction) and *bayanihan* (a common struggle), one can understand the strong sense of interaction in the community. In the interviews conducted in the Philippines, the nature of creating associations is affected by the activities of the *barangay* (minimal political unit) and by Catholic action, though far less through direct political life.

Second factor: labour market conditions

We will here try to provide the main explanations for mobility stagnancy in the domestic service sector through the analysis of qualitative data obtained from fieldwork involving Filipino, women carried out in Madrid during 1996 and 1997 (Oso, 1998), which has been extended and updated for this contribution.

The arrival of Filipino women and their incorporation in the Madrid domestic service market in the 1970s constituted the first migratory wave of women arriving alone, and which would develop further during the 1980s and 1990s in the Madrid metropolitan area. Those feminized flows would also include other women from various countries such as Morocco, the Dominican Republic, Peru, Ecuador and Colombia. Female migratory flows are attracted by a demand for reproductive services (especially domestic work, care of children and the elderly). This demand for domestic service, especially for live-in help, was developed due to a number of reasons: (1) the mass incorporation of young Spanish women to secondary and university education and to the skilled job market (Garrido, 1992), (2) weak public services offering a solution to the need for social reproduction in Spanish homes (nurseries, old people's homes, etc.) and the lack of a family policy, (3) the problems experienced by Spanish males in adapting to the need to share domestic tasks (Oso, 1998), (4) The development of the detached home in the areas surrounding cities such as Madrid, which provides enough space in the home for live-in domestic help (Herranz, 1996).

Apart from the traditional demand for domestic workers by the wealthiest classes, as a symbol of their status and social position, the 1970s saw an increase in this demand by the middle classes, many of them young professional couples, who needed domestic help to replace the absent figure of the woman in the home (housework, looking after the children). To this, we must also add the on-going demographic change in the population, requiring more and more assistance for the elderly.

It must be noted that the figure of the maid from the countryside, who traditionally filled this social need, disappeared from the Madrid scene to become a permanent day time domestic worker paid by the hour, or an employee of cleaning companies, thereby rejecting the low prestige job, the least valued form of cleaning and social care employment, namely live-in domestic service (Catarino and Oso, 2000). Consequently, a niche was created in the labour market, which, as we have already mentioned, was first discovered and filled by Filipino women.

Fieldwork carried out in Madrid brought to light the fact that it was the wealthy Spanish families, with businesses in the Philippines, who brought the first Filipino women to Spain to work as domestic servants in their Spanish homes (Oso, 1998). This form of recruitment created in turn a series of employment networks between employers and immigrants which would later feed migratory trends, filling the demand for domestic service in Madrid among the wealthiest sectors of the population with Filipino women. This demand for domestic service was, as has been pointed out earlier, linked with a function of social differentiation rather than with a survival strategy, as occurred with middle-class women, subjected to the pressure resulting from work in and outside the home. The excellent reputation of the Filipino domestic employee – obedient, Catholic, English-speaking, exotic and a symbol of social status – highlighted in a number of studies of the phenomenon of female immigration in southern Europe, also favoured the consolidation of this group in the more stable sectors, offering them good working conditions and the highest salaries on the domestic service market, enabling them to occupy a clearly differentiated position in comparison with other groups on an ethnically defined professional scale.

Indeed, the middle classes can devote less of their salary to pay for domestic help. Furthermore, variations in family income among this class have a direct impact on the consumption of services; during an economic crisis of expenditure, domestic service is the first to be reduced, directly affecting job stability. Moreover, the middle classes usually employ only one person, who has to carry out all the household tasks on her own, whilst the wealthier classes tend to hire more than one employee, who specialize also in certain functions, leading to a lower work load. Similarly, the middle classes have a greater tendency to contract women in an irregular legal situation, thereby saving on social security costs. The result of this was that the Filipino woman, thanks to her excellent reputation and consolidated employer–employee social networks, gradually penetrated the wealthy social classes, occupying an optimum employment niche for new arrivals.

As we have already underlined, a number of studies have highlighted the way in which ethnic entrepreneurs create a mechanism for tackling the obstacles hindering integration into the job market, in the absence of a network facilitating their absorption into the national labour market. The fact that Filipino migration to Madrid moves towards the domestic service niche among the wealthy classes and the existence of an employer–employee network on the national market channelling the jobs to these immigrants results in that, unlike other groups of Asian immigrants, the ethnic economy among the Filipino community remains practically not developed. Moreover, live-in domestic work meets the migratory objectives of Filipino women, in as much as it provides them with the opportunity to save most of their wages, due to the fact that their board and lodging expenses are covered.

One of the features of Filipino immigration is the influence brought to bear by the collective component on immigration, in the sense that immigration mostly responds to a strategy implemented by the family staying behind in the Philippines. Indeed, many single Filipino women leave their country to provide their families with the resources they lack in order to provide their brothers and sisters, nieces and nephews with a good education, which implies considerable responsibility and a sense of sacrifice towards their relatives in the Philippines. There are also married women who leave their husbands behind. Some authors have suggested that one of the reasons why some immigrants fail to develop an ethnic entrepreneurial spirit is that the migratory project is a temporary and not a long-term one. This means that the immigrant prefers to work as a waged employee instead of facing the financial risks involved in setting up a business. Likewise, the fact that immigrants have to send money back to their country of origin means that the money they earn cannot be used for investing in a business (Waldinger *et al.*, 1990). It is precisely the financial burden represented by the family in the Philippines which explains why female migrants feel obliged to continue sending money home and which prevents them from saving and business investment (Table 9.3). Live-in domestic help is therefore the job which is best suited to migrant women, in enabling them to succeed in a collective strategy for social mobility. Therefore, it is precisely the fact that social mobility through migration is focused on the society in their country of origin and not on the country of destination which determines the overwhelming decision by women not to opt for the ethnic economy.

The importance of gender weighs heavily on the women's family strategy, backed up by their feelings of responsibility towards the family and their sense of sacrifice, which is part of the traditional Filipino female culture. In fact, after the mother of the nuclear family, the role of the emigrant is often

Table 9.3 Means to exit the domestic service sector

Care service employment
Leaving live-in domestic help and obtaining cleaning jobs paid on an hourly basis
Leaving cleaning services and doing other type of private family services (care of the sick, care of the elderly, as planned in the future EU offer of nursery services)

The care service business
Care services which arrange contracts between employer and employee. They organize live-in domestic help, cleaning services, babysitters, care of the elderly, substitutes for catering workers, shop attendants and clerical work.

Catering industry employment
Working as waiters in Chinese and other types of restaurants

Catering industry entrepreneurship
Opening bars and restaurants, a strategy adopted by Filipino couples in particular

Small shop entrepreneurship
Family-owned and not women-owned
Directed at an ethnic clientele or small older Spanish shops taken over by Filipinos
Food shops, small supermarkets, hairdressing salons
Combined goods, for example: telephone centres with hairdressers

Peddler work
Network business based on remittance services
They include a variety of countries and have offices in the main areas of migrant concentrations in Spain
Clandestine textile warehouses entrepreneurship

Informal income strategies
Women's informal economy among the community: buying and selling cosmetics, clothes, take-away food. These strategies are usually combined with other formal arrangements

Leaving productive work completely

---

adopted first by the eldest sister, which would explain the large exodus of single women travelling alone, yet those who do do so within the context of their (own) family project. Fieldwork has also shown that some of these single female migrants opt for 'permanent celibacy', due to the fact that marrying in Spain may place the collective strategy for social mobility in the place of origin at risk. For instance, in the case of the husband's objection to the sending of money back to the woman's home, it appears that they value the collective – family – livelihood strategy more then being a married woman. Besides, marriage in Spain implies giving up live-in domestic service, with the corresponding loss of savings capacity and transfer of money back to the country of origin. The result is that many Filipino women tend not to abandon live-in domestic service, driven by this sense of responsibility towards the family and personal sacrifice, despite the fact that this type of employment brings with it strong paternalistic relationships of domination (Catarino and Oso, 2000). They behave unlike other migrant women, such as those from Ecuador and Colombia, the majority of whom, after a spell as live-in domestic workers, opt to live out, within the context of their social mobility strategy in Spain. In fact, we have come across cases of Filipino women who have been working as live-in domestic servants for the same family for more than thirty years.

On the other hand, family reunion, especially in the case of men, implies considerable expense for the transnational home, right from the moment when the woman is forced to leave her live-in employment and spend her wages on keeping the reunited family members and home, thereby slowing down the transfer of money back home. As Lever-Tracy *et al.* (1991) point out, family reunion facilitates the development of the ethnic economy. Consequently, the difficulties experienced by the Filipino woman employed in live-in domestic service in reuniting her family hinder the creation of ethnic entrepreneurship and businesses. Moreover, the situation of the job market in Madrid and Barcelona presents various obstacles for the male migrant, as, unlike the case of women immigrants, there is no niche for men enabling them the join the work force (Ribas-Mateos, 2002). While Filipino men tend to work with their wives, in positions offered for couples in live-in domestic service, Filipino men also work as waiters or chefs, many of them in Chinese restaurants. It is interesting to see how working in ethnic businesses owned by Chinese immigrants is a common practice adopted by Filipino men and women alike.

Conclusion

We have selected the case of Filipino women in Spain as a paradigmatic case (labelled as an anti-entrepreneurial example). It has enabled us to go in depth through a rich interpretation of business start-up conditions, and to ask why some others do and others do not, when considering ethnic, migrant origins as well as gender divisions. By opening up the debate on the difficulties of starting up female migrant business, and by showing the limited opportunities migrants have in rising up the occupational ladder, for example characterized by tight exit barriers leading to blocked social mobility, we decided to utilize the mixed-embeddedness approach. Indeed, the presented pioneer example, Filipino migration to Spain, is made up of a feminized migratory flow of domestic service employees and, unlike other Asian communities, such as the Chinese, did not develop an ethnic economy. The female Filipino migration flow is considered as a relevant case in this chapter because of its strong presence in southern European urban sector services, its original type of residential concentration and as being a clear example of the internationalization of reproductive services in global genderized labour migration. All these factors contrast with the situation of other Asian communities in Spain. Our aim in this chapter has been to describe our specific case, that of Filipino women and entrepreneurship, within the framework of some of the more relevant theoretical thoughts on ethnic entrepreneurship.

The mixed embeddedness approach has helped us to evidence the general context in which to understand the tight barriers that exist to domestic help in an ethnic and gendered urban segmented market. In other words, it has helped us to establish factors based on the type of economy and the type of welfare and group characteristics, which are supposed to interact in the development of immigrant business. Indeed, we have defined Filipino women's immigration to Spain, their position in the domestic service market and their failure to develop an ethnic economy against a background composed of: the influence of structural and politico-institutional factors, the conditioning factors of urban economies (in Barcelona and Madrid), and social networks and women's strategies. Within the analysis of this latter factor, that of women's strategies, and within the framework of a gender approach, we have found how stagnant mobility must be understood in the context of the feminization of reproductive services. The gender approach has also enabled us to reconsider the importance of informal networks among women, and the invisible role of female migrants in the public and economic space. The explanatory factors in tables 9.1–3 summarize our explanatory argument on the conditions of entrepreneurship.

Notes

1 We have basically used the following secondary sources: (1) data analysis (statistics from the Spanish Interior Ministry, the Ministry for Employment and Social Affairs, and special requested data regarding work permits for foreign workers, Izquierdo, 2000), (2) qualitative studies, based on in-depth interviews with Filipino women in Madrid (Oso, 1998) and qualitative fieldwork in the country of origin (Ribas, 1993).

2 The feminization of migratory flows is evident in the Autonomous Region of Madrid (51 per cent of foreigners with a residence permit are women) (*Anuario Estadístico de Extranjería*, 2001) and similar also for the city of Barcelona (Ribas-Mateos, 2002).

3 Chinese are often characterized by their entrepreneurial ethics, which make it easy for them to set up a small business (facilities, information, know-how), their use of labour costs (family and community resources, gender divisions) and by a labour system based on hard work (Beltrán, 2003).

References

Anderson, B. (1996) 'Living and working conditions of overseas domestic workers in the European Union. A report for Stichting Tegen Vrouwenhandel'. July.

Anthias, F. (1992) *Ethnicity, Class, Gender and Migration*. Aldershot: Ashgate.

Anuario Estadístico de Extranjería Año 2001. Madrid: Delegación del gobierno para la extranjería y la inmigración.

Barcelona City Council (2001) *La poblacion estrangera de Barcelona*. Barcelona: Ajuntament de Barcelona, Departament d'Estadistica.

Beltrán, J. (2003) *Los ocho inmortales cruzan el mar. Chinos en el Extremo Occidente.* Barcelona: Bellaterra.

Beltrán Antolín, J. (1998) 'The Chinese in Spain', in Gregor Benton and Frank F. Pieke (eds) *The Chinese in Europe.* Basingstoke: Macmillan.

Bonacich, E. (1973) 'A theory of middleman minorities', *American Sociological Review*, 38: 583–94.

Catarino, C. and Oso, L. (2000) 'La inmigración femenina en Madrid y Lisboa: hacia una etnización del servicio doméstico y de las empresas de limpieza', *Revista Papers* monograph, *Inmigración femenina en el Sur de Europa*, 60: 183–207.

Estadística sobre permisos de Trabajo a Extranjeros (1999) Madrid: Ministerio de Trabajo y Asuntos Sociales.

Fundació CIDOB (2001) *Observatori de la immigració estrangera a Ciutat Vella*, Informe 6.

Garrido, L. (1992) *Las dos biografías de la mujer en España.* Madrid: Instituto de la Mujer.

Herranz Gómez, Y. (1996) *'Formas de incorporación laboral de la inmigración latinoamericana en Madrid. Importancia del contexto de recepción'*, Ph.D. thesis, Departamento de Sociología y Antropología, Universidad Autónoma de Madrid.

Hillmann, F. (1999) 'A look at the hidden side: Turkish women in Berlin's ethnic labour market', *International Journal of Urban and Regional Research*, 23 (2): 267–82.

Izquierdo, A. (ed.) (2001) *Mujeres inmigrantes en la irregularidad. Pobreza, marginación laboral y prostitución.* Unpublished research report, Instituto de la Mujer, Ministerio de Trabajo y Asuntos Sociales.

King, R., Fielding, A.J. and Black, R. (1997) 'The international migration turnaround in southern Europe', in R. King and R. Black (eds) *Southern Europe and the New Immigrations.* Brighton: Sussex Academic Press.

Lever-Tracy, C., Ip, D., Kitay, J., Phillips, I. and Tracy, N. (1991) *Asian Entrepreneurs in Australia. Ethnic Small Business in the Chinese and Indian Communities of Brisbane and Sydney.* Report to the Office of Multicultural Affairs, Department of the Prime Minister and Cambine. Canberra: Australian Government Publishing Service.

López de Lera, D. (2004) 'Análisis cuantitativo sobre género y empresariado étnico en España', in L. Oso (ed.) *El empresariado étnico como una estrategia de movilidad social para las mujeres inmigrantes en España.* Unpublished research report. Madrid: Instituto de la Mujer.

Lora-Tamoyo d'Ocon, G. (1999) *Extranjeros en la comunidad de Madrid.* Madrid: Delegación Diocesana de Migraciones and ASTI.

Mingione, E. (1999) 'Immigrants and the informal economy in European cities'. *International Journal of Urban and Regional Research*, 23 (2): 209–11.

Morokvasic, M. (1991) 'Roads to independence: self-employed immigrants and minority women in five European states', *International Migration*, 29 (3).

Oso, L. (1998) *La migración hacia España de mujeres jefas de hogar.* Madrid: Instituto de la Mujer.

Oso, L. (ed.) (2004) 'El empresariado étnico como una estrategia de movilidad social para las mujeres inmigrantes en España'. Unpublished research report. Madrid: Instituto de la Mujer.

Rath, J. (2000) *Immigrant Business. The Economic, Political and Social Environment.* Basingstoke and London: Macmillan.

Ribas-Mateos, N. (1993) 'Des de Manila sense mantó. Un estudi sobre l'origen del procés emigratori de la dona filipina a Catalunya'. Barcelona: Institut d'Estudis Catalans.

Ribas-Mateos, N. (2000) 'Female birds of passage: leaving and settling in Spain', in Anthias Floya and Gabriella Lazaridis (eds) *Women in the Diaspora. Gender and Migration in Southern Europe*. Oxford: Berg.

Ribas-Mateos, N. (2001) 'Revising migratory contexts: the Mediterranean caravanserai', in Russell King (ed.) *The Mediterranean Passage. Migration and the New Cultural Encounters in Southern Europe*. Liverpool: Liverpool University Press.

Ribas-Mateos, N. (2002) 'Migrant women in southern European cities', in M.L. Fonseca, J. Malheiros, N. Ribas, P. White and A. Esteves (eds) *Immigration and Place in Mediterranean Metropolises*. Lisbon: Luso-American Foundation.

Ribas-Mateos, N. and Díaz, F (2001) *Migrants and Ethnic Minorities on the Margins (MEMM). The Case of Spain*. Report for the Vienna Observatory on Racism.

Solé, C., Ribas, N., Bergalli, V. and Parella, N. (1998) 'Irregular employment amongst migrants in Spanish cities', *JEMS*, 24 (2): 333–46.

Teixeira, A. (1998) 'Entrepreneurs of the Chinese community in Portugal', in Gregor Benton and Frank Pieke (eds) *The Chinese in Europe*. Basingstoke: Macmillan.

Waldinger, R. (1995) 'The other side of embeddedness: a case study of the interplay of economy and ethnicity', *Ethnic and Racial Studies*, 18: 555–79.

Waldinger, R., Aldrich, H. and Ward, R. (1990) *Ethnic Entrepreneurs. Immigrant Business in Industrial Societies*. Race and Ethnic Relations series 1. Newbury Park, CA: Sage.

Wilson, K. and Portes, A. (1980) 'Immigrant enclaves: a comparison of the Cuban and black economies in Miami', *American Journal of Sociology*, 78 (May): 135.

10 Filipina women as domestic workers in Paris

A transnational labour market enabling the fulfilment of a life project?

Liane Mozère

Among Asian migrant workers in Paris, Filipino men and especially women occupy a specific economic niche. They are known predominantly as domestic workers, where they are highly valued by employers. As a Filipino joke puts it, they are reputed to be 'the Mercedes-Benz of domestic workers'. Filipinas in Paris employed in domestic service are, as in other parts of the world, skilled, educated and English-speaking. Furthermore, they often accept downward social mobility in order to send remittances back home. This is a paradox that a sociologist is eager to analyse. This chapter will tackle two major issues:

1   The first concerns the question why these Filipinas migrate. Seen from the supply side, the labour market in the Philippines offers them few, and usually poorly paid, opportunities. Moreover, the Filipino government has favoured emigration in order to solve domestic economic and social difficulties. On the other hand, this migration is linked with an international demand for domestic services Filipinas are able to offer.
2   International labour migration usually means exile, living far from home, with a heavy burden of sorrow, isolation and, in some cases, ill treatment. However, in this chapter, I would like to highlight the way in which the careers of Filipina international domestic workers are patterned and structured as life projects, during which they become more empowered. In other words, I would like to suggest that migration should be viewed in the long term. Many of these women have been in Paris over ten years and they act as entrepreneurs of a special type, which we will define as *entrepreneurs of themselves*.

In order to develop these two points, I will first briefly present the status of women in the Philippines from a historical perspective. Using life narratives I collected, I will then analyse why and how these women migrated and describe the emigration system the Filipino government organized. After that, I will focus on some elements of the life trajectories of Filipinas I interviewed in Paris, in order to show the way they exemplify careers that I choose to analyse in entrepreneurial terms.

A traditional and a modern society coexisting?

Before the Spanish conquest, Filipino society was based on traditional forms of familialism, where mutual assistance and solidarity patterned relations among members of the group. But the traditional society also allowed women to maintain responsibility for a great deal of family and economic matters and to develop personal autonomy (e.g. concerning sexual behaviour). Since the 'debt of gratitude' always prevailed in private and social relations (Hunt, 1966; Lacar, 1995; Tacoli, 1999), conflicts between social demands and individual autonomy possibly occurred. So it is worth stressing that this society accepted the coexistence of conflicting interests.

After the seventeenth century, the Spanish christianized the Filipino people; today, more than 90 per cent of Filipinos are Christians. The christianization process partly tended to reduce women's autonomy but they maintained their social and economic vitality (Feliciano, 1994: 552). At the end of the nineteenth century, in 1898, the United States took over the Philippines when they defeated the Spaniards. American influence led to an increase in the level of education, especially for women. Education policies favoured women's autonomy, due to primary schooling for girls and boys (1901) and free admission to college and university for both girls and boys (1908) (Feliciano, 1994). Schooling of girls also probably transformed family structures and social behaviour and thus lessened the weight of traditional social constraints.

Today, Filipino society is still structured by influences that could be analysed as conflicting – autonomy versus constraining solidarity ties – that the Filipino people and especially Filipinas seemingly manage to reconcile. Thus, the idea of a traditional pattern of socialization, shaped by subjection to patriarchal rules and behaviours, and generally reinforced by Christian precepts, must be deconstructed and reconceptualized: Filipinas do not adopt a mere passive position, but are rather, at the same time, both submissive as well as autonomous. In other words, Filipinas lead life projects that enable them to gain empowerment that is already at stake thanks to their good training (Tria Kerkvliet, 1990). As we shall see, the migration process will enhance this empowerment. But some difficult questions emerge. Why emigrate when one has graduated from school or when one has a job? And why emigrate so far away when one has a family and children that one leaves behind in the archipelago?

My fieldwork[1] gave some hints to answer these riddles. Most of my respondents already had jobs in the archipelago, and they migrated for a variety of reasons. Some vignettes from fieldwork will show this. Vivian's husband emigrated to Saudi Arabia to earn as much money as possible to be able to send their five children to private schools. Unfortunately, because of his health, he had to return to the Philippines. Through available information and friends, the couple found out about job opportunities in Europe for Filipino domestic workers, especially women. So Vivian left her five children and husband and went to Paris on a three-month tourist visa. Unable to survive alone ('I was crying all the time. I couldn't stand not being with my husband'), she convinced her husband to join her in Paris. The five children were left in the Philippines, with the eldest daughter, aged fifteen, in charge of the whole bunch. In the Philippines, Vivian used to do clerical work but she is paid much better in Paris as a domestic worker than at home. The interview revealed that the dramatic situation of having to leave her children back home on their own has not prevented Vivian and her husband from making a living with this painful decision. As of today, Vivian has been living in Paris for sixteen years.

Stella was a religion teacher in a private school after having completed her studies. When her father died suddenly, the family could not live on her salary. The 'household strategy' that the family developed implied that girls were 'naturally' obligated to sacrifice their position in order to support the rest of the family. So Stella emigrated to Paris when she was eighteen, although she had an elder brother who was married and who could have emigrated. Luizia, a trained midwife, migrated to France eight years ago. In the Philippines she had hoped to earn a better salary. Her brother convinced her to enter the emigration business where, he assured her, she would be better off. She became a go-between, supposed to recruit women wishing to emigrate. She was paid according to the number of clients she found. Then, when her boss ran away with the money, she had to hide because she couldn't refund the stolen money to the swindled clients. Getting out of the country was a way to escape her fate.

Most of the women I interviewed earned far more in Paris than their previous training would have allowed them to earn in the archipelago. But this opportunity could be seized only if there was a demand at stake, at the other end of the chain, overseas. And somehow all the women I met knew about this possibility and made use of all their serendipity in order to be able to emigrate. For example, Suzie had a neighbour who had emigrated to Canada, Bella had a sister and two sisters-in-law in Italy; Melanie heard about Paris through colleagues in the office. As many other researchers throughout the world have indicated, information and tips spread quickly in the archipelago, as some emigrants come home for visits and their stories inspire dreams among others, imagined projects in search of an opportunity to become reality.

Rhacel Parrenas, among others, argues that an entire immaterial diasporic world has developed around the Filipino migration process. It has become familiar through narratives, telephone calls and e-mails, as well as through specific media such as *Tinig Filipino*, a magazine created and run by a former domestic worker, which connects and informs migrants throughout the world by providing information, addresses, and contacts and by publishing poems, letters to the editor and photographs of domestic workers. 'Migrant Filipina domestic workers in Rome and Los Angeles also share similar experiences with their counterparts in other regions of the world. Based on writings of domestic workers, in the multinational monthly magazine *Tinig Filipino*, Filipina domestic workers in most other countries maintain transnational household structures' (Parrenas, 2001: 246).

Thus, the possibility of emigrating becomes part of everyday life in the archipelago for educated women or others who really wish to make a better living. In my survey, there were only two women who had not graduated from high school. They had the necessary qualifications, having gone through a socialization process that enabled them to enter domestic service. They are used to gender segmentation in everyday chores; they can adopt the submissive behaviour required in this type of job; and they are English-speaking. To jump the step becomes part of picture although everybody fears to leave so far. Most of my respondents remember how scared they were before departing, and all of them had to encounter numerous difficulties.[2] But they nevertheless also experienced the call of adventure and the dream of future fulfilment. But this migration process must not be analysed only at an individual level, since it is embedded in a structural setting shaped by public policy.

How migrants leave the Philippines

Since the economic situation in the Philippines is disastrous, most educated young people have to cope with low salaries or are unemployed. Since it is well known in the archipelago that there are job opportunities abroad, many wish to migrate. It became the policy of Ferdinand Marcos and his successors to encourage migration in order to obtain foreign currency through migrants' remittances. The Labor Export Policy (LEP) was developed to meet the World Bank's requirements and thus to entitle the Philippines to receive international loans. The emigrants were hailed as the 'heroes of the modern times' of the country (Gonzalez, 1992: 24, Jackson *et al.*, 1999: 44). This policy has made emigration attractive. Today the Philippines are the number one labour-export country in the world, and migrants depart for practically all parts of the world: East Asia, the Gulf countries, Western Europe and North America. In 1991 it was already estimated that 1.6 million Filipinos had been living abroad for at least fifteen years (Rodriguez, 1998). It is estimated today that about 10 per cent of the Filipino population work abroad.

There are two sorts of emigration processes: the official one and the undocumented one. The official system is co-ordinated by a State agency where potential emigrants must apply and where demands are collected from different countries overseas that have established formal agreements with the Philippines. These agreements establish the type of labour wanted, the work conditions and the length of the stay. One finds, for instance, that nurses are required in certain countries, such as in the Gulf area or the Netherlands. The agency then establishes contacts and helps people to emigrate to the countries concerned. Furthermore, the agency is supposed to guarantee the terms of exchange between the partners (employers and emigrants). The migrants are supposed to send back their remittances through official financial channels, thus helping to sustain the Philippine economy. These 'overseas contract workers' (OCWs) are documented workers who migrate legally, in an official framework (Gonzalez, 1992; Jackson *et al.*, 1999). This type of migration is, of course, the result of an individual choice, but it is controlled and supervised by public authorities in both countries, although it seems that the supervision is loose.

However, there exists significant undocumented ('illegal') migration, regardless of the existence or not of such agreements. In countries, such as France, where no official agreement exists, immigration results only from an individual process. France is also, like other European countries, under the Schengen restrictive regulations that are supposed to prevent undocumented immigration. Yet, as in other countries of the North, such as the United States or Canada, Filipino migrants manage to enter domestic employment. In my research most of the respondents had been, at one time or another, in an illegal situation. Some respondents had arrived in Paris long enough ago (around 1975) to be entitled to get documents on their arrival. For all the others one finds two types of situations. In one case, the Filipina enters the country with a visa valid for three months, and in the other, she enters illegally without any valid documents.

The valid visa may very well have been 'bought' illegally, through an informal private business; this was the case for most of the women I met who had gone through a lot of trouble getting in touch with networks around the migration business. These agencies have contacts in embassies and airports in the Philippines and, in certain cases, in the host country, which can be mobilized and used to organize the migration process. A lot of my respondents had fake visas or fake passports and nevertheless entered France legally, without any trouble. They sometimes also used false identities.[3] These Filipinas only became undocumented after three months, when the tourist visa expired. In some cases, some of the Filipinas were first sent to a transit country (Belgium or the Netherlands) to avoid suspicion and entered France by train. In other, more dangerous, situations Filipinas had first to enter a non-Schengen country and then get into France illegally. Suzie first flew to Budapest, then was driven with a group of other Filipinas to the Italian border, hidden in a train to cross Italy, where the customs officers discovered her but they were not sending her back. She finally managed to find the contact she was supposed to meet in Rome where an Italian lady put her up. She then had to walk with a group of migrants across the Alps before reaching France. Her epic voyage may be a heroic one, but other Filipinas had to undergo similar journeys. All these Filipinas came to France and found employment in the domestic service sector, where a specific demand exists.

An international demand for domestic services

Saskia Sassen, studying 'global cities', shows how the new global profession-als working and living in these cities need all sorts of labour-intensive, ser-vices cheap, domestic services in particular, thus leading to employment of poor, often undocumented, migrants. She also posits that the same process occurs in all big cities and not only in the 'global cities' (*Libération*, 28 July 2003). If one turns to empirical research one can see that Filipinas are engaged in domestic work in practically all parts of the world except for Africa and South America, where domestic work is done by native women (Lan, 2001; Parrenas, 2001; Jackson *et al.*, 1999; Pertierra, 2000). I shall give some examples showing how Filipinas settled in European countries.

Italy is one of the main destinations for Filipinas, because of close links between the Catholic Church and migration. According to Cecilia Tacoli (1999) in the 1990s the Philippine community accounted for half the total number of immigrants in the country. In 1995 it was estimated that in Rome alone there were 50,000 Philippine migrants, half of them undocu-mented. Domestic work is a 'natural' outcome for migrant Filipinas since Italian law forbids foreigners to engage in business. For the past decade, half of all Philippine migrants have been women migrating through the contract system, whereby official recruiting agencies in Rome hire Filipinas. In this legal migration process, but also in the informal, undocumented, migration trajectories of women, the Catholic Philippine clergy provide support and act as a placement agency. Cecilia Tacoli (1999) even refers to a 'safety net'. Since then, Italy has organized two waves of sorting out (regularization)[4] of undocumented migrants. As in other countries, in Tacoli's sample 86 per cent of the migrant Filipinas ($n = 154$) had members of their family or their friends already settled in Italy.

In Spain the first Philippine migrant women arrived in the 1970s. This migration trend continued to develop and today many Filipino domestic workers are employed by wealthy Spanish families (Oso Casas, 1998, 2002). Contract workers, especially nurses, migrated in the early 1960s to the Netherlands. They were mostly recruited through Church institutions. Later came political refugees belonging to the National Democratic Front of the Philippines. During and after the 1980s many 'mail-order brides' and undocumented immigrants arrived (Spaan *et al.*, 2001). Although paid domestic workers are not widespread in the Netherlands compared with southern European countries, in the 1990s many Filipinas were engaged in domestic work.

In France there are no contract workers, and the regulations of the French immigration policy are very strict. However, these facts don't prevent Filip-ina women from migrating to Paris. In my research I found that almost all the Filipinas worked in very wealthy families. Very few are live-ins,[5] many are employed full time with very flexible work hours, and some clean house for different employers. All are domestic workers and stay in these jobs even when they have attained documents. Fieldwork helped me to understand the reason why Filipinas so easily found employment in this sector. Why were they especially chosen? And why were they as popular as they are in the rest of the world? As said above, why are they the 'Mercedes-Benz of all domestic workers'?

What can a Filipina offer?

Filipinas engaged in domestic work are highly valued by families who employ them. What was said above concerning the coexistence of tradition and modernity in the socialization process they embodied – first as young women and later as married women – must be kept in mind at this stage. Skilled, educated, behaving well, Christian, moral, English-speaking, Filipinas are polite, silent, submissive and deferential. Because of the socialization patterns for girls in the Philippines, the 'self-presentation' (Goffman, 1973) they adopt is a 'deferent' (Goffman, 1963), if not 'humble' one (Hughes, 1996). That's also why women are generally preferred to men.[6] As such, their behaviour is valued by their wealthy employers since they apparently accept willingly their subordinate position. They consent to call their employers 'Madam' and 'Sir' (Goffman, 1956, Rollins, 1990), thus embodying the social importance of the employer as Veblen described it a century ago (1899). The Filipina domestic worker thus becomes a source of 'distinction' (Bourdieu, 1979) for her employers and offers them numerous side advantages. Being often mothers themselves their maternal competences are particularly valued, and they can also coach children in English, and pass on values of decency and morality, without stepping out of their social realm, which their employers appreciate. Even when the servant/employer arrangement is more informal, Filipinas never forget their social position and continue referring to the employer as 'ma'am'.[7] Although it is common courtesy in the Philippines to say 'ma'am' and does not necessarily indicate the employer's class position, in Paris this infers a subordinate position that values the employer who in general is not addressed so by cleaning women. This labour market is highly specialized and invisibilized;[8] it cannot be compared to the usual domestic services offered by other migrant workers employed as cleaning women (Africans, North Africans, and more recently Poles, Russians and other East European migrants).

They stay in domestic employment because their diplomas cannot be used in France. But as we have just shown it is also because the demand seems to provide a sufficient supply of jobs. Never have I met or heard of a Filipina jobless for long.[9] By making instrumental use of incorporated behaviour and manners, Filipinas created a niche at the very top rank of domestic services. This counts for France but also for other European countries. For example, Laura Oso Casas shows that Philippine domestic workers in Madrid are the best paid among all domestic workers (2002). Mutual support is widespread. For example, in many cases I could notice that, in each specific network I was close to, an informal employment bureau existed. When Vivian's husband died, a neighbour helped her to recover, settled things and accompanied her to the police station where Vivian, unconscious of being endangered by her illegal situation as an undocumented immigrant, registered his death. Her employer and the Philippine embassy paid for the trip back to the Philippines to bury him and she had to leave very promptly. The moment she left, another Filipina belonging to the same network replaced her at her employers' home. Employing a Philippine domestic worker is, in a way, engaging in a 'collective arrangement' where each Filipina is interchangeable in a market where all available jobs are occupied, which in return means that they control this 'closed' market as a form of monopoly. Thus the different segments of the Philippine community in Paris draw the outlines of a market for domestic workers that addresses upper-class employers, to whom this type of servant is particularly attractive.

But what does this all mean for the life trajectories of the individual emigrant? To answer this set of questions one needs first to analyse their careers as transnational and also as life projects.

Transnational careers

Among my respondents most of the women had transnational links in at least two ways. First, they often migrated with the help of relatives or friends who were already in Paris; second, many Filipinas return to the archipelago quite often. Yolanda migrated to Paris, after a two-year stay in Singapore, because she knew her aunt Suzie was already there, although she and her family had never met her. Raphaëlla fleeing Iran heads for Paris where her sister Carol is living already. But these relations can also spread further. Melanie, who heard about job opportunities in Paris through an office colleague, also knows about Europe, because she has close relatives in Italy. Helen first migrated to Spain and decided to leave for Paris, where her cousin found her a job. Luizia was a contract worker in Lebanon and came to Paris with her employers; she ran away, not wanting to return to Beirut, where she was unhappy, in order to stay in Paris, where she had friends.

The ties with the Philippines are maintained often a long period. Vivian has returned three times since she has been in Paris, when her husband died, to fetch Eiffel-François and send him back. Teresa, who now has documents, flies back from time to time. Suzie flies back to see her family a first time and then a second to look after her ill daughter. These transnational links are strengthened by numerous telephone calls, e-mails, but also other information channels such as visitors bringing back news, photographs and videotapes. A lot of goods are also conveyed through these networks, strengthening and renewing social, cultural and economic ties in the Paris community. When one shares meals and gatherings with Filipinas one really feels that all this connecting activity requires competences and skills which

Tarrius calls 'migrating competences' (2003). This means these Filipinas are able to transfer overseas, in other situations and contexts, their know-how and experience from 'home'. Not only must they transfer, they must reorganize them, 'translate' them so they can be useful elsewhere (i.e. in the host country). This means to be able to make use of 'management competences': to link networks, to develop serendipity, to negotiate new roles in unexpected situations and contexts that may occur. 'Migration competences' as well as 'management competences' must coexist to enable each Filipina to analyse and face situations, which means to mobilize cognitive as well as subjective resources such as involvement and tenacity. Filipinas employed in domestic employment are not amateurs; their careers are built professionally.

As Peraldi has shown with migrant sellers who develop businesses transnationally, informal activities grow in 'territories that are outside of the State boundaries, on the roads, on borders and towns – or parts of the city – that are extraterritorial *enclaves* . . . [where] the State is not absent but suspended [*Etat suspendu*]' (2003: 34). These activities are, in a certain sense, *beside* or adjacent to the nation State. As many researchers have shown, capitalist, developed economies need a reserve of unskilled, flexible work in many sectors, such as in the building and construction industry, the confection industry and the agricultural sector (Mozère and Maury, 1999). In a more surprising way, the demand for domestic workers seems to answer the same type of need.

Could one then argue that the Filipinas are part of a proliferating or 'rhizomatic' (Deleuze and Guattari, 1972, 1980) international labour market? Could one consider it being embedded in what Arjun Appadurai calls an 'ethnoscape' (Appadurai, 1996)? By using this word, he wants to stress the importance of cultural phenomena in creating what he calls 'communities without a place'.[10] Basing his work on findings of cultural studies, Appadurai uses Benedict Anderson's concept of 'imagined worlds'[11] (1996) to analyse the way the identities, kin, friendship and work networks are influenced and affected by the permanent transnational mobility of tourists, migrants, refugees, academics and businessmen. In Appadurai's words 'ethnoscapes' provide people all over the world with new images and life scripts. These 'ethnoscapes' are full of media information (newspapers, magazines, television, movies) as well as political ideas ('ideoscapes') such as 'democracy'. As such, he develops a theory of global flows today where the 'identity landscape' is cosmopolitan and not territorialized in a nation-state setting. In this sense, can one argue that the Filipinas pattern their migration in the realm of this type of ethnoscape and are, as such, equipped with a stock of knowledge, pictures and ideas that become part of their selves? Aren't they living transnational lives only because, even if they travel to and from, they will finally settle back home, in a rather traditional pattern of migration?[12]

During my research, once the usual reason to migrate had been worded (i.e. economic necessity), I could feel that the respondents had not told the whole story. Some slips of the tongue, some ways of expressing their satisfaction at being in Paris, encouraged me to deepen the questioning. In the case of Ginnie, whose husband had left her with four children in Manila, some of these reasons, omitted in the beginning, clearly appeared. Her in-laws put her up and supported her and her children, but she felt ashamed that she had to depend on them. So she decided to pay them back and leave for Paris. Her mother-in-law asked her if she didn't feel bad becoming a domestic worker while she had maids herself in the Philippines. Ginnie answered that she felt independent and free, able to support her own children since their father didn't.[13] Stella has been in Paris for more than eight years now and hopes to obtain documents[14] and a work permit. She feels homesick, but as I discovered, she had built her life around the fulfilment of another desire which has to do with her own life interests and eventual life project. In the Philippines, she was a paid religion teacher; in Paris, in the American church, she teaches in Sunday school, probably without pay, since she is undocumented. But thanks to this job in Paris she increases her competences; she gains extra qualifications (learning French, learning through the contact with an American community, gaining metropolitan habits and competences). Vivian has paid for the education of all her children, but since her young husband wants to return to the Philippines, she stays on.

The women in my sample seem quite happy to be in Paris, although, a seeming paradox, they have downward social mobility: becoming maids and really missing their family. Seeing them frequently, joining them in different settings and in different social 'frames' (Goffman, 1973), it soon appeared that migration, which in a way they considered – using Becker's (1985) wording – as a 'career', was less about economic survival alone and more about what one could call a 'life project'. The migration decision is a result of a cluster of reasons where the economic level is certainly determining but can nevertheless hide or legitimate other, maybe less 'politically correct' or socially admitted, reasons.[15]

So it is necessary to find out more about this cluster of motivations, the puzzling coexistence of apparently contradictory feelings and choices that were mingled, mixed and entangled in the interviews. This complicated picture does not match the familiar one, where the process of emigrating is traditionally linked to a clear-cut opposition between home ('I leave and I wish to return') and the host country ('where I am supposed to live temporarily'). Once the migrant has sent back sufficient remittances to support the family and to educate the children isn't the trip home on the agenda? And, especially in the case of feminine migration, women are supposed to want to return home to their children as soon as possible. For the Filipinas shouldn't migration be a parenthesis, a digression on a home-centred life?

Filipinas as 'entrepreneurs of themselves'

Fieldwork offered a clue that might give meaning to the trajectories of these women and to what I call their life projects. Similar to what we have seen for Stella, other women also pursue goals that are not univocal. Teresa, who now has a work permit and has also paid her 'debt of gratitude', could return to the Philippines. During an interview she finally says, 'My family has a small poultry business. I continue sending money. It helps them. The business works well now, and I can even get some returns. Actually I can't go back to the Philippines, it's too backward a country. I'm used to my life in Paris.' Migration thus creates new living standards, new habits, and leaves space for new desires Teresa doesn't want to give up. But she nevertheless is closely linked with her home country; she sends remittances, and she travels to the Philippines to visit her family. Luizia is a midwife. She tells me that she wants to use the money she earns to buy a small clinic in the Philippines, but neither does she want to emigrate back. Melanie wants to build a hotel in the Philippines and make a living out of it, but at the same time she wants to stay in Paris. Actually, her husband and children have joined her and they all have documents. 'Here we have good education for the children, good medical care, and now that we have our documents, we have a decent flat. Of course I miss home, but we can travel back and forth. We are building this hotel in Manila.'

Migrating, one can now see, is embedded in broader life projects, where micro-economic individual ventures link the migrants with the archipelago and tend to favour a transnational relation between home and overseas. This new pattern seems to indicate that these women think of their migration as part of a process which they build pragmatically and which evolves in the time–space of migration. None of our interviewees had a pre-established plan when leaving the Philippines. They build a life project that starts with goals they establish at their departure and that they complete later, in the host country. And this goal is not only a family-oriented goal; it encompasses what migration has embodied in their own desires. It is a goal where their selves and their lives are transfigured and transvalued by new ethnoscapes and ideascapes. They build themselves anew in a sense, which is why they are 'entrepreneurs of themselves'. They are entrepreneurs, not only in the sense that their trajectory is grounded on financial and other resources – they devote some of their time and energy to small trade activities linked with their back-and-forth trips to the archipelago (for instance, small jewellery and native products) – but on a more moral level. To show this leads us to make use of other theoretical tools.

Michel Peraldi has researched young 'second generation' immigrants from North Africa, usually born in France and thus French, in so-called 'disadvantaged' neighbourhoods. He has shown how their bad curriculum (many are dropouts) and racial stigma can lead many of these youngsters (especially boys) to illegal activities such as dealing drugs. At the same time, they often construct and pattern their own occupation, using network resources (which may be ethnic or not) as well as other social competences. They usually don't create very small enterprises, but pattern their activity as 'a personal adventure', creating for themselves a social status. Their work consists in connecting, through 'weak ties' (Granovetter, 1985, 2000), different commercial 'worlds'. There are no offices (and thus no costs), and 'work' is grounded on reciprocal confidence. Their network can always develop. The capital is their own capital of contacts, their serendipity, and their capacity to seize opportunities and deals that are available.

But another way of analysing what seems at stake is to turn to the paradigm of the *political entrepreneur* Toni Negri uses when he analyses the new pattern of entrepreneurship in a postmodern immaterial and material economy that he considers as a 'second transformation', reminding us of Polanyi's 'great transformation'. Studying the case of Benetton in Italy, Negri and his co-authors show that the factory is no longer at the centre of the production process. As the economists who had previously analysed 'industrial districts', production in a district, in their view, depends on the possibility of building a successful tension between competition and solidarity between different economic agents, who in this framework are at the same time opposing as well as collaborating. In this sense the entrepreneur occupies a 'political' position in so far he 'favours . . . political exchanges by developing clan-like, reciprocal, and solidarity-oriented practices in the work organization and the circulation of goods' (Lazzaratto *et al.*, 1993, Negri, 1996). Social co-operation is then essential, and what Negri calls 'social aggregation competences', which enable one to switch from one world to another and to develop 'translation' competences that help cross-situational borders, are now necessary to link heterogeneous groups and frameworks. This means subjectivity has to be actively mobilized in these new socio-economic procedures. The immaterial firm needs the city, but it also needs the subjective 'entrepreneurial' competence of its workers.

Why should these ideas be used in the case of the often undocumented Philippine women migrating and working as domestic workers? Similar to the young men Peraldi discusses or the political entrepreneurs of Toni Negri, the Filipinas use their position in an economic *niche* to develop entrepreneur-type behaviour. They act as agents, actors – one might say 'authors' of their own story (Peraldi, 1997), switching from one world to another, in the host country as well as on a transnational level. Migration changes their relation to home in so far it transforms the 'desire for home' by creating new life standards, new tastes and new desires borne by the host society. It increases migrants' empowerment in due process. Crossing social, economic and geographical borders,[16] mobilizing the social aggregate, connecting competences Negri analysed in the case of the 'political entrepreneur', they enhance their social and economic prestige in the Philippines and retain the advantages the host country can offer. This process should be understood as a process of *affirmation*, asserting the will to build one's own life. Whether they have their family join them or not, they certainly do gain autonomy.

Three years of field work, much of it done by participant observation, have convinced me that these women who were particularly full of courage and guts were 'entrepreneurs of themselves', migrating to fulfil a life-project and made it doubtful whether one should consider them as mere victims of a system to which they would be confined. Given the existing demand for this type of domestic work in the countries of the North, they seize the opportunity. Couldn't we posit that the Filipinas make instrumental use of their competences as well as their social and subjective skills to attain their goal? In Alain Tarrius's words this crossing [of borders] movement has a meaning, has sense for these migrants. In this in-between space [the migrants] say 'project' when we hear 'exile' (2003). It is in this sense that the Philippine domestic workers are 'political entrepreneurs of themselves' in a transnational world market.

Saying this does not mean their entrepreneurial 'career' is only an economic one, but refers to the fact that it combines economic as well as social interests,[17] cultural and moral advantages which build a more acceptable and suitable self in their eyes. Through migration the Filipinas I interviewed had gained self-esteem, a socially enviable position in home and community, and increasing gender equity. Becoming the main breadwinners, women in the Philippines on trips back or abroad gain prestige in the home community as well as social and symbolic consideration. When the whole family has joined her in Paris, the Filipina still pursues family-oriented goals in the Philippines such as house building, supporting family businesses that are still embedded in family and kin networks. If the family is still in the archipelago, by staying in Paris, she will guarantee them a better life by more education or support in periods of unemployment, as did Vivian when her son was jobless. Moreover, men left behind, in many cases, have to mind the children[18] and consequently 'act' as women, according to Pingol and Diliman (2001),[19] even when grandmothers or sisters take care of the children. Without hiding the negative side effects of their migration on the children,[20] many women of my sample were of opinion that what they were doing in Paris would, in the end, be useful and helpful to them. In this way they manage to reconcile the apparently contradictory goals of 'home' and their life project.

Conclusion

Making use of an socio-anthropological analysis has enabled me to draw the outlines of the migration of Filipinas to Paris today as well as the ambiguities that shape them. These ambiguities, I have tried to point out, tend to highlight the importance of gender and of transnational connections to explain how and why Northern countries attract Filipinas. Providing them with a *niche*, this international market of domestic services offers the women an opportunity, through migration, to paradoxically develop, although through 'partial citizenship' (Parrenas, 2001), empowerment in terms of gender and life projects. Notwithstanding suffering, exile and difficulties, as well as a deficit of care for their children (Parrenas, 2001; Tronto, 2001) they are perhaps emblematic of new migrants who patiently build and set up a *bricolage* to live here and there, across borders, trying perhaps to open routes for their children, their families and their communities in a globalized world.

Notes

1   This socio-anthropological empirical research, funded by the Mission du Patrimoine Ethnologique of the French Ministry of Culture, was led during three years in Paris among three different networks of Filipino migrants, dominantly composed of women. One is based in the American Church Quai d'Orsay in Paris, the other in a group called Babaylan which was first a political group in the 1970s and now is engaged in cultural activities and offering assistance to migrant domestic workers, especially undocumented ones, and the third group which developed around Suzie, a charismatic domestic worker who sued her employer and now has become a member of a French trade union (CFDT). Over twenty-five in-depth interviews were led, participant observations were led as well as what I call urban strolling, accompanying these women in everyday occupations and/or helping them in institutional steps. I also participated in parties and informal meetings, encountering a high number of Filipino people. The latter usually lead a very intensive social life and in other countries such as Hong Kong are used to gathering in public parks on weekends, as one can see in the Chinese film *Yiyi* (Fruit Chan, 2001).

2   I would like to say that fear, dread, dangers as well as, in some cases, ill treatment were stated in the interviews. I do not wish to minimize these facts or pretend they are not morally unacceptable and/or a source of infinite distress. But this chapter wishes to focus on these women's courage, endurance that enables them precisely to engage in life projects, becoming, as I shall show later, 'entrepreneurs of themselves'.

3   Raul Pertierra shows that identities in the Philippines can be 'multiple', thus dedramatizing this type of illegal behaviour (2000).

4   Tacoli also shows how the Filipino domestic workers organized to fight for their rights, struggle against invisibilization and obtained documents. Parrenas also shows this (2001).

5   Some had been so when they first arrived in Paris but all tried to escape this situation, preferring to be on their own, even in tiny, uncomfortable rooms, mostly under the roofs, called *chambres de bonnes* (servants' rooms).

6   Some men are nevertheless employed as servants in Paris. It is then often the case of husbands who join their wives or who have come with them. Some colleagues from Nice said they had seen advertisements of employers wishing to hire a couple of Philippines.

7   When they pick up the phone and hear a woman's voice that they recognize as non-Filipina, they immediately say 'ma'am' and it took months before they would call me Liane instead of 'ma'am'. It's because of this general behaviour that Filipinas are preferred to Filipinos in general. One must also add that in the Philippines one prefers to see girls and women migrate because they spend less overseas and send more remittances home than men.

8   Because of the social invisibilization due to the presence of undocumented and often not declared workers which take part in what Portes calls 'the lower glob-

alization' (1999) and because of the muffled, discreet and unassuming social worlds they serve in (Pinçon and Pinçon-Charlot, 1985).

9 On the contrary I could notice the existence of an informal job agency organized in each network. Each time a Filipina was ill or returned home, another Filipina was immediately replacing her at her employer's. Most of the respondents had first started working this way, temporarily replacing another Filipina. This also favours long trips back home (which generally means to buy a new visa unless an employer sends an invitation officially which entitles a three-month temporary visa). Suzie was worried about her daughter's health, she returned two months to the archipelago without worrying, as she had set up everything to be replaced and find a job once she was back in Paris.

10 The concept of ethnoscape can be linked to the idea of diaspora (Ma Mung, 2000) in so far it encompasses space, time and moral spheres. What it maybe adds is the fact that not only members of diasporas are mobile, but that in a globalized world, human, cultural, technical, financial and ideological flows affect practically everybody.

11 In other words the multiple worlds that result from the historically situated imagination of people and groups scattered around the world.

12 The existence of such a 'traditional' pattern can well be questioned. Take the North African migrants who migrated to France between 1950 and 1974. They were supposed to emigrate to make money in France and return home. But in 1974 a new policy enabled them to have their families join them. So the return journey was delayed for decades: children born in France becoming French and brought up in French schools were soon strangers in Algeria, Morocco or Tunisia. Nowadays practically no return trips, except for vacations, do really occur. And in her documentary *Mémoires d'immigrés*, Yamina Benguigui shows that even the Last Voyage is often delayed for ever: more and more of these oldest migrants are now buried in France and no longer in the home village (*bled*).

13 Which never meant they were not, at the same time, unhappy not to be home, with their children, for example.

14 The immigration in France at the moment allows undocumented migrants to apply for documents and a work permit if they can prove they have been there for ten years. So Stella stays on, although she has overwhelmingly repaid the 'debt of gratitude'.

15 Divorce being illegal in the Philippines, a certain number of women choose (or prefer) to emigrate to escape an unhappy marriage.

16 One must remember that they are educated and have to 'pretend' they are servants, adopting a submissive behaviour: this is border crossing. So is the fact they may have led a middle-class position in the Philippines and now are economically at the bottom of the scale in France, although they earn far more than they did in the archipelago.

17 Such as a good private education for the children.

18 Not all men, however: some will openly choose a new partner, as Suzie discovered on a trip back home. Some also choose to reconcile two, apparently contradictory, statuses: on one side being a 'real' man in the community (i.e. engaging in community responsibilities) and on the other taking care of the children, which belongs to the feminine activity sphere.

19 Some women teasingly call their husbands 'housebands' (Pingol, 2000; Pertierra, 2000).

20 Filipinas migrating overseas leaving their own children in the Philippines take care of children of wealthy families abroad, thus creating an international division of care. Researchers have shown the negative impact of migration on children left behind (Battistella and Conaco, 1998; Parrenas, 2001). Parents, and

especially mothers, try, by all means, to keep contact with their children through telephone calls, letters or e-mails, but often state the fading of emotion as Ginnie says in the interviews. This means that if the Filipinas in Paris do act as 'entrepreneurs of themselves', this also means they act in a setting of constraints that entrench and diminish empowerment. But there again, if this may lead to subjection, one must keep in mind the ambivalence that can nevertheless exist in their minds and behaviours.

References

Appadurai, A. (1996) *Modernity at large. Cultural Dimensions of Globalization*, Minneapolis MN: University of Minnesota Press.

Battistella, G. and Conaco, M.G. (1998) 'The impact of labor migration on the children left behind: a study of elementary school children in the Philippines', *Sojourn*, (13) 2: 1–22.

Becker, H. (1985) *Outsiders*. Paris: Métaillié.

Bourdieu, P. (1979) *La Distinction. Critique sociale du jugement*. Paris: Minuit.

Castells, M. (1996) *The Rise of the Network Society. The Information Age* I. Oxford: Blackwell.

Castells, M., Portes, A. and Benton, L. (1989) *The Informal Economy. Studies in Advanced and Less Developed Countries*. Baltimore MD and London: Johns Hopkins University Press.

Cock, J. (1980) *Maids and Madams*. Johannesburg: Ravan Press.

Deleuze, G. and Guattari, F. (1972) *L'Anti-Œdipe*. Paris: Minuit.

Deleuze, G. and Guattari, F. (1980) *Mille plateaux*. Paris: Minuit.

Eviota, E.U. (1992) *The Political Economy of Gender. Women and the Sexual Division of Labor in the Philippines*. London: Zed Books.

Fassin, D. and Morice, A. (2001) 'Les épreuves de l'irrégulraité: les sans-papiers entre déni d'existence et reconquête d'un statut', in Dominique Schnapper (ed.) *Exclusions au cœur de la cité*. Paris: Anthropos.

Feliciano, M.S. (1994) 'Law, gender, and the family in the Philippines', *Law and Society Review*, 28: 547–60.

Giddens, A. (1984) *The Construction of Society*. London: Polity Press.

Goffman, E. (1959) *The Presentation of Self in Everyday Life*. New York: Anchor Books.

Goffman, E. (1963) *Behavior in Public Places. Notes on the Social Organization of Gatherings*. New York: Free Press.

Goffman, E. (1973) *Les Rites d'interaction*, Paris: Minuit.

Goffman, E. (1991) *Les Cadres de l'expérience*, Paris: Minuit.

Gonzalez, A. (1992) 'Higher education, brain drain and overseas employment in the Philippines: towards a different set of solutions', *Higher Education*, 23: 21–31.

Granovetter, M. (1985) 'Economic action and social structure: the problem of embeddedness', *American Journal of Sociology*, 91 (3): 481–510.

Granovetter, M. (2000) *Le marché autrement*. Paris: Desclée de Brouwer.

Hillman, F. (1999) 'A look at the hidden side: Turkish women in Berlin's labor market', *International Journal of Urban and Regional Research*, (23) 2: 262–87.

Hughes, E. (1996) *Le Regard sociologique*, Paris: EHESS.

Hunt, Ch.L. (1966) *Social Aspects of Economic Development*, New York: McGraw-Hill.

Jackson, T.R., Huang, S. and Yeoh, B. (1999) 'Les migrations internationales des domestiques philippines', *Revue européenne des migrations internationales*, 15 (2): 37–67.

Laacher, S. (2002) *Après Sangatte*, Paris: la Découverte.

Lacar, L.Q. (1995) 'Familism among Muslims and Christians in the Philippines', *Philippine Studies*, 43: 42–65.

Lan, P.-C. (2001) 'Doing gender in the continuum of domestic labour Filipina migrant domestic workers and Taiwanese employers', unpublished.

Lazzaratto, M., Moulier-Boutang, Y. and Negri, T. (1993) *Des entreprises pas comme les autres.* Paris: Edisud.

Ma Mung, E. (2000) *La diaspora chinoise. Géorgraphie d'une migration.* Paris: Ophrys.

Morokvasic, M. (1984) 'Birds of passage are also women', *International Migration Review*, 18 (68): 886–907.

Mozère, L. and Maury, H. (1999) *Petits métiers urbains au féminin. Le cas des assistantes maternelles et des nourrices.* Paris: CNAF, FAS, Plan urbain.

Mozère, L. (1999) *Travail au noir et informalité. Liberté ou sujétion?* Logiques sociales. Paris: L'Harmattan.

Mozère, L. (2001) 'La philippine ou la "Mercédès Benz" des domestiques. Entre archaïsme et mondialisation. Carrières de femmes dans l'informalité', *Sextant* (Brussels), 15–16: 297–317.

Mozère, L. (2003) 'Des domestiques philippines à Paris: un marché mondial de la domesticité?', *Revue Tiers Monde*, 43 (170): 373–96.

Negri, T. (1996) 'L'entrepreneur politique', in Michel Peraldi and Evelyne Perrin (eds) *Réseaux productifs et territoires urbains, pour le Plan Urbain.* Toulouse: Presses universitaires du Mirail.

Oso, L. and Catarino, Ch. (1996) 'Femmes chefs de ménage et migration', in J. Bisilliat (ed.) *Femmes du Sud. Chefs de famille.* Paris: Karthala.

Oso Casas, L. (1998) *La migración hacia España de mujeres jefas de hogar.* Madrid: Instituto de la Mujer.

Oso Casas, L. (2002) *'Domestiques, concierges et prostituées. Migration et mobilité sociale des femmes immigrées, espagnoles à Paris, colombiennes et équatoriennes à Madrid'*, Thesis, Paris: Université de Paris I, Sorbonne.

Parrenas, R. (2001) *Servants of Globalization. Women Migration and Domestic Work.* Palo Alto CA: Stanford University Press.

Peraldi, M. (1997) 'Portraits d'entrepreneurs', in *En marge de la ville, au coeur de la société. Ces quartiers dont on parle*, La Tour D'Aigues: l'Aube.

Peraldi, M. (2003) *La Fin des norias.* Paris: Maisonneuve & Larose.

Pertierra, R. (2000) 'Multiple Identities, Overseas Labour and a Diasporal Consciousness in a Local Community'. Presented to the second conference of the South European Network of Asian Studies (SENAS), Marseilles 14–15 October.

Pinçon, M. and Pinçon-Charlot, M. (1985) *Les beaux quartiers.* Paris: Seuil.

Pingol, A. (2000) 'Ilocano masculinities' *Asian Studies* 36 (1): 123–33.

Pingol, A. and Diliman, Q.C. (2001) *Remaking Masculinities. Identity, Power and Gender Dynamics in Families with Migrant Wives and Househusbands.* University Center for Women's Studies.

Piore, M. (1983) *Birds of Passage.* Cambridge University Press.

Portes (1999) 'La mondialisation par le bas', *Actes de la Recherche en Sciences sociales*, Paris.

Rodriguez, E.R. (1998) 'International migration and income distribution in the Philippines', *Economic Development and Cultural Change*, 46 (2): 15–26.

Rollins, J. (1985) *Between Women. Domestics and their Employers*. Philadelphia: Temple University Press.

Rollins, J. (1990) 'Ideology and servitude', in Roger Sanjek and Shellee Colen (eds) *At Work in Homes. Household Workers in World Perspective*. American Ethnological Society Monographs 3. Washington DC: American Anthropological Association.

Rotkirch, A. (2001) 'The Internationalization of Intimacy. A Study of the Chains of Care'. Communication to the fifth conference of the Association Européenne de Sociologie, Helsinki, 28: 8–1.9.

Spaan, E., Naerssen, T. van and Nieling, M. (2001) 'Asian immigrants in the Netherlands. Between the general labour market and entrepreneurship'. Paper presented at the ESF workshop 'Asian Immigrants and Entrepreneurs in the European Community', Nijmegen, 10–11 May.

Tacoli, C. (1999) 'International Migration and the Restructuring of Gender Asymmetries: Continuity and Change Among Philippino Labor Migrants in Rome', *International Migration Review*, 33 (3): 649–82.

Tarrius, A. (2003) *La Mondialisation par le bas*. Paris: Balland.

Tria Kerkvliet, B.J. (1990) *Everyday Politics in the Philippines*. Berkeley CA: University of California Press.

Tronto, J. (2001) 'Multicultural Care'. Paper for the annual session of the American Political Science Association, San Francisco, 30 August–2 September.

Veblen, Th. (1899) *An Economic Study of Institutions*, rev. edn. London: Macmillan; New York: Viking Press, 1960; French trans., *Théorie de la classe de loisirs*, Paris: Gallimard.

Part III

Self-employment and entrepreneurship among Asian immigrants in Europe

11 Asians on the Norwegian labour market

Industrial concentration and self-employment careers

Geir Inge Orderud and Knut Onsager

For quite some time, empirical and theoretical research on entrepreneurship among immigrants has focused on the concentration of the phenomenon in certain industries. Empirically speaking, the tendency to cluster has been observed over a long period of time, in different countries and against the backdrop of different theories. The most comprehensive corpus of material comes from the United States, but European research picked up the gauntlet in the late 1990s and in the run-up to the millennium.[1] One general characteristic is that immigrant entrepreneurs from poor countries have tended to go into businesses with low barriers to entry and which involve highly liquid assets, making it equally easy for the immigrant entrepreneurs to get out again. Food retailing and catering[2] are two examples. Other types of services are also common, e.g. businesses that meet the demands of particular groups of immigrants.

With a view to this pattern, efforts have been made to examine the mechanisms that lead immigrants into or enable them to go into particular trades. One discovery is the tendency for immigrants to go into businesses that other groups, commonly the majority population, have abandoned owing to their low status and income.[3] Certain urban commercial services have been identified as examples of such industries. However, research also indicates that immigrants go into trades in which they have previous experience,[4] through family ties, previous employment or both. This makes career paths the focal point.

In other words, the focus of the present study[5] is on immigrants' career paths in the job market and business world. In addition to determining whether the patterns that prevail in countries with an immigration history and institutional system different from Norway's also prevail in Norway, our goal is to identify the scope of and distinctions in recruitment and career paths for entrepreneurs. However, such a perspective calls for theories that can deal with the choices made by the players: immigrants and natives, minority(ies) and majority. That being said, it is equally important that the theoretical basis takes structural conditions into account. In the authors' opinion, theory development in this field made great strides in the right direction in the 1990s, e.g. Waldinger *et al.* (1990)

introduced the categories or concepts of opportunity structures, comprising market conditions and access to ownership, and group characteristics, comprising predisposing factors and resource mobilization, thereby combining structure – agency; Waldinger (1996) linked networking and structural forces with context and a specific historical process, wherein ethnic queuing regarding ethnic niches is central; after having focused on middleman minorities and structural gap theories (Light and Bonacich, 1988) Light during the 1990s develops a theory based on demand and supply side interaction (Light and Rosenstein, 1995; Light and Gold, 2000) with a resource and network perspective, wherein class and ethnic resources are central; and lastly Kloosterman and Rath (2000, 2001) and Rath (2002) have brought in the mixed embeddedness perspective comprising a socio-economic dimension and a political-institutional dimension. Overall, theories on ethnic entrepreneurs have been linked with wider political and economic structures. We are of the opinion that the mixed embeddedness postulated by Kloosterman and Rath (2001), along with the resource perspective of Light and Gold (2000), facilitates understanding of the mechanisms and processes that underlie and create career paths among groups of immigrants.

The investigation was aimed at immigrants from countries in southern and eastern Asia. Generally speaking, immigrants from these countries are known for having a high percentage of entrepreneurs. This means the groups are well suited to a career analysis. By the same token, there are differences among immigrants from different Asian countries, as evidenced by the figures for the individual countries. However, it will also be useful to compare the patterns of immigrants from southern and eastern Asia with those of immigrants from affluent countries.[6] Danish immigrants stand out as a relevant group, partly due to linguistic kinship and partly due to the high percentage of entrepreneurs.

Immigrants from southern and eastern Asia[7] constitute about 25 per cent of Norway's immigrant population.[8] The percentage has been more or less stable in recent years at the same time as the percentage of immigrants has increased, meaning immigration from southern and eastern Asia has kept pace with this growth. The various Asian groups have different immigration histories in Norway. Pakistanis and Indians began to arrive in Norway during the labour immigration phase in the 1960s and early 1970s. Since then, members of these groups have continued to arrive for the purpose of family (re)unification. On the other hand, many Vietnamese and Sri Lankans came to Norway during the refugee phase, meaning they have not been in Norway as long, but that their residence time has none the less varied. In brief, the demographics[9] can be summarized as follows:

1   The 25,000+ Pakistanis constitute about 31 per cent of the immigrants from eastern and southern Asia, making them the largest group of immigrants in Norway, ahead of Swedish and Danish immigrants.[10] The

Vietnamese are the next largest group, accounting for roughly 21 per cent.

2   Seven countries – Pakistan, Vietnam, Sri Lanka, the Philippines, India, China and Thailand – constitute about 95 per cent of the immigrant group, making their mark on this ranking.

3   About half the South and East Asian immigrants live in the municipality of Oslo, and the percentage is on the rise. About 73 per cent of Pakistani immigrants live in Oslo, compared with only about 25 per cent of Vietnamese and Thai immigrants.[11]

4   The age structure is generally young, but it also varies among groups from different countries. A relatively high percentage of Pakistanis have children under the age of fifteen, although the figure has been diminishing. At the same time, a growing percentage of Pakistanis are joining the ranks of senior citizens. The percentage of immigrants of working age is higher among those from the other Asian countries.

Thus the percentage of immigrants in Norway is not especially high by international standards, and the number of immigrants is correspondingly low. Of today's approximately 350,000 immigrants,[12] approximately 55 per cent come from non-Western countries, and it is reasonable to postulate that this group, comprising about 190,000 immigrants, 85,000 or so of whom are from southern and eastern Asia, has a significant impact on the scope of and type of enterprises in which these immigrants end up.

The empirical basis for the analysis is registered figures adapted by Statistics Norway for the purpose of the present analysis, and interviews with immigrants from a selection of Asian countries, living in Oslo and actively involved in food retailing or catering. The *registered figures* include all immigrants during the years 1989, 1993 and 1997, and their data.[13] This makes it possible to follow individuals over time if they have lived in Norway during at least two of the relevant years. Accordingly, this offers a good chance to analyse career opportunities. Like all register-based statistics, reliability depends on the quality of the figures received by Statistics Norway. The self-employment data have been burdened by a number of weaknesses, but we are of the opinion that this has not had a decisive impact on the findings presented in this context.[14]

The *informant analysis* comprised twenty-four semi-structured interviews with self-employed immigrants (Onsager and Sæther, 2001). Most of the visits took place at the workplace, but it was necessary to book appointments in advance at some larger enterprises. The length of the interviews varied, partly because some were less open than others, but also because some had been selected as case studies for broader presentation. The informants came from Turkey, Pakistan, China, Iran and Iraq,[15] and they were engaged in food retailing, mainly grocery stores, snack bars and restaurants, including one fast-food chain.[16] Although the number of interviews is limited, the material facilitated in-depth studies of topics illustrated quantitatively in the register-based analysis. Our ambition was to combine finds derived from the two analytical methods to increase insight into the topics at hand.

Industrial concentrations

We analysed the distribution of industries among self-employed immigrants from southern and eastern Asia along the following axes: two industrial classification systems, that is, a conventional ISIC classification and a classification based on the wage level in different industries; gender, age and residence time; educational level; municipality of residence; and the years 1989, 1993 and 1997. The most important finds from this analysis can be summarized as follows in the light of the traditional industrial classification for business activities (ISIC):

1   Immigrant self-employment from countries in southern and eastern Asia and poor countries[17] is generally dominated by food retailing and catering.[18] Concentration in this area increased from 1989 to 1997. The concentration on eating-places is especially strong among Chinese immigrants[19] (in 1997 about 90 per cent of the males and 79 per cent of the females). Pakistanis, on the other hand, are to a large extent involved in food retailing (in 1997 about 45 per cent of males and 59 per cent of the females) but Pakistani men are also considerably involved in restaurants (31 per cent in 1997). The Vietnamese resemble the Pakistanis but with a reverse order; for example, a higher proportion of eating places than food stores, while the share of food stores is fairly high among Indians (36 per cent among men and 56 per cent among women). As the shares show, the pattern is fairly similar for both genders, although the levels vary somewhat. Women are generally more involved in food retailing. The industrial distribution is generally more widespread among immigrants from affluent countries.

2   The percentage of *male* immigrants from countries in southern and eastern Asia and poor countries in general involved in food retailing is especially high in Oslo and adjacent municipalities. The figure increased during the economic boom from 1993 to 1997. On the other hand, eating places play a greater role, relatively speaking, in smaller municipalities. Regarding Oslo, immigrants from southern and eastern Asia account for a fairly high percentage of total firms engaged in food related services, as food retailing, snack bars and kiosks, and restaurants. For food retailing, and snack bars and kiosks, the share is nearly a third, while it is around 15 per cent for restaurants.

3   Focusing on immigrants from poor countries the figures for 1997 reveal that the percentages of self-employed males involved in food retailing and restaurants are inversely proportional to *age*, while the engagement in manufacturing industries, construction work, wholesale trading, other retailing and business services increases with age.[20] The tendency

for business services also applies for immigrants from rich countries but the increase was most pronounced among immigrants from poor countries, while the percentage of immigrants from rich countries engaged in construction industries decreases with age. This structural pattern dismantles somewhat when the focus is restricted to immigrants from southern and eastern Asia, not due to the fact that the region also includes richer countries but because the Pakistanis make up a large share. The pattern for food stores disappears, probably as a result of this being an industrial niche for the Pakistanis. Nevertheless, for males we still find the percentage engaged in restaurants inversely proportionate to age, while wholesale trading, construction work and business services increase with age.[21] The number of females in the eldest demographic group is too low to give reliable figures for self-employed women from southern and eastern Asia.

4   Although *age* and *residence time in Norway* are, to some extent, overlapping variables, the trend is less pronounced for residence time than for age, both for immigrants from poor countries and from southern and eastern Asia. However, when looking at male immigrants from poor countries, the percentages involved in wholesale trading, other retailing and business services increases with residence time, while restaurants and food retailing show a more mixed pattern, although there are some signs of a inverse pattern relative to residence time.[22] Among the male immigrants from southern and eastern Asia we find a clear pattern for wholesale trading; for example, the percentage increases with residence time. The same is true for some other industries too, such as manufacturing, construction, business services and private services, but the differences are rather small; for example, in 1997 from zero to about two percentage points.[23] On the other hand the pattern for food stores is not clear, although the percentage is far larger among immigrants in the category with lowest residence time, a pattern which happens to be largely caused by the Pakistanis.

5   We should also expect *educational level* to impact on the selection of industry to be self-employed in. But for a long time there have been problems with these data due to lack of information on educational level among immigrants; for example, a quite large number of immigrants have been categorized as unknown, or with no education. Nevertheless, for male immigrants from poor countries there is a tendency to engage in wholesale trading, other retailing or business services to increase with educational level, while the opposite is true for food retailing and restaurants.[24] Turning to male immigrants from southern and eastern Asia, the picture again gets more mixed. We find a tendency for food retailing and restaurants to diminish as the educational level increases, while an opposite pattern emerges for other retailing and business services.[25] The pattern for business services is true for immigrants from both rich countries and poor countries.

By wage level,[26] industries show a clear pattern. Generally speaking, immigrants from poor countries are often found in the lowest pay grades, while immigrants from affluent countries earn higher wages. Between 80 per cent and 95 per cent of immigrants from southern and eastern Asia are involved in industries in the two lowest pay grades, ranging from Indians at 80 per cent to Chinese at 95 per cent. There is a tendency for the percentage in the lowest wage categories to decrease as residence time and age rise, while the opposite applies to residence time and the two highest wage categories.

The pattern of industrial concentrations detected thus far indicates that immigrants from countries in southern and eastern Asia and poor countries in general are concentrated in a handful of industries, generally those featuring low wage levels. Age, residence time and education influence this pattern somewhat, but the main tendencies are the same. There are also some country-specific variations within the framework of the main pattern, e.g. the strong concentration of Chinese in the operation of eating places; the strong tendency among Pakistanis to be involved in food retailing, with restaurants as a good No. 2; the strong tendency for Vietnamese to be engaged in restaurants, with food retailing as a good No. 2; and the tendency for Indians to be engaged in food retailing. Additionally, by Norwegian standards, the Chinese, Pakistanis and Indians are immigrant groups with a high proportion of self-employment (10–20 per cent).

A quantitative analysis of career paths into and out of self-employment

By focusing on career paths, the present study sought to determine the impact of training and experience, defined as experience from the same business, on self-employment among immigrants. This was done by examining the course of immigrants' career development from 1989 to 1993 and 1993 to 1997, respectively. The various types of career development paths or streams are illustrated in Figure 11.1. First, we have identified the industries in which wage earners in certain industries are registered as self-employed four years later, for example, the pattern of distribution (*a* in Figure 11.1). Second, we have examined which industries self-employed immigrants in certain industries have been recruited from, for example, the recruitment pattern (*b* in Figure 11.1). The same two-pronged approach was then applied in the other direction, from self-employment to wage earning. Naturally, some people come from or move to unemployment or something else, for instance, education or social security benefits, but we maintain that this does not impact our conclusions about career paths based on training and experience.[27] In addition to these two main trends, we will analyse immigrants who became self-employed during the period and then moved on to something else. This means they were self-employed only in 1993, and that their career development consisted of a recruitment pattern and then a dis-

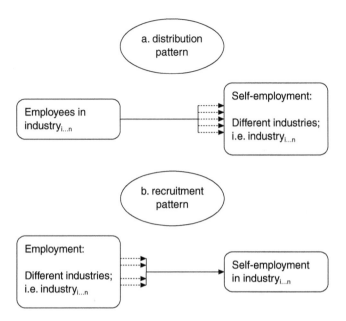

Figure 11.1 Model of career paths.

tribution pattern. Finally, we will look at business careers among those who were self-employed at two different points in time.

For career paths *from wage earning to self-employment*, the following main characteristics were observed:

1   With a view to *the pattern of distribution*, the main pattern is either to establish a business in the same industry as the one in which one had paid work, or to establish a grocery store, eventually a snack bar. The percentage with experience from grocery stores and restaurants who establish themselves in the same industry is especially high, and this tendency is also prominent in industries such as snack bars and convenience stores. On the other hand, the percentages that choose the same business is virtually insignificant in industries such as transport (taxi), cleaning, food production and the healthcare sector.

The percentage that start their own business in the same industry in which they had paid employment increased from the first period (1989 to 1993) to the second period (1993 to 1997). This was especially true for grocery stores, restaurants and snack bars. This indicates that financial progress ensues when several people start businesses in the same industry.

In Table 11.1 and Table 11.2, the seven wage-earner industries for 1989 and 1993, respectively, with the highest number of immigrants moving into self-employment are shown. These seven industries

Table 11.1 Distribution pattern: a set of wage-earner industries in 1989 distributed on self-employment industries in 1993, male immigrants from southern and eastern Asia (%)

To:	From: Wage earner industry in 1989							
	Food processing[a]	Grocery stores[b]	Restaurants[c]	Transport[d]	Public sector[e]	Cleaning[f]	Health sector[g]	Other industries
Self-employment industry in 1993	Grocery 33.3 Kiosk 23.8 Snack bars 19.0	Grocery 64.3	Restaurants 56.9 Snack bars 12.8 Grocery 9.2 Kiosks 8.3	Grocery 28.8 Snack bars 13.5	Grocery 30.4	Snack bars 19.4 Grocery 16.7 Kiosks 13.9 Cleaning 2.8	Grocery 33.3 Kiosks 16.7	Grocery 16.6 Snack bars 10.4 Restaurants 8.3 Kiosks 8.3
No.	21	28	109	52	23	36	24	181

Notes
a Isic code 31 Manufacture of food, beverages and tobacco.
b Isic code 6222 Retailing of grocery goods.
c Isic code 6311 Operation of authorized restaurants.
d Isic code 711 Land transport.
e Isic code 911 Local government administration.
f Isic code 952 Laundries, laundry services and cleaning and dyeing plants.
g Isic code 933 Medical, dental, other health and veterinary services.

Table 11.2 Distribution pattern: a set of wage-earner industries in 1993 distributed on self-employment industries in 1997, male immigrants from southern and eastern Asia (%)

	From: Wage earner industry in 1993							
To:	Food process	Grocery stores	Kiosks[a]	Restaurants	Snack bars[b]	Cleaning	Health sector	Other industries
Self-employment industry in 1997	Grocery 34.6	Grocery 74.2	Kiosks 58.6	Restaurants 74.5	Snack bars 54.8	Grocery 30.2	Grocery 27.3	Grocery 13.8
	Restaurants 26.9	Kiosk 6.2	Grocery 20.7	Snack bars 14.0	Restaurants 19.4	Snack bars 18.6	Cleaning 13.6	Snack bars 10.4
	Kiosks 15.4			Grocery 4.0	Grocery 12.9	Restaurants 16.3		Restaurants 9.6
	Food processing 3.8							Kiosks 8.5
No.	26	97	29	200	62	43	22	260

Notes
a Isic code 6226 Retailing of tobacco, chocolate, fruit and ice cream.
b Isic code 6312 Operation of snack bars, salad bars and hot-dog bars.

comprise about two-thirds of the wage-earner flow into self-employment. Furthermore, the wage earners in the category 'other industries' most frequently start up in grocery stores and snack bars, followed by restaurants and kiosks, but generally the percentages are lower than they are for the selected wage-earner industries, indicating a more dispersed pattern, probably caused by the fact that grocery stores, restaurants, etc., are among the wage-earner industries listed in the two tables.

2   *The recruitment perspective* demonstrates another pattern. During the first period, from 1989 to 1993, it was only in the restaurant business where internal recruitment dominated, and the restaurant industry was also an important contributor to the recruitment mix in kiosks and snack bars. During the second period, from 1993 to 1997, internal recruitment was still most prominent in the restaurant sector, but also in grocery stores, kiosks, and snack bars the internal recruitment constituted the largest single recruitment industry, even though the proportion was generally well below 50 per cent.

As Table 11.3 and Table 11.4 show, the four selected self-employment industries comprise about two-thirds of the recruitment into self-employment, a little lower in 1993 and a little higher in 1997. The pattern for the category 'other industries' reveals a mixed pattern; for example, varying wage earner industries, but restaurant also makes its way into this category from the first to the second period. Furthermore, the percentages are low and decrease from the first to the second period.

For the career path *from self-employment to wage earning*, it is only *the pattern of distribution* that is of interest. The pattern is clear: the vast majority end up being wage earners in the same industry in which they were self-employed. Once again, the closest links are in the restaurant industry, with figures well over 50 per cent, while the proportion is well below 50 per cent in the other industries. There were no clear distinctions between the two periods.

Although the evidence did not allow us to examine an equal number of industries among women, the same general patterns also appear to apply for women. However, among women there was a higher percentage of wage earners in the grocery business than in the restaurant industry setting up a business in the same industry. The proportion was greater than 50 per cent in both industries.

The main pattern as described above also applies to immigrants from Pakistan. On the other hand, the restaurant industry dominates almost completely among Chinese immigrants. Almost 90 per cent of those who leave paid employment in the restaurant industry to start for themselves do so in the same industry. And a comparable percentage returns to paid employment in the restaurant industry when they are no longer self-employed in the restaurant industry. This is indicative of a strong bias in career paths among Chinese immigrants in Norway.

Table 11.3 Recruitment pattern: wage earners from different industries in 1989 to a set of self-employment industries in 1993, male immigrants from southern and eastern Asia (%)

From:	To: Self-employment industry in 1993				
	Grocery stores	Kiosks	Restaurants	Snack bars	Other industries
Wage-earner industry in 1989	Grocery stores 17.8	Restaurants 19.8	Restaurants 69.7	Restaurants 26.4	Transport 7.6
	Transport 14.9	Transport 10.9	Transport 5.6	Transport 13.2	Restaurants 7.6
	Health 7.9	Cleaning 10.9	Hotels[a] 4.5	Cleaning 13.2	Cleaning 7.6
	Food processing 6.9	Food processing 10.9	Cleaning 4.5	Snack bars 7.5	Public sector 6.5
	Public sector 6.9	Health sector 8.7		Fabricated metal[b] 7.5	Health sector 5.9
	Cleaning 5.9	Public sector 6.5		Food processing 7.5	Education sector 4.9
No.	101	46	89	53	185

Notes
a Isic code 632 Operation of hotels.
b Isic code 38 Manufacture of fabricated metal products, machinery and equipment.

Table 11.4 Recruitment pattern: wage earners from different industries in 1993 to a set of self-employment industries in 1997, male immigrants from southern and eastern Asia (%)

From:	To: Self-employment industry in 1997				
	Grocery stores	Kiosks	Restaurant	Snack bars	Other industries
Wage-earner industry in 1993	Grocery stores 45.6	Kiosks 30.9	Restaurants 71.6	Snack bars 32.4	Restaurants 6.6
	Cleaning 8.2	Grocery 10.9	Snack bars 5.8	Restaurants 26.7	Transport 6.6
	Food processing 5.7	Transport 7.3	Hotels 3.8	Cleaning 7.6	Retail wearing 6.6
	Restaurants 5.1	Social 7.3	Cleaning 3.4	Grocery stores 3.8	Cleaning 5.6
	Snack bars 5.1	Food processing 7.3	Food processing 3.4	School sector 3.8	Health sector 5.6
No.	158	55	208	105	213

The data also enabled us to examine the career path *from paid employment to self-employment and back to paid employment again*. The large degree of internal recruitment detected for restaurants was corroborated. Most of those who quit after attempting to set up their own eating place return to paid employment in the industry. In the grocery business the distribution remains broader, but a higher proportion than those who came from the grocery industry remains in the industry once they abandon self-employment. It is also possible to talk about an *industrial career path* for those who are self-employed during two consecutive years. Our corpus indicates that the vast majority (about 90 per cent) were registered in the same industry in two consecutive years. In the restaurant industry, the figure was as high as 97 per cent.

The quantitative analysis has identified an obvious learning effect based on industry, but the effect varies from one industry to another. For eating places, especially restaurants, the learning effect is strong, and the industry is characterized by robust internal career paths, both from wage earning to self-employment and vice versa. There is also a learning effect in the grocery business, but the industry recruits from a far broader range of industries than the restaurant business. On the other hand, during the period under review, the learning effect had little influence on the transition between paid employment and self-employment in industries such as transport and cleaning.

Expanding the analysis to immigrant groups other than those from southern and eastern Asia corroborates this pattern, as also when selected other industries are taken into account. Such an expansion also demonstrates that the learning effect in industries such as building and construction, transport and cleaning is more prominent for immigrants from affluent countries. This points in the direction of other factors and mechanisms having an impact on the importance of the learning effect, and on the extent to which immigrants from different countries are able to take advantage of such a learning effect.

A qualitative analysis of careers and acclimatization strategies in the food industry in Oslo

Generally speaking, the semi-structured interviews, conducted by Onsager and Sæther (2001), and their analysis revealed that immigrants are integrated into an employment system and, over time, occupy different positions and roles as wage earners and entrepreneurs in various food-related activities such as grocery stores, snack bars, convenience stores, cafés, restaurants, fast-food places and import businesses. The interviews disclosed more sporadic transitions between other industries such as taxi driving and cleaning. Another general trait is that immigrants tend to be wage earners in low-status occupations for quite some time before establishing their own businesses. They often hold down multiple jobs to earn enough money and gain sufficient expertise to start up on their own account.

Retailing

As indicated in Figure 11.2, most started their businesses without prior experience of self-employment, either in their country of origin or in Norway, but many have experience from paid employment in grocery stores, and they have seen how acquaintances have managed. Hence one common career in retailing is to build up capital through hard work and a frugal lifestyle, then to start one's own business to improve one's working situation and finances. In addition to accumulated capital from earnings, it is common to borrow money from relatives and acquaintances in Norway, and possibly also abroad. Some immigrant entrepreneurs had also taken out bank loans. Most had purchased shops from Norwegians, but some had started from scratch. Both ethnic and class resources are essential for starting and operating businesses, for example, ethnic bonds and acquaintances, mutual trust and a common language play a decisive role in getting started right, as well as for sharing knowledge. The career path may take the direction of consolidating the store, often leading into a subsistence situation; expansion;

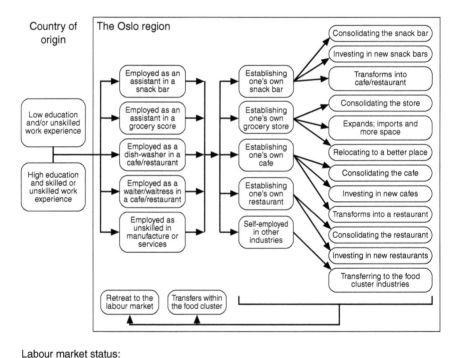

Figure 11.2 Illustration of career paths among immigrants in the food-related industrial system in the Oslo region (source: NIBR report (2001: 6)).

and relocating to a better site; or the owner may return back to the wage-earner situation.

Snack bars, convenience stores and cafes[28]

In contrast to the general picture of Figure 11.2, this group of entrepreneurs had little relevant education or experience prior to starting their own businesses. However, the informants who had been in business the longest had some experience of running a shop and bookkeeping in their country of origin. Several had been employed in food and service industries before striking out on their own. Another trait is that some started snack bars after having tried other types of enterprises such as grocery stores or restaurants. It was common for owners to have had other employment during the initial phase, and that they kept those jobs until their snack bar began to show a profit. In general, the threshold for starting a snack bar was considered low. For example, snack bars require considerably less stock than grocery stores and the pace of their turnover is significantly faster. Less capital is required to get a snack bar up and running. These entrepreneurs' subsequent careers can move in several directions. Attempting to upgrade a snack bar to a café is one alternative and expanding into a chain of operations is a second. However, most stay put in the snack-bar business and the business becomes a subsistence enterprise.

Restaurants

Figure 11.3 indicates that the paths into operating a restaurant are many, both directly from a wage-earner position and via snack bars and cafes. The range between marginal subsistence enterprises and capital-forming growth enterprises appears to be vast in the restaurant industry. Small restaurants are similar to cafés in many respects, although the restaurants offer a higher level of service, must meet certain requirements for a liquor licence and

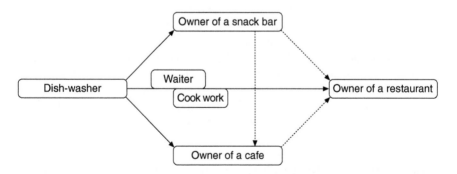

Figure 11.3 Illustration of career path within the catering trade (source: NIBR report (2001: 6)).

feature dinner menus, meaning they call for a higher level of expertise and greater ability to deal with Norwegian legislation and regulations. The higher barriers to entry mean that many hesitate to enter the restaurant business until they have gained experience and expertise as a restaurant employee or as an owner of a snack bar or café.

Some of the informants started their businesses based on informal competence and a small amount of start-up capital acquired through hard work as an employee in one or more jobs at a restaurant, snack bar or café. Another group of the informants based their entrepreneurship on prior experience of company ownership and operation in similar activities, often accumulated in family businesses and possibly through training. One common feature is thus several years of experience in the restaurant industry as a dishwasher, cook or server. But occupational experience is broader, and specialization in certain industries is weaker among those who run subsistence enterprises, than is the case for owners of expansive growth enterprises. Another common feature is that entrepreneurs draw upon class and ethnic resources, especially if they have experience of establishing and operating restaurants in Oslo. Moreover, labour is often recruited among people from the same country of origin, thus reducing communication problems and ensuring improved stability and control. Some have bought restaurants, while others have started them from scratch.

The conclusion is that self-employment stands out as a project with a more or less deliberate preparatory phase, followed by an establishment phase with trial (and error) and avenues of retreat to paid employment, at the same time as many have paid employment parallel to setting up a business of their own. Finally, they reach an operational phase. Most of these businesses will be subsistence enterprises, but a handful of immigrants create expansive growth enterprises. The interviews also revealed that some had changed their line of business.

The Onsager and Sæther (2001) study also concludes that food-related immigrant businesses are embodied in a multi-ethnic food cluster, wherein enterprises are linked to each other through an economic and social network, comprising different ethnic minorities and the majority population. Vertically in the value chain goods are bought from Norwegian and foreign producers, suppliers and import firms located in Oslo. The market is generally multi-ethnic but overall dominated by the majority population, although some are more focused on their own ethnic group. Horizontally, drawing on certain co-ethnic industry-related resources and networks is more common. On the other hand, Krogstad (2002), also studying the food sector in Oslo, concluded that co-operation between firms within food-related industries is as frequent across ethnic groups as within them, e.g. some co-operate in focusing on different areas, some on the time they operate, some on procurement. Furthermore, Krogstad's survey, covering a broader range of immigrants than the study by Onsager and Sæther, revealed that about half the respondents had helped others through the start-up phase, and of those 20 per cent were relatives, while 39 per cent were co-ethnic or from the same country of origin, but more than a third were immigrants from other countries or with a mixed background. But also Krogstad found that Asian enterprises are strongly based on an extended household, comprising strong internal bonds and a tradition of internal organization.

Empirical finds in a theoretical perspective

We maintain that Light and Gold's focus on resources and Kloosterman and Rath's focus on embeddedness and the balance between supply and demand represent a fruitful theoretical basis for the pattern discussed above. The structure that applies to this perspective is demonstrated in Figure 11.4. The supply side represents the players and the resources they draw on in order to establish and operate their own businesses, while the demand side represents the structural embeddedness they operate within as entrepreneurs.

Above, we have demonstrated that immigrants from southern and eastern Asia, like immigrants from other poor countries, focus on food-related industries, and that these industries feature a generally low wage level and correspondingly low entrepreneurial income. This bears witness to the fact that the opportunity structure, including availability and growth potential in the market, establishes a clear pattern for the industries in which immigrants set up in business, indicating a concentration mechanism inherent in

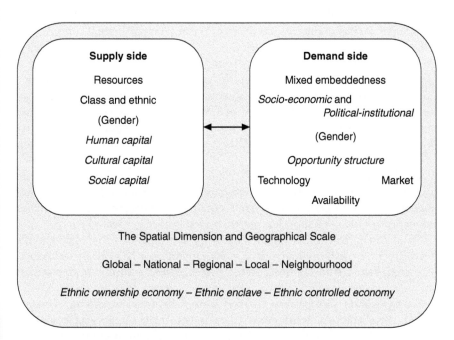

Figure 11.4 Theoretical approach (source: NIBR report (2001: 15)).

the opportunity structure; for example, a demand resulting in a market niche to be met, together with economic and technological premises favouring certain industries more than others. But this concentration also has consequences regarding the financial conditions self-employed immigrants operate within; for example, the income potential is generally modest, combined with long working hours.

By the same token, *inter alia*, residence time and education reduce the differences in entrepreneurial income between groups of immigrants. Hence this reflects interaction between resources on the one hand and opportunity structure on the other. This may be deemed a newcomer's disadvantage, and it is reflected in the industries chosen by immigrants from poor countries and countries that differ greatly from Norwegian society in terms of culture. This may be related to institutional conditions; for example, the industries immigrants from poor countries go into may have lower entry barriers than industries immigrants from rich countries choose, as is exemplified by the strong concentration of immigrants from other Nordic countries in the building industry; for example, to establish a building firm certain requirements regarding education and professional background must be met. Nevertheless, variations also exist between industries immigrants mostly enter. For instance, starting a restaurant is professionally more demanding than starting a grocery store, and furthermore the restaurant industry is subject to a more stringent control system than, for instance, grocery stores, due to regulations enforced by the municipal planning and building authorities, the municipal branch office of the Norwegian labour inspectorate, the municipal Fire and Emergency Service, the municipal branch office of the Norwegian Food Control Authority, and the municipal business authority, comprising the restaurant and permit department. Although starting a grocery store requires approval from the same authorities, the regulations are stronger for those establishing a restaurant, due for instance to the prescriptions regarding emergency in case of fire; regulations regarding preparing meals in a kitchen, both equipment and qualifications; regulations regarding hygiene standards and sanitary facilities; and passing a test regarding liquor serving regulations. This may serve to intensify the barriers for those who do not fulfil the requirements. Such conditions help explain differences in recruitment streams within different industries. Meanwhile, to a greater extent than immigrants from poor countries, immigrants from affluent countries are involved in industries with even higher barriers to entry.

This career pattern is characterized by the fact that industry-specific recruitment to self-employment is predominant in a handful of industries, e.g. restaurants, while in other industries, e.g. food retailing, entrepreneurs are to a greater extent recruited from and transferred to other industries. In order to explain this pattern there is a need to draw on several elements under the theoretical framework. Class is one main category of resources, and in the form of human capital helps some of the immigrants to go into business activities because of their own experience or inherited experience, either through the petty bourgeois or upper-class business people. This tendency is reinforced by socio-economic embeddedness, where ties to a profession and a social community make some options more obvious than others, as is the case for immigrants getting a job in certain industries thereby becoming a member of network of information and knowledge. And, furthermore, the institutional-political embeddedness makes it easier to go into certain industries than others, as exemplified above by grocery stores, restaurants and the building industry. Next, the ethnic dimension is linked to socio-economic embeddedness; for example, an interaction between ethnic preferences and producer and consumer network advantages, or some may characterize it as disadvantages because it contributes to low-pay industry lock-in, thereby making it preferable to draw on co-ethnics and simultaneously building a pool of ethnic resources to draw on later. Because of the tendency for immigrants from different countries to be concentrated in certain industries the ethnic resources channel the entrepreneurship in the direction of some industries rather than others. Put more simply: class resources, socio-economic embeddedness, institutional-political embeddedness and ethnic resources working together (Figure 11.4).

A similar chain may be attributed to immigrants without business class experience and business class heritage. The class resources are restricted to the wage-earner domain: for example, not including the experience and culture of doing business, and dominated by industry-related knowledge. Following Hardman Smith's (1996) distinction between a wage-earner life form and a self-employment life form, former wage earners may face bigger problems due to a lack of a proper habitus, to use Bourdieu's term. Nevertheless, the career path will be shaped through the same interaction as described above, although the outcome may vary, a matter so far not answered by Norwegian research.

In terms of regions, it appears that immigrants from poor countries who live in Oslo have a greater propensity than immigrants living elsewhere in the country to be involved in food retailing, while the pattern in catering is, relatively speaking, characterized by a broader regional distribution. This raises the question of the significance of geography to self-employment. Oslo, and other big cities by Norwegian standards, represent the largest markets, with the highest numbers of positions open to newcomers. In spite of this, the larger percentage of immigrants in catering is outside Oslo, in far smaller markets. Of course, this may be caused by stronger competition from native Norwegians in the Oslo market, especially in certain attractive areas of the city, due to the fact that Oslo is the capital, serving a large and varied market. The competition in certain segments in the upper parts of the market is therefore strong. On smaller places the market is smaller and the economic margins are correspondingly smaller, making it easier for immigrants to enter and take over positions left by Norwegians. At the same time, we know that food retailing is relatively strong among Pakistanis, and Pakistani immigrants generally live in Oslo. On the other hand, the

Chinese are more strongly represented in catering, especially restaurants, and this group has a more widespread pattern of settlement. Hence the regional pattern is to a certain degree ascribable to the immigrants' embeddedness in birth-country-specific socio-economic conditions. But there is more to it than that. For one thing we should consider the immigrant history of Pakistanis and Chinese, the first from the labour immigrant period, and the latter a more recent history, but none of them contain refugees at any rate. On the other hand, Pieke (1998), in a study of the Chinese in Europe, underlined the role self-employment played as a symbol of success, and owning a restaurant was for many a central aim. But the flocking into the restaurant sector caused market niches to saturate geographically, and according to Pieke, a dispersed pattern emerged rather than cut-throat competition in a given area. So, if this strategy applies to Norway, the differences regarding regional self-employment patterns for food retailing and restaurants may be due to a combination of ethnic resources and socio-economic embeddedness.

There is no ethnic enclave economy in Oslo, but there is an ethnic ownership economy and a tendency to have an ethnic-controlled economy in catering, as indicated by the figures above. This is important for understanding the interaction between supply-side resources and embeddedness, together with the opportunity structure on the demand side. Barriers diminish, and it becomes easier to set up in business or get a job in certain industries for co-ethnics or ethnics possessing a similar cultural and religious background. Thus interaction between resources and opportunity structure is mutually enhanced. However, it also represents a lock-in of career paths and may contribute to the creation of new barriers. Immigrants remain in a handful of industries, and it is conceivable that immigrant entrepreneurs might compete each other right out of business, causing employee wage levels to be depressed even further. However, as Waldinger (1996) asserts, over time, groups of immigrants move into new industries, and this takes place as described by Kloosterman and Rath (2001), for example, that innovators break new ground, paving the way for their successors. New parameters are established for the opportunity structure and the mixed embeddedness.

Thus far, this review has not treated gender systematically, although reference has been made to males and females, but the numbers of females were often too low to give reliable figures for immigrants from southern and eastern Asia. That does not mean that gender is not an issue. In Figure 11.4, gender is placed in an intermediate position on the demand and supply sides alike, without any straightforward effect vectors. This is because women mostly follow the same pattern as men; for example, when we look at larger groups of immigrant women, the effects of for instance age and time of residence have the same direction but the degree of change according to categories of age and time of residence is smaller. Female immigrants from countries in southern and eastern Asia are concentrated in the same industries as men, although women are engaged more often in food retailing and less often in the restaurant industry. In addition female immigrants in general also follow the same pattern as men when it comes to training and careers, but internal recruitment is lower in the restaurant business and higher in food retailing. To conclude, the overall pattern is much the same but the effect varies between males and females.

Common features indicate that the driving forces are the same for both genders, while the differences are at the very least related to the extent of the effect. This may be because women react differently and make different choices. Or perhaps women face another type of mixed embeddedness than men, so that the same outcomes do not have the same outcomes as for men. Or perhaps the options have a different composition, channelling a higher proportion of women into other types of acclimatization strategies. In conclusion, self-employment among immigrants from southern and eastern Asia is dominated by Pakistanis, Indians and Chinese, who show high percentages of self-employment. Nevertheless, self-employed people from other countries also are visible because of concentration in a few industries, mainly food retailing, and catering industries, which also is dominant among Pakistanis, Chinese and Indians. An analysis of career paths revealed connections between certain wage-earner industries and corresponding self-employment industries. This is especially true for the restaurant sector but also evident in other catering industries and food retailing, although the recruitment pattern for the last mentioned industry is more dispersed. Above, this pattern is theorized through a focus on resources on the supply side and the mixed embeddedness on the demand side, and the interactions between dimensions have been highlighted; for example, as exemplified by class resources – socio-economic embeddedness – institutional-political embeddedness – ethnic resources.

Notes

1 Relevant contributions from the United States include Bonacich (1973), Light and Bonacich (1988), Waldinger *et al.* (1990), Min (1993), Waldinger (1996), Yoon (1997), Bates (1997) and Camarota (2000). As regards European contributions, Pollins (1984), Ladbury (1984), Waldinger *et al.* (1990), Blaschke *et al.* (1990), Brune (1996), Najib (1999), Wilpert (1999), Hillmann (2000), Bager and Rezaei (2000) and Hjarnø (2000) are all relevant.

2 Catering is here meant as an overall concept covering restaurants, snack bars, cafés, etc., and should not be mistaken as the NACE category 55.52 Catering.

3 Waldinger (1996) has done empirical and theoretical research in this field. Sassen (1991) places such development characteristics in a broader context.

4 Waldinger (1996) and Light and Gold (2000)

5 We are grateful for the funding received for this research from the Norwegian Research Council and the Norwegian Institute for Urban and Regional Research.

6 As a group, immigrants from affluent countries have a greater propensity for entrepreneurship than immigrants from poor countries. However, if we look beyond the general pattern, several Asian countries have a relatively high proportion of entrepreneurs. In fact, their percentages are approximately equal to

that of Danish immigrants (Orderud, 2001), as Denmark is very similar to Norway in terms of culture.

7 Southern and Eastern Asia refers to countries east and south of Pakistan and Afghanistan, and including China in the north-east direction.

8 At the beginning of the twenty-first century, immigrants accounted for about 7.5 per cent of the overall population of Norway (home page of Statistics Norway, 2004).

9 The data are from 2004 (home page of Statistics Norway, 2004).

10 This is the situation for first-generation immigrants and children born in Norway of foreign-born parents. If we restrict the figures to first-generation immigrants, the Swedes are the largest group, ahead of the Danes, while the Pakistanis are in fifth place, behind, but close to, immigrants from Iraq and Bosnia-Hercegovina.

11 The differing regional pattern can to a large degree be attributed to the practise of assigning refugees and people granted asylum to live in different places in Norway, although many of them as time passes migrate to the larger cities and especially Oslo (Sørlie, 1994). The Pakistanis, as mentioned in the text above, started arriving in the labour immigration phase, and therefore from the start had a more centralized residential pattern.

12 This includes first- and second-generation immigrants, defined according to the Statistics Norway definition, respectively as individuals living in Norway and a child of two foreign-born parents and individuals born in Norway of parents who immigrated to Norway. Had the criterion been one foreign-born parent the share would have increased markedly. Vassenden (1997) showed that the percentage of immigrants in 1996 would have increased from about 6 per cent to roughly 9 per cent, but this last group had a rather heterogeneous background, according to Vassenden (1997).

13 The following variables are linked to the individual: gender, year of birth, country of origin, year of immigration, year of emigration if applicable, municipality of residence, highest education, employment income, entrepreneurial income, financial income, social security income, assets, and ties to the job market and industry as a wage earner. In addition, there are corporate data linked to individuals registered as owners of enterprises, including branch of industry, municipal affiliation, number of employees, employees with same country of origin as the owner, or other foreign country of origin, and sales.

14 The statistical material was improved in the 1990s for a number of service industries requiring professional qualifications. However, the present analysis (Orderud, 2001) discovered that data on industrial classification and educational level are often lacking for immigrants who come from poor countries and have been in Norway less than four years.

15 Turkey, Iran and Iraq are not part of the region traditionally defined as southern and eastern Asia but they are part of a larger Asian region.

16 Distribution by country: ten Turks, ten Pakistanis, one Chinese, one Kurd, one Iranian and one Iraqi. Distribution by business: fourteen retail outlets (mainly grocery stores), five snack bars and five restaurants, including a fast-food chain.

17 The distinction between poor countries, moderately affluent/poor countries and affluent countries is based on a ranking of countries by purchasing power parities in US dollars (http://www.cia.gov/cia/publications/factbook). The interval between the lowest and highest values is split in three categories (in US dollars below $11,000; between $11,300 and $22,000,; and above $22.000).

18 Catering includes approved restaurants and snack bars, salad bars and hot-dog stands, as well as other eating places.

19 The Chinese here comprise immigrants from the Republic of China, Taiwan and Hong Kong.

20 In 1997 the percentages were these for immigrant males from poor countries, listed consecutively for the three demographic groups, twenty to thirty-four years, thirty-five to forty-nine years and fifty to sixty-nine years: manufacturing 3.1, 6.1, 10.0; construction 4.0,5.6, 7.7; wholesale trading 6.2, 9.3, 12.3; other retailing 9.4, 12.8, 15.4; business services 1.1, 3.6, 12.3; food retailing 30.3, 25.5, 19.2; catering 35.6, 27.8, 9.2.

21 In 1997 the percentages for the three categories twenty to thirty-four years, thirty-five to forty-nine years and fifty to sixty-nine years, of males from southern and eastern Asian countries were these, respectively: construction 1.1, 1.8, 6.1; wholesale trading 4.8, 8.9, 16.3; business services 1.1, 0.8, 4.1; catering 40.7, 36.7, 16.3.

22 Among immigrants from poor countries the 1997 percentages for the three residence time categories up to four years; five to fourteen years and fifteen or more years, respectively, were: wholesale trading 4.0, 5.4, 9.8; other retailing 6.0, 9.9, 11.8; business services 0.0, 1.2, 3.9; food retailing 36.0, 26.7, 29.4; catering 30.0, 40.2, 23.5.

23 Among immigrants from southern and eastern Asia the 1997 percentages for three residence time categories up to four years; five to fourteen years and fifteen years, respectively, were: wholesale trading 0.0, 3.5, 8.8; food retailing 61.9, 28.3, 38.8; catering 23.8, 53.4, 29.5.

24 Among immigrants from poor countries the 1997 percentages for the following four categories of educational level: no education, elementary school, secondary school, college/university, respectively, were: wholesale trading 12.2, 4.7, 9.8, 11.0; other retailing 9.8, 10.3, 11.4, 16.3; business services 1.8, 1.3, 1.8, 11.3; food retailing 33.5, 32.7, 23.5, 20.7; restaurant 28.0, 36.3, 24.6, 19.7.

25 Among immigrants from southern and eastern Asia the 1997 percentages for the following four categories of educational level: no education, elementary school, secondary school, college/university, respectively, were: other retailing 9.6, 9.7, 9.6, 20.4; business services 0.0, 0.9, 1.5, 3.5; food retailing 34.0, 38.1, 31.7, 29.2; catering 37.2, 40.6, 30.3, 31.0.

26 Four categories have been used, based on hourly wage level at the end of the period, i.e. in 1997: (1) less than NOK 95; (2) NOK 95–110; (3) NOK 111–139: and (4) NOK 140 or more. Converted to euros (NOK 8.30 to one euro) and US dollars (NOK 7.40 to one dollar), this amounts to: (1) less than €11.30, or US$12.80; (2) €11.30–€13.20 or US$12.80–$14.80; (3) €13.21–€16.80, or US$14.81–$18.80; and (4) above €16.80, and above US$18.80.

27 Of those who were self-employed from 1989 to 1993 or from 1983 to 1997, respectively.

28 The distinction between snack bars and cafés is rather fuzzy, but for the purpose of the present study the term 'snack bar' includes establishments that offer limited space but where it is possible to stand or sit on stools at counters along the wall and dine on hamburgers, kebabs, chips, etc. Owners who want to furnish such premises with tables and chairs have to apply to the municipality of Oslo for permission.

References

Bager, T. and Rezaei, S. (2000) 'Immigrant Businesses in Denmark: Captured in Marginal Business Fields?' Revised version of paper submitted to the eleventh Nordic Conference on Small Business Research, Aarhus, Denmark.

Bates, T. (1997) *Race, Self-employment and Upward Mobility*. Baltimore MD: Johns Hopkins University Press.

Blaschke, J., Boissevain, J., Grotenbreg, H., Joseph, I., Morokvasic, M. and Ward, R. (1990) 'European trends in ethnic business', in R. Waldinger, H. Aldrich and R. Ward *et al.* (eds) *Ethnic Entrepreneurs. Immigrant Business in Industrial Societies*. London: Sage.

Bonacich, E. (1973) 'A Theory of Middleman Minorities', *American Sociological Review*, 38 (5): 583–94.

Brune, Y. (1996) *Invandrare som egenföretagare*. Stockholm: Arbetsmarknadsstyrelsen, NUTEK and Statens Invandrarverk.

Camarota, S.A. (2000) *Reconsidering Immigrant Entrepreneurship. An Examination of Self-employment among Natives and the Foreign-born*. Washington DC: Center for Immigration Studies.

Hillmann, F. (2000) 'Are Ethnic Economies the Revolving Doors of Urban Labour Markets in Transition?' Paper presented at the conference 'The Economic Embeddedness of Immigrant Enterprises', Jerusalem, 18–20 June.

Hjarnø, J. (2000) 'Innvandrere som selverhvervende: en sammenlignende analyse av udbredelsen af selverhverv hos danske pakistanere, tyrkere og eksjugoslaver', *Dansk Sosiologi*, 11 (3): 95–112.

Kloosterman, R. and Rath, J. (2000) 'Mixed Embeddedness: Markets and Immigrant Entrepreneurs. Towards a Framework for comparative Research'. Paper presented at the conference 'The Economic Embeddedness of Immigrant Enterprises', Jerusalem, 17–20 June.

Kloosterman, R. and Rath, J. (2001) 'Immigrant entrepreneurs in advanced economies: mixed embeddedness further explored', *Journal of Ethnic and Migration Studies*, 27 (2): 1–10.

Krogstad, A. (2002) *En stillferdig revolusjon i matveien. Etniske minoriteter og kulinarisk entreprenørskap*. ISF report 2002:7. Oslo: Institute of Social Research.

Ladbury, S. (1984) 'Choice, chance or no alternative? Turkish Cypriots in businesses in London', in Robin Ward and Richard Jenkins (eds) *Ethnic Communities in Business. Strategies for Economic Survival*. Cambridge: Cambridge University Press.

Light, I. and Bonacich, E. (1988) *Immigrant Entrepreneurs. Koreans in Los Angeles, 1965–1982*. Berkeley CA: University of California Press.

Light, I. and Gold, S.J. (2000) *Ethnic Economies*. San Diego CA: Academic Press.

Light, I. and Rosenstein, C. (1995) *Race, Ethnicity, and Entrepreneurships in Urban America*. New York: Aldine de Gruyter.

Min, P.G. (1993) 'Korean Immigrants in Los Angeles', in Ivan Light and Parminder Bhachu (eds) *Immigration and Entrepreneurship. Culture, Capital, and Ethnic Networks*. New Brunswick NJ: Transaction Publishers.

Najib, A.B. (1999) Invandrarföretagande: några grundläggande fakta. Bilaga 2 i SOU 1999: 49 *Invandrare som företagare. För lika möjligheter och ökad tillväxt*. Stockholm: Inrikesdepartementet.

Onsager, K. and Sæther, B. (2001) *Etniske entreprenører og selvsysselsetting*. NIBR report 2001:6. Oslo: Norwegian Institute of Urban and Regional Research.

Orderud, G.I. (2001) *Innvandring og selvsysselsetting*. NIBR report 2001:15. Oslo: Norwegian Institute of Urban and Regional Research.

Pieke, F.N. (1998) Introduction, in Gregor Benton and Frank N. Pieke (eds) *The Chinese in Europe*. London: Macmillan.

Pollins, H. (1984) 'The development of Jewish business in the United Kingdom', in

Robin Ward and Richard Jenkins (eds) *Ethnic Communities in Business. Strategies for Economic Survival.* Cambridge: Cambridge University Press.

Rath, J. (2002) 'Needle games: a discussion of mixed embeddedness', in Jan Rath (ed.) *Unravelling the Rag Trade. Immigrant Entrepreneurship in Seven World Cities.* Oxford: Berg.

Sassen, S. (1991) *The Global City. New York, London, Tokyo.* Princeton NJ: Princeton University Press.

Smith, L.H. (1996) *Indvandrere med egen forretning: en kulturanalytisk undersøgelse af selverhvervende indvandrere i København og deres livsformer.* Esbjerg: Sydjysk Universitetsforlag.

Sørlie, K. (1994) *Unge innvandreres flytting i Norge.* NIBR report 1994:5. Oslo: Norwegian Institute of Urban and Regional Research.

Vassenden, K. (ed.) (1997) *Innvandrere i Norge. Hvem er de, hva gjør de og hvordan lever de?* Statistiske analyser 20. Kongsvinger: Statistics Norway.

Waldinger, R. (1996) *Still the Promised City? African-Americans and New Immigrants in Postindustrial New York.* Cambridge MA: Harvard University Press.

Waldinger, R., Aldrich, H. and Ward, R. (1990) *Ethnic Entrepreneurs. Immigrant Business in Industrial Societies.* London: Sage.

Wilpert, C. (1999) 'A Review of Research on Immigrant Business in Germany'. Paper presented at the 'Working on the Fringes: Immigrant Businesses, Economic Integration and Informal Practices' conference, Amsterdam, 7–9. October.

Yoon, I.-J. (1997) *On my Own. Korean Business and Race Relations in America.* Chicago: Chicago University of Chicago Press.

12 Asian business strategies in the United Kingdom

A qualitative assessment

Monder Ram and Trevor Jones

Within the literature on South Asian entrepreneurship, one persistent theme is the seemingly particularistic nature of social ties within the Bangladeshi, Indian and Pakistani communities, which are believed to fuel entrepreneurial activity and encourage co-operative work relations (Werbner, 1990). Crucial to entrepreneurial success is the 'heritage of networks' (Gidoomal, 1997), often argued to be the hallmark of these ethnic groups (Srinivasan, 1995). Often implicitly, studies in this tradition suggest that the 'strategies' that South Asian entrepreneurs adopt in developing and managing their businesses are distinctive, shaped largely by the specific internal characteristics of the community (Basu, 1998; Metcalf *et al.*, 1996).

For the purposes of this chapter, which aims to understand the concrete ways in which South Asian firms are managed, such approaches are problematic on several counts. First, the actual dynamics of the workplace, particularly in relation to the management of people, have rarely been granted an explicit analytical focus with existing studies. An appreciation of these dynamics is necessary in order to assess the extent to which South Asian business 'strategies' are 'distinctive'. Second, theoretical approaches to the study of ethnic minority entrepreneurship – from structural as well as cultural traditions – tend to be insufficiently integrated with insights from other relevant disciplines, notably the growing literature on industrial relations in small firms (see Scase, 1995, and Wilkinson, 1999, for review). Finally, the quantitative orientation of much extant research (Aldrich *et al.*, 1981; Basu, 1998; Metcalf *et al.*, 1996) arguably fails to develop an adequate account of how the actual *processes* of management in South Asian firms are operationalized.

Aiming to plug these gaps, the present chapter conducts a direct examination of how employee relations in South Asians firms are managed. Insights from the developing 'mixed embeddedness' perspective (Kloosterman *et al.*, 1999) are drawn upon to demonstrate that labour management strategy is conditioned by a variety of factors, of which ethnicity is but one. There are three main aspects to the chapter. First, studies of South Asian businesses from different economic sectors are drawn upon to illustrate that ethnicity alone does not have a determinate influence on business practices, which in reality are mediated by a range of influences, particularly market pressures. Second, the studies all deploy intensive qualitative case history approaches, designed to illuminate the dynamics of people management. Finally, the focus is on concrete work practices – recruitment, remuneration and supervision – in different market contexts. It is argued that the management of these practices relies upon the 'negotiation of order' within the workplace.

Employee relations in ethnic minority firms

A note on strategy

Before we come to consider the special case of ethnic minority enterprise, we note that the management process itself has been the subject of considerable speculation in the general small business literature. Particular attention has been paid to the problem of securing consent on the shop floor, with much debate on whether or not this is achieved through deliberate management strategy. Among the various typologies of management control are R. Edwards's (1979) three-stage simple–technical–bureaucratic model and Friedman's (1977) continuum ranging from 'responsible autonomy' to 'direct control'. In relation to the clothing industry, Rainnie (1989) has suggested a regime of managerial autocracy, produced by the imperatives of sheer survival in a cut-throat market dominated by the buying power of High Street retailers.

However, the very phrase 'negotiation of order' implies that idealized notions of a coherent strategy on the part of management are bound to be problematic. In practice, it is doubtful whether firms have explicit well defined strategies of management control. Rather they will respond to situations in a pragmatic, often *ad hoc* way. There are likely to be a variety of means of controlling labour and securing commitment, and this will rarely conform to an ideal type, even if the firm does have clear policies (Edwards, 1986).

Rather than a simple managerial imposition, control is better seen as the product of a number of different influences. There are many ways in which control can be secured, ranging from close supervision to the favourable treatment of key workers, methods often operating simultaneously within the same firm. Systems of control emerge from the complexities of managerial practices and worker activity. Rarely sensitive to this, control typologies generally suggest that particular strategies are devised and implemented, whereas in reality control is diverse and complex, the product of past struggles within the social relations of work (Edwards, 1986). Recognizing this, Edwards (1986) and Hyman (1987) offer a way of conceptualizing relations at work that acknowledges the importance of structural conditions without being deterministic, and allows for action but within certain constraints. This view accepts the classical belief that inherent within the employment relationship is a basic antagonism between capital and labour. Together with this basic antagonism, however, are elements of co-operation, since employers need to secure workers' willingness to work while workers rely on firms for their livelihoods. This fundamental antagonism does not actually determine events, since they have to be interpreted in action. A negotiation of order occurs but within a definite material context.

Increasingly, the discourse of labour management in small firms has recognized these nuanced interpretations of the multi-dimensional and contested nature of management control. Conceptually, the trend has been towards the recognition that small-firm employee relations are characterized by complexity and heterogeneity (see Curran, 1991, and Scase, 1995, for further comment). As such, the approach small firms adopt when managing labour should not just be seen simply as a function of size (Wilkinson, 1999). Paralleling this recognition of complexity and heterogeneity, there is also growing acknowledgement of the importance of informal regulation in the workplace, individualized negotiation, and the distinct nature of labour and product market arrangements (Curran and Blackburn, 1993; Ram, 1994; Holliday, 1995; Kitching, 1997).

Employee relations in South Asian firms

Little of this theoretical development is in evidence in the burgeoning literature on South Asian entrepreneurship, a parallel intellectual universe often isolated from the mainstream of entrepreneurial thought (Ram *et al.*, 2000). Much of this literature is conditioned by the manner in which ethnic minority businesses have been theorized. For example, culturalist approaches portray belief that ethnic minorities are competitively advantaged by their privileged insider access to the productive resources offered by close-knit familial and communal networks (Metcalf *et al.*, 1996; Srinivasan, 1995; Werbner, 1990), a body of resources now referred to as 'ethnic social capital' (Flap *et al.*, 2000). Especially germane to the present argument is the assumption that the strength of co-ethnic ties and the ensuing trust between community members mediates the class distinctions between owners and workers, producing in effect a conflict-free workplace. In this rather idyllic scenario, class identity is decisively overridden by ethnic identity, another instance of the way in which the urge to exoticize ethnic minorities presents them as immune from the normal rules of social life. Not surprisingly, such mysticism has not gone unchallenged and there are well established critiques of the cultural approach, pointing out above all its neglect of the role that the wider economic context plays in shaping the fortunes of ethnic minority firms (Phizacklea, 1990; Ram and Jones, 1998; Rath, 2002).

A fuller appreciation of the economic context is discernible in the accounts of Mitter (1986), Jones *et al.* (1992) and Phizacklea (1990). Considerable explanatory importance is attached to the disadvantages that minority ethnic communities face in a deindustrialized labour market with inbuilt racist biases. Rather than stemming from a culturally specific predisposition for entrepreneurship, recourse to self-employment is more a reflection of limited alternative opportunities. Within self-employment itself, opportunities tend to be limited to some of the least rewarding sectors of the economy (Jones *et al.*, 2000), creating large numbers of marginal employers, who are often obliged to exploit an even more marginalized co-ethnic work force in order to survive (see Hoel, 1984; Mitter, 1986; Phizacklea, 1990). Recent developments in immigrant entrepreneurship theory have attempted to reassert the importance of the economic context in which ethnic minority firms operate, whilst not losing sight of the socio-cultural dimension (Kloosterman *et al.*, 1999; Rath, 2002).

According to this view, the well established concept of embeddedness (Granovetter, 1985) is acknowledged as a useful device for understanding immigrant entrepreneurship. However, a one-sided deployment of embeddedness as purely socio-cultural lends undue emphasis to ethnic ties at the expense of the wider structures in which minorities are embedded. To counter this tendency, Kloosterman *et al.* (1999: 254) propose 'the more comprehensive concept of *mixed embeddedness* ... encompassing the crucial interplay between the social, economic and institutional contexts'. For present purposes, this broad framework is useful, since it recognizes that, notwithstanding their ethnic identity, entrepreneurs are also shaped by the nature of the economic activities in which they are engaged. Significantly, this acknowledgement of the influence of market environment and economic sector resonates with mainstream debates on the influence of customer relations on working practices within the small firm (Rainnie, 1989). Building upon this insight, we are now able to examine more closely the dynamics involved in the managing labour in South Asian firms operating in different market contexts.

Data sources: market contexts

Three studies are drawn upon to examine how the employment relationship is managed in South Asian firms. These are referred to as the consultancy project (Ram, 1999a, b), the clothing project (Ram *et al.*, 2001a) and the catering project (Ram *et al.*, 2000, 2001b).[1] Each of the projects had different objectives. For example, a key issue of the consultancy research was to examine how market relations impinged upon the workplace; the clothing study was part of a multi-sector investigation on the impact of the National Minimum Wage (introduced in Britain in April, 1999) on small firms; and the catering project examined how independent restaurants owned by different (but mainly South Asian) ethnic minority groups operated in the city of Birmingham. However, there were key commonalities. First, the analytic focus was on the management of the employment relationship. Second, as outlined in the preceding section, the theoretical approach was based on the centrality of negotiated order, which was shaped by product and labour market circumstances as well dynamics within the workplace. Finally, each project deployed intensive qualitative research methods. Hence, the consultancy project was based upon a year-long ethnography of three small consultancy firms (one of which was owned by a South Asian Muslim); eighteen case studies of South Asian clothing firms were undertaken, which involved repeat interviews with employers and employees; and twenty-three South Asian restaurants were examined over a year, with repeat interviews with owners and workers also featuring in this study.

A key argument is that economic sector is absolutely central in shaping work relations, so it is important to elaborate briefly the market context in each study. IsCo, the South Asian firm researched as part of the consultancy project, belongs to a segment of the small business population (business services) that has been conspicuous by its rapid expansion during the 1980s (Keeble *et al.*, 1992) and early 1990s (Bryson, 1997). This growth continues. In 1996, nearly one in three VAT registered enterprises (52,000) was in the business services sector. During 1996 there was a net increase of 21,000 VAT registered businesses in this sector. Most other sectors saw a net fall in the stock of registered enterprises (Business Monitor, 1996). IsCo comprises two distinct commercial activities. There is a training arm that provides training services for the transport industry. This is the main business and employs six training staff as well as administrators. The other part of IsCo is a consultancy operation, which provides business development services to a public-sector client base. This organization, which commenced trading in 1995, usually draws upon the services of associates to expedite its contracts.

The eighteen clothing firms were drawn from the South Asian-dominated West Midland clothing industry, and from a broader sector that according to the British government's Textile and Clothing Strategy Group (TCSG) is 'facing the greatest challenges in its history' (2000: 5). Among these challenges are a reduction in domestic sourcing and a corresponding fall in production levels, the imminent removal of the protectionist Multi-fibre arrangement, a relocation of production to Central Europe, the collapse of overseas markets and the strength of the pound (TCSG, 2000). This global restructuring of garment manufacture has fundamentally shaped the operating context of the West Midland clothing sector. It has intensified, encouraged informalization of working practices and increased reliance on cheap labour.

Although the restaurant industry has enjoyed spectacular growth in Britain, participants in the catering project were also seen to be in a highly competitive market place where supply outlets are multiplying even faster than demand (Ram *et al.*, 2001b). Ethnic restaurants and take-aways have been increasingly vulnerable to competition from other forms of eating out, including traditional forms like fish and chips and pubs and chain restaurants, which are able to compete on price with many ethnic eat-in restaurants. Competition from such outlets increased by 27 per cent between 1993 and 1997 (Mintel, 1998). Despite the unavailability of accurate figures on

Birmingham's ethnic restaurants, these broad sectoral trends were discernible in employers' assessments of their market environment. Most identified the intensification of competition as the main factor impinging upon their business. In the words of one respondent, restaurateurs in the locality were 'at each others throats'. Hyper-competition often led to extreme price and cost cutting – in itself a major factor in the employment of family and co-ethnic labour and 'informal' work practices.

Recruitment

Regularly noted in studies of South Asian business is the widespread tendency to recruit labour from within the ranks of the co-ethnic community (Basu, 1998; Metcalf *et al.*, 1996; Werbner, 1990). Co-ethnic employees are often deemed to be more 'trustworthy' than other employees, thereby easing the imperative of labour control (Ward, 1991). However, it is doubtful if their significance can be assessed without an appreciation of the range of considerations that shape employers' choice of workers. As Jenkins (1986) in particular has shown, employers seek workers who are 'suitable', that is, able to expedite the technical requirements of the job, as well as 'acceptable' in the sense of being able to fit into the prevailing pattern of social relations in the workplace. Kitching (1994) draws upon Jenkins's framework in his elaboration of a 'work force construction' approach to the recruitment process in small enterprises. Accordingly, employers' choice of workers is contingent upon a number of interrelated requirements, including the aforementioned technical skills; non-specific job attributes consistent with acceptable behaviour in the workplace; time, particularly when demand for labour is variable; and labour supply factors.

When such an approach is adopted, the recruitment of workers to ethnic minority firms is clearly more complex than a simple invocation of ethnic ties. Rather, it rests upon the complex array of influences that shape the labour process, as well as workers' ethnic backgrounds. For example, in the catering study, chefs were often of the same ethnic group, a practice usually justified on the grounds of their mastery of 'authentic' food. However, given the importance of the chef to the production process, technical expertise and the capacity to prepare 'authentic' food were not the only criteria that were important. Employers needed to ensure that chefs would remain with the firm, hence comparatively better terms and conditions were offered to such workers. A greater mix of ethnicities was to be found among waiting staff in the restaurant. Nonetheless, there was a predominance of Bangladeshi waiters employed in the South Asian restaurants (but less so for Pakistanis, who were more reliant on direct co-ethnics). This was due, in the words of one respondent, to 'plenty of Bangladeshi workers being available'. Indeed, 60 per cent of all Bangladeshi working-age males are employed in the catering sector (Sly *et al.*, 1998: 608); levels of unemployment are also particularly high in this community.

Co-ethnic recruitment is also a feature of the South Asian-dominated garment sector in the West Midlands; but again, it is more complex than simply drawing on ethnic ties. Sectoral, regulatory and labour-market pressures interact, thus influencing the terms upon which social networks are utilized. As a recent review of the industry noted:

> it is clear that Asian domination of the clothing niche in the West Midlands is explained by limited opportunities in the wider labour market, the availability of family and co-ethnic labour, and legislation that could be exploited to channel women into low paid work.
>
> (Ram *et al.*, 2002: 76)

The clothing project further illustrates the changing shape of mixed embeddedness by examining how changes in regulations, in particular the National Minimum Wage, influenced working practices. The eighteen case study firms were interviewed just before the introduction of the Minimum Wage, and a year later. Virtually all the clothing firms said that they were worse off than a year before. While two had shut down completely, employment levels were also reduced to some extent in almost all the rest. This was blamed primarily on deteriorating market conditions, though the new regulations were sometimes reported to have significantly increased costs. As Rath (2002) has observed, there is little that the mobilization of social networks can do to thwart the impact of such drastic changes in the market. One response by firms was to move into the *grey* market (that is, engage in the production of counterfeit goods). Because of the illegal nature of the product, the company relied heavily on particular home workers (usually extended family members) or trusted owners of other small manufacturing firms to produce these goods. In effect, it was retreating further into family networks in order to sustain this way of operating.

Co-ethnic recruitment figured prominently in IsCo, the case study firm involved in the consultancy project. However, the distinctive nature of product and labour market circumstances created a rather different approach to recruitment. It is characteristic for small consultancy firms like IsCo to draw on the services of associates rather than just in-house staff. They are deployed in order to meet specialist client demand to enable the firm to compete for a wider variety of work than if operating on their own. Enlisting someone as an associate can also serve as a means of assessing their suitability for a full-time position in the firm. As the owner of IsCo explained, it was a process that had to be carefully assessed:

> Well, I think that track record is very important. I followed him [the associate] for at least a year before I offered him anything. I just literally followed him. I enquired of companies that he worked for to see whether they were happy.

This was confirmed in discussions with the associate. He met the proprietor of IsCo some three years ago, when they first discussed the prospect of working together. A process that the associate termed 'checking each other out' began, after they met by chance at the premises of a mutual client. The associate recalled:

> We met three or four times and then he checked me out and I checked him out. I felt that his nature and my nature were sort of very similar ... His view of me initially was looking at me assisting him as an associate, but with a view ... if things work out in the long term ... to join him. So that was a longer-term objective.

The careful scrutiny of potential employees is explained by consultants' direct relationships with clients. They would be involved in cultivating 'demand-side networks' (Bryson *et al.*, 1993). Such networks are thought to be particularly distinctive in firms like those in the research, because of the importance of 'reputation' and personal contacts (Clark, 1995) and the significance attached to 'managing' the interface with clients (Alvesson, 1993). Occupants of such strategically important positions are often required to spend much of their time developing relationships with clients and the co-ordination of such activities is usually achieved by nurturing high-trust relations (Goffee and Scase, 1995). 'Repeat business' between the parties (i.e. between consultant and associate) was integral to this process in the case of IsCo.

From this assessment of the recruitment in South Asian firms in three contrasting sectors, we gather that the process is clearly more complex and multi-faceted than commonly portrayed. Despite the undoubted importance of ethnic social networks, they cannot be detached from other elements of labour process or from the wider operating environment of the enterprise. It is the social relations of production and the mediated impact of 'external' factors that will 'ultimately shape and underpin the existing structure of employment, the job opportunities available to particular groups within the labour market, and the path and development of industrial relations in particular companies' (Nolan, 1983: 309). Hence, the significance of ethnic ties needs to be evaluated in the light of these basic work processes rather than isolated as a particular 'variable' in the recruitment decision.

Remuneration

Drawing once again from the general small business literature, we find that another hallmark feature of small firms is their informal and individualized mode of pay determination (Curran and Blackburn, 1993; Kitching, 1997). When it comes to ethnic minority firms, however, wages appear to be a taboo subject for all but a few writers (Mitter, 1986; Phizacklea, 1990). Doubtless this neglect stems from reluctance to confront the possible exploitation of ethnic labour by co-ethnic employers, a classic instance of a clash between minority rights and individual rights, often female (Deveaux, 2000). Hence the present findings can shed some much needed light to this vital question.

From the outset, we stress that reality is not a simple one-dimensional matter. In the first instance, it is difficult even to pin down accurate estimates of workers' earnings, as the evidence from our own respondents makes only too clear. Pay determination in the three projects[2] reviewed here was largely an individualistic and *ad hoc* affair, with no regular pay reviews, no recognized pay structure and with individual bargaining for pay increases. In the catering project, workers reported being paid per shift rather than per hour and the following case of a waiter is not untypical: 'I get £70 for twenty hours. Sometimes he [the boss] calls me four hours later, sometimes I start at ten o'clock [in the morning].' Further blurring the picture is the prevalence of informal payment practices (cash in hand) and the provision of extras like free meals, transport and, in some cases, accommodation. Such payments in kind partly compensate for low money wages, as with the respondent who received £150 a week; however, he declared himself satisfied because he was living rent-free with the proprietor, who had previously helped him to get work in Britain. For all its comforts, however, such profound paternalism quite clearly develops bonds of dependence and obligation, resting as it does entirely on the continuing goodwill of the employer.

Ostensibly the British state's attempt to regulate low wages ought to have exerted a major impact on the remuneration regimes operating in firms such as these. On the evidence of the present research, however, the advent of a statutory National Minimum Wage in the United Kingdom has encouraged neither greater formality nor higher wages. Indeed, few owners in the clothing and catering projects had taken any steps to prepare for the Minimum Wage and at least a third of employers paid less than the rate. Moreover there appeared to be an absence of concern about the consequences of continuing to (under)pay workers in such an informal manner. Revealingly one of the chefs interviewed claimed that all the kitchen workers and waiters in his establishment were claiming benefit. Because of this practice, the National Minimum Wage would not have any real impact, since 'most of the people working here will always be … signing on … Whatever they'll get, they're better off … so a minimum wage would not in any sense affect them.' He maintained that most of the restaurants in the locality operated in this manner.

Once again it would be unwise to attribute these practices to essentially 'Asian' or any other 'ethnic' values. On the contrary, they are to be attributed to the over-concentration of Asians in sectors of the economy where entrepreneurial viability is sustainable only by recourse to desperate survival strategies shading into illegality. While catering is notorious in this respect, the same strictures apply equally to the clothing industry, where low pay is also an endemic feature of the industry in general (Phizacklea, 1990) and the

West Midland clothing sector in particular (West Midlands Low Pay Unit, 1991). Having recognized this, however, the prevailing circumstances do not preclude fruitful negotiations around pay, especially, where workers feel they enjoy some leverage. The clothing project, along with other shopfloor studies of the sector (Ram, 1994; Ram *et al.*, 2002) provide many examples of skilled workers taking advantage of labour shortages and the seasonal nature of demand to boost their bargaining power with employers.

However, this leeway is not available to all workers, particularly those working from home (Mitter, 1986). Indeed, the clothing project throws up cases of sheer powerlessness, with all the leverage in the employer's hands. One of the home workers interviewed is paid around £150 per week. To earn this sub-living wage, she works around nine hours a day, but this can fluctuate between seven and ten hours. Poverty and insecurity are compounded by the absence of any pension, holiday pay or overtime entitlement. Moreover, in relation to the negotiation of pay:

> I have very difficult negotiations with the employers. I have to fight for a decent wage but if you negotiate too much, for example more than three times then they tell me to go elsewhere. I feel like I am exploited and used because I don't know English and cannot go elsewhere. I am not paid on a regular basis and I have to keep asking for my wages. I am unaware of the minimum wage ... But who is going to pay us that? We get paid according to the number of pieces we complete ... I work long hours and for such a low pay. I have also been left with a string of medical problems.

It is clear that the regulation of remuneration is contingent upon the balancing of a range of influences. Even in sectors like clothing and catering, where the structure of the industry is such that management are often thought to have little discretion over pay, levels of remuneration were subject to significant variation. As Herman (1979) and Bailey (1985) suggest in their accounts of the ethnic minority restaurant sector, there is a certain flexibility incorporated in the wage structure that provides some advantages for both employers and employees. The catering and clothing projects offer some support for this view but, arguably, they go further than extant studies by highlighting that not *all* types of labour benefit to the same degree, as well as the range of contradictory pressures that owners face in managing the effort bargain.

Patterns of control

Rather than the prevailing categorizations of consensus (Werbner, 1990) or autocracy (Hoel, 1984), the pattern of control was shaped by a number of processes that are more consistent with the concept of mixed embeddedness. For example, in the case of the catering sector, the influence of the customers, the particular skill mix in the firms (typically, the position of chefs in relation to the waiters, and the presence of family members) combined to create a surprisingly complex and negotiated pattern of control in the workplace.

Though curiously neglected in existing studies of ethnic minority-owned restaurants (Herman, 1979; Bailey, 1985), the general literature on the restaurant trade has much to say on the role of customers in regulating workers (Gabriel, 1988; Wood, 1992; Kitching, 1997). It was clear that restaurant owners in the catering study actively used customers to ascertain satisfaction with the level of service and quality of food; to this end, one restaurateur commented, 'I go and ask every customer if they're well enough or if they've got problems, let me know ... then I apologize and then I'll personally talk to the waiter.' In addition to the imperatives deriving from customers, owners also had to handle the dynamics emanating from the different skills of workers. For instance, owners had little scope for imposing their authority on chefs; illustrating this point, one of the chefs remarked, 'No, no, he doesn't have to [supervise], we know ... We teach him what to do when he comes here. I know what to do.'

To understand the dynamics of employee management in the consultancy firm, IsCo, it is again necessary to examine the pressures exerted by the market, recognizing the extent to which clients tend to exert leverage over social relations within the firm. Immersion in client affairs is a common characteristic of this sector, so that significant aspects of a firm's internal management process are shaped by the nature of market relations (Ferner *et al.*, 1995). On the question of the evaluation of 'professional' and service work, Zucker (1991) argues that the exercise of authority does not usually rest within the boundaries of the firm. Rather, it 'migrates to the market' (1991: 164). Clients often shape authority relations in the firm by prescribing incentives, procedures, and adjusting costs, quantities and prices. Contracts obtained by IsCo usually contained quite detailed guidelines on the mode of delivery, the personnel to be deployed and the nature of evaluation. The owner of IsCo claimed that 'the criteria that most funding agencies use nowadays are that you have to meet certain standards. Unless you meet those standards they will not contract with you.' He maintained that 60 per cent of the firm's internal operations were 'dictated' by clients.

In clothing firms, it is well established that control is exercised through the carrot and stick of the payment system. Most machinists are paid piece rates, that is, on the basis of the number of garments (or parts of garments) that they sew (Phizacklea, 1990; Ram *et al.*, 2002), thus obviating the need for close control over machinists. Yet the clothing project shows once again how different aspects of a firm's environment can influence patterns of work relations in the firm. As noted earlier, some companies responded to the arrival of the Minimum Wage by adopting 'grey' market practices. However, when others did attempt to comply with the new regulations it had implications for patterns of control. For example, one manufacturer said that they 'have had to change everything' as a result of the Minimum Wage, compliance with which necessitated shifting most workers to day rates (although a few better workers still had a piece-rate component), resulting in a 30 per cent increase in pay costs. It was also commonly claimed that productivity was down as a result: 'The minimum wage is a bad thing, not so much in actual pay terms but because there are no incentives any more. For the borderline cases it has destroyed incentives, as they will be paid £3.50 (*sic*) whether they work hard or not'; 'they have decided that they will get paid anyway so performance is down ... the only way to improve productivity is to stand there and watch them'. From this last quotation we gather that tighter control is one unintended consequence of the Minimum Wage.

In essence, these illustrations from the three projects highlight the fluidity and complexity of control in the different research settings. They demonstrate that the process of supervision and discipline involves the mediation of pressures emanating from customers, regulations, the differing skill and authority levels of staff, and the influence of social networks.

Conclusion

In examining the dynamics of labour management in South Asian firms operating in three different market contexts, this chapter has argued for a balanced perspective. Our findings highlight the need to approach employment relations by drawing on the political-economic as well as the sociocultural aspects of embeddedness. Unlike culturalist accounts, our aim is to break with the exclusive focus on internal dynamics and so escape the risk of exaggerating the importance of ethnic influences on social relations within the firm. This is not to advocate a switch to an economically deterministic approach which, while useful in highlighting disadvantage and marginality, fails to capture the nuanced manner in which employment relations in such firms are constituted. Much more fruitful is the mixed embeddedness framework, with its insistence that the kinds of outcomes discussed in this chapter are shaped by the interplay of internal and external forces across a multitude of dimensions.

As well as the need to escape from a real world contextual vacuum, there is also a need to escape the discourse vacuum in which ethnic business studies often operates. Since by definition Asian entrepreneurs are members of a small entrepreneurial community (class) as well as of an ethnic community, they are not to be fully understood without input from the general literature on that community. In particular, the illumination of workplace dynamics in South Asian firms necessitates the incorporation of insights from wider debates on employment relations in small firms. As noted earlier, notions of cohesiveness (Werbner, 1990) or autocracy (Hoel, 1984; Mitter, 1986) do not reflect the multi-faceted ways in which social relations are managed in diverse small-firm contexts, irrespective of the ethnicity of the owners. The importance of informal processes (Scott *et al.*,

1989), conflict (Curran and Stanworth, 1979) and particular market environments (Rainnie, 1989 and Scase, 1995) to the trajectory of workplace relations have been argued elsewhere but rarely integrated within empirical studies in the field of ethnic minority entrepreneurship. The studies reviewed here can be seen as contributions to this process.

Finally, the qualitative approaches in the three studies have lent empirical weight to the mixed embeddedness perspective on ethnic minority entrepreneurship, and conceptualizations of employment relations that point to the inevitability of tensions arising from employers' attempts to manage a range of conflicting pressures (Edwards, 1986; Hyman, 1987; Ram, 1994; Kitching, 1997). The qualitative methods used have helped to detect the largely informal negotiations that shaped workplace relations in these settings; hence they have contributed to the illumination of the processes of labour management in action.

Notes

1 The writers wish to thank the Economic and Social Research Council for funding this study (catering: Ref. L130241049); (clothing: Ref: L212252031); (consultancy: Ref. R000221861). The clothing project (which did include other sectors) was jointly undertaken with Professor Paul Edwards, James Arrowsmith and Mark Gilman. Tahir Abbas, Gerald Barlow and Balihar Sanghera contributed to the catering project. The usual disclaimers apply.
2 The regulation of pay was not investigated in the same depth in the consultancy project as in the other two. However, in the case firm, IsCo, salary levels were acknowledged by management and workers alike to be markedly less than competitors'. For example, Younis, an IsCo employee claimed that he received 'less than a foundry worker'. Another agreed, and then added, 'I've looked in the papers, but the jobs just aren't out there.' The owner claimed that this was inevitable in 'small companies'.

References

Aldrich, H., Cater, J., Jones, T. and McEvoy, D. (1981) 'Business development and self-segregation: Asian enterprise in three British cities', in C. Peach, V. Robinson and S. Smith (eds) *Ethnic Segregation in Cities*. London: Croom Helm, 170–90.

Alvesson, M. (1993) 'Organizations as rhetoric: knowledge-intensive firms and the struggle with ambiguity', *Journal of Management Studies*, 30 (6): 997–1015.

Back, L. and Solomos, J. (1992) 'Black politics and social change in Birmingham, UK: an analysis of recent trends', *Ethnic and Racial Studies*, 15 (2): 327–51.

Bailey, T. (1985) 'A case study of the immigrants in the restaurant industry', *Industrial Relations*, 24 (2): 205–21.

Basu, A. (1998) 'An exploration of entrepreneurial activity among Asian small businesses in Britain', *Small Business Economics*, 10: 313–26.

Braverman, H. 1974. *Labor and Monopoly Capital. The Degradation of Work in the Twentieth Century*. New York: Monthly Review Press.

Bryson, J. (1997) Business service firms, service space and the management of change, *Entrepreneurship and Regional Development*, 9: 93–111.

Bryson, J., Wood, P. and Keeble, D. (1993) 'Business networks, small firm flexibility and regional development in UK business services', *Entrepreneurship and Regional Development*, 5: 265–77.

Business Monitor (1996) *Size Analysis of UK Businesses* (PA 1003), Sheffield: Department of Trade and Industry.

Clark, Timothy (1995) *Managing Consultants. Consultancy as the Management of Impressions*. Buckingham: Open University Press.

Curran, J. (1991) 'Employment and employment relations in the small enterprise', in J. Stanworth and C. Gray (eds) *Bolton Twenty Years On. The Small Firm in the 1990s*. London: Paul Chapman Press, 190–208.

Curran, J. and Blackburn, R. (1993) *Ethnic Enterprise and the High Street Bank*. Kingston on Thames: Kingston University Business School/ESRC Centre for Research on Small Service Sector Enterprises.

Curran, J. and Stanworth, J. (1979) 'Self-selection and the small firm worker: a critique and alternative view', *Sociology*, 13 (3): 427–44.

Curran, J., Kitching, J.A. and Mills, V. (1993) *Employment and Employment Relations in the Small Service Sector Enterprise*. Kingston on Thames: Small Business Research Centre, Kingston Business School, Kingston University.

Deveaux, M. (2000) 'Conflicting equalities? Cultural group rights versus sex equality', *Political Studies*, 48 (3): 522–39.

Edwards, P.K. (1986) *Conflict at Work*. Oxford: Blackwell.

Edwards, R. (1979) *Contested Terrain: The Transformation of Work in the Twentieth Century*. London: Heinemann.

Ferner, A., Edwards, P. and Sisson, K. (1995) 'Coming unstuck? In search of the "corporate glue" in an international professional service firm', *Human Resource Management*, fall: 343–61.

Flap, H., Kumcu, A. and Bulder, B. (2000) 'The social capital of ethnic entrepreneurs and their business success', in J. Rath (ed.) *Immigrant Business: The Economic, Political and Social Environment*. London: Macmillan.

Friedman, A. (1977) *Industry and Labour*. London: Macmillan.

Gabriel, Y. (1988) *Working Lives in Catering*. London: Routledge.

Gidoomal, R. (1997) *The UK Maharajahs. Inside the South Asian Success Story*. London: Brealey.

Goffee, R. and Scase, R. (1995) *Corporate Realities: The Dynamics of Large and Small Organisations*. London: Routledge.

Granovetter, M. (1985) 'Economic action and social structure: the problem of embeddedness', *American Journal of Sociology*, 91 (3): 481–510.

Herman, H. (1979) Dishwashers and proprietors: Macedonians in Toronto's restaurant trade, in S. Wallman (ed.) *Ethnicity at Work*. London: Macmillan, 71–90.

Hoel, B. (1984) 'Contemporary clothing sweatshops: Asian female labour and collective organisation', in J. West (ed.) *Work, Women and the Labour Market*. London: Routledge.

Holliday, R. (1995) *Investigating Small Firms. Nice Work?* London: Routledge.

Hyman, R. (1987) 'Strategy or structure? Capital, labour and control', *Work, Employment and Society*, 1 (1): 25–55.

Jenkins, R. (1986) *Racism and Recruitment*. Cambridge: Cambridge University Press.

Jones, T., Barrett, G. and McEvoy, D. (2000) 'Market potential as a decisive influence on the performance of ethnic minority business', In J. Rath (ed.) *Immigrant*

Businesses. The Economic, Political and Social Environment. London: Macmillan, 37–53.

Jones, T., McEvoy, D. and Barrett, G. (1992) *Small Business Initiative. Ethnic Minority Component.* Swindon: ESRC.

Jones, T., McEvoy, D. and Barrett, G. (1994) 'Labour-intensive practices in the ethnic minority firm', in J. Atkinson and D. Storey (eds) *Employment, the Small Firm and the Labour Market.* London: Routledge, 172–205.

Keeble, D., Bryson, J. and Wood, P. (1992) 'The rise and role of small service firms in the United Kingdom', *International Small Business Journal,* 11 (2): 11–22.

Kitching, J. (1994) 'Employers' work-force construction policies in the small service sector enterprise', in J. Atkinson and D. Storey (eds) *Employment, the Small Firm and the Labour Market.* London: Routledge, 103–46.

Kitching, J. (1997) 'Labour Regulation in the Small Service Sector Enterprise'. Ph.D. dissertation, Kingston on Thames: Small Business Research Centre: Kingston University.

Kloosterman, R., van der Leun, J. and Rath, J. (1999) 'Mixed embeddedness: (in)formal economic activities and immigrant businesses in the Netherlands', *International Journal of Urban and Regional Research,* 23: 252–66.

Metcalf, H., Modood, T. and Virdee, S. (1996) *Asian Self-employment. The Interaction of Culture and Economics.* London: Policy Studies Institute.

Mintel (1998) *Ethnic Restaurants, Leisure Intelligence.* London: Mintel International.

Mitter, S. (1986) 'Industrial restructuring and manufacturing home work: immigrant women in the UK clothing industry', *Capital and Class,* 27: 37–80.

Mulholland, K. (1997) 'The family enterprise and business strategies', *Work, Employment and Society,* 11 (4): 685–711.

Nolan, P. (1983) 'The firm and labour market behaviour', in G. Bain (ed.) *Industrial Relations in Britain*: 291–310.

Phizacklea, A. (1990) *Unpacking the Fashion Industry.* London: Routledge.

Rainnie, A. (1989) *Industrial Relations in Small Firms: Small isn't Beautiful.* London: Routledge.

Ram, M. (1994) *Managing to Survive.* Oxford: Blackwell.

Ram, M. (1999a) 'Managing professional service firms in a multi-ethnic context: an ethnographic study', *Ethnic and Racial Studies,* 22 (4): 679–701.

Ram, M. (1999b) 'Management by association: interpreting small firm–associate linkages in the business services sector', *Employee Relations,* 21 (3): 267–84.

Ram, M. and Jones, T. (1998) *Ethnic Minority Enterprise in Britain.* London: Small Business Research Trust.

Ram, M., Jerrard, R. and Husbands, J. (2002) 'Still managing to survive: Asians in the West Midlands clothing industry, in J. Rath (ed.) *Unravelling the Rag Trade. Immigrant Entrepreneurship in Seven World Cities.* Oxford: Berg.

Ram, M., Arrowsmith, J., Gilman, M. and Edwards, P. (2001a) 'Once more into the Sunset? Asian Clothing Firms after the National Minimum Wage'. Paper prepared for the third conference of the International Thematic Network on Public Policy and the Institutional Context of Immigrant Businesses, 'Working on the Fringes: Immigrant Businesses, Economic Integration, and Informal Practices', Liverpool, 22–25 March.

Ram, M., Abbas, T., Sanghera, B., Barlow, G. and Jones, T. (2001b) 'Apprentice entrepreneurs? Ethnic minority workers in the independent restaurant sector', *Work, Employment and Society,* 15 (2): 353–72.

Ram, M., Sanghera, B., Abbas, T., Barlow, G. and Jones, T. (2000) 'Ethnic minority business in comparative perspective: the case of the independent restaurant sector', *Journal of Ethnic and Migration Studies*, 26 (3): 495–510.

Rath, J. (2002) 'Needle games: mixed embeddedness of immigrant entrepreneurs', in J. Rath (ed.) *Unravelling the Rag Trade. Immigrant Entrepreneurship in Seven World Cities*. Oxford: Berg.

Scase, R. (1995) 'Employment relations in small firms', in P. Edwards (ed.) *Industrial Relations in Britain*. Oxford: Blackwell, 569–95.

Scott, M., Roberts, I., Holroyd, G. and Sawbridge, D. (1989) *Management and Industrial Relations in Small Firms*. Research Paper 70. London: Department of Employment.

Sly, F., Thair, T. and Risdon, A. (1998) 'Labour market participation of ethnic groups', *Labour Market Trends*, 106 (12): 601–15.

Srinavasan, S. (1995) *The South Asian Petty Bourgeoisie in Britain*. Aldershot: Avebury.

Storey, D. (1994) *Understanding the Small Business Sector*. London: Routledge.

Textile and Clothing Strategy Group (TCSG) (2000) *A National Strategy for the UK Textile and Clothing Industry*. London: DTI.

Ward, R. (1987) 'Ethnic entrepreneurs in Britain and Europe', in R. Goffee and R. Scase (eds) *Entrepreneurs in Europe*. Beckenham: Croom Helm, 83–105.

Ward, R. (1991) 'Economic development and ethnic business', in.J. Curran and R. Blackburn (eds) *Paths of Enterprise*. London: Routledge, 51–67.

Werbner, Pnina (1990) *The Migration Process. Capital, Gifts and Offerings among British Pakistanis*. Oxford: Berg.

West Midlands Low Pay Unit (1991) Report *The Clothes Showdown. The Future of the West Midlands Clothing Industry*.

Wilkinson, A. (1999) 'Employment relations in SME's', *Employee Relations*, 21 (3): 206–17.

Wood, P.A. (1990) 'Conceptualising the role of services in economic change,' *Area*, 23: 66–72.

Zimmer, C. and Aldrich, H. (1987) 'Resource mobilization through ethnic networks: kinship and friendship ties of shopkeepers in England', *Sociological Perspectives*, 30 (4): 422–55.

Zucker, L. (1991) 'Markets for bureaucratic authority and control: conformation quality in professions and services', *Research in the Sociology of Organizations*, 8: 157–90.

13 Asian immigrants and entrepreneurs in the Netherlands

Ernst Spaan, Ton van Naerssen and Harry van den Tillaart

The migration of Asians to the Netherlands has a long history.[1] It includes such groups as the Chinese, repatriates from the former Dutch East Indies (including Moluccans), refugees from Vietnam and, more recently, such diverse groups as Iranians, Filipinos, Sri Lankan Tamils and Chinese from Hong Kong. Since the late 1960s Asian immigration has increased significantly and has become more diverse in terms of composition of the migration flows and the socio-economic characteristics of the immigrants. The (potential) contribution of immigrant workers and immigrant business to the economy has been recognized by academics and policy makers. However, the position of the various Asian groups in terms of labour market insertion varies and the elements contributing to their position on the labour market and the development of immigrant businesses are quite diverse.

The different immigrant groups have particular migration histories and differ in terms of demographic composition and geographical dispersal, which has had a bearing on their specific incorporation trajectories and the emergence of incipient entrepreneurship. The theories that have been developed to explain this process have identified certain (clusters of) factors at different levels, subsumed under the headings of ethnic-cultural approaches versus structural approaches. Culturalist explanations point to the ethnic inter-group properties and resources, in particular the close-knit family and communities that can be tapped for capital and labour at low cost. The immigrants' social capital in the form of social networks can be put to use in generating income (Waldinger, 1995). The socio-cultural background, i.e. ethnic and class resources (Light and Rosenstein, 1995), of the specific Asian immigrant groups clearly plays a role in their mode of labour market incorporation and the drive towards entrepreneurial activities. The structural approach stresses the socio-economic context of receiving countries that determines the position of immigrants on the labour market. The more recent 'mixed embeddedness approach' (Kloosterman and Rath, 2001; Rath *et al.*, 2000; Light and Gold, 2000) is an attempt to merge the two approaches. The embeddedness in the economic, politico-institutional and social environments are determinant of the opportunities open to (potential) immigrant entrepreneurs (Rath, 2000: 10). Thus, both cultural and structural factors should receive due attention in the explanation of immigrant labour market insertion.

Some observers have argued that immigrants are forced into self-employment as a result of their lack of economic opportunities on the labour market, together with racism (Light, 1980). The development of immigrant business can be seen as a 'survival strategy'. The question is whether the barriers to participation in the labour market fosters the development of immigrant businesses and entrepreneurs, as has been suggested by (cultural and) structural theories of immigrant entrepreneurship (Light, 1972, 1980; Bonacich, 1973; Waldinger *et al.*, 1990), or whether also other factors play a role. The degree to which the immigrants are oriented towards the recipient society at the outset will play a role. Bonacich (1973) argues that immigrant groups considering themselves to be transients ('sojourners') rather than settlers will display an instrumental attitude toward the recipient society. Furthermore, this could lead to small business development through which they aim at accumulating capital in a short time through a life of hard work, self-exploitation and thrift. As is argued here, the specific immigration histories (and the migration regimes in the recipient countries) are important factors to be taken into consideration.

In general, Asian migrants have the reputation of being fairly active in the labour market and they are supposed to be well represented in small business and in self-employment in particular. In this chapter, however, we will consider two Asian groups in the Netherlands who are *under*represented in these sectors and for specific but different reasons are not intensively engaged in entrepreneurship. The two groups are the Moluccan group originating from east Indonesia (in the Netherlands often known as Ambonese) and the Philippine migrant group. It will be shown that the policy settings for these different ethnic groups are different. For example, contrary to the relatively small group of Philippine immigrants, the Moluccan group is formally recognized as a minority group in the Netherlands and specific policies aimed at improving their labour market position have been devised and implemented. Besides differences in the policy setting, there are social specialities, as in the case of Philippine migrants: for many of them integration into the Dutch society has been facilitated by marriage. However, it will also be shown that the lack of an entrepreneurial tradition and cultural preferences also play a role. Before going into more detail on these specific case studies of immigrant groups in the Netherlands a general picture of Asian immigration, with a specific focus on the labour market and the development of Asian immigrant entrepreneurship, is outlined.

Asian immigrants

While the world population reached the six billion mark in the year 2000, the population of the Netherlands reached sixteen million persons. Of these sixteen million, 2.8 million were non-nationals or *allochtonen* (native Dutch are called *autochtonen*) (CBS, 2001a).[2] Although older immigrant groups such as Turks, Moroccans and Surinamese are larger, during the 1990s there was a notable increasing trend in the number of Asians in the Netherlands. Turks, Moroccans and Surinamese represented 11 per cent of all immigrants in the year 2000, down from 25 per cent a decade earlier (Heering *et al.*, 2002). The Netherlands Central Statistical Bureau (CBS) predicts that Asians will form the largest *non-Western* foreign population in 2010 at an estimated 397,000 persons, thereby superseding long-standing non-Western immigrant groups such as Turks and Moroccans. An estimated one million Asians will reside in the Netherlands by 2050. This is mainly due to the immigration of Asians but also as a result of the higher than average fertility of the growing number of Asian women (Alders, 2003).

However, the above cited figures of Asians in the CBS statistics do not include the quite considerable part of the population (and their descendants) that shortly after the Second World War migrated from Indonesia, the former Dutch East Indies, to the Netherlands. They are not considered as an ethnic minority in the Netherlands, and most of them are of mixed European-Indonesian origin bearing European family names. Also, the Dutch Eurasians (or Indo-Europeans, 'Indos') in general don't consider themselves as such (de Vries, 1999) and they have Dutch nationality. On 1 January 1990 there were 404,000 people in the Netherlands of Indonesian origin, which increased slightly to 407,000 in 1999. Due to the fact that this population group is ageing, it is predicted that the group will decrease to around 369,000 by 2015 (Prins and Verhoef, 2000: 88). A specific group is demobilized soldiers and their dependants (most of Moluccan origin; about 12,500 people) who came to the Netherlands after the dismantling of the former Dutch colonial army and Indonesian independence in 1949.[3] Of the Indonesian group,[4] the second generation is now larger (about two-thirds) than the first generation, due to the fact that the majority of this group came to the Netherlands between 1945 and 1962 and concerned whole families. These immigrants opted for Dutch citizenship and were naturalized in large majority. In the period 1990–95 the average yearly increase of the Indonesian group was a low 1 per cent and in the following period the rate of growth was negative (−1.0 per cent) (Folkerts, 1999: 11).

In 2003 there were slightly over 690,000 *allochtonen* in the Netherlands with an Asian background (Table 13.1).[5] This was an increase of 15 per cent as compared with 1998. Of these, 58 per cent (or about 401,000) originated from Indonesia (including the former Dutch East Indies), among them approximately 42,000 people of Moluccan descent.[6] The other population groups of Asian origin amounted to 289,895 people on 1 January 2003. Leaving aside the Indonesian group, the largest groups of Asian immigrants at the start of the new millennium in the Netherlands were from China (and Hong Kong), Afghanistan, followed by Vietnam, Thailand, Pakistan and India (Table 13.1).

The ethnic minorities are not evenly distributed in the Netherlands. The four largest groups (Turks, Moroccans, Surinamese and Arubans/Antilleans)

Table 13.1 Asian populations in the Netherlands, by country of origin, 1998 and 2003

Country of origin	1998	2003
Afghanistan	11,551	34,249
China	26,191	38,815
Philippines	8,868	11,755
Hong Kong	17,304	17,923
India	10,302	12,971
Indonesia/Dutch East Indies	407,885	400,622
Iran	20,685	28,043
Malaysia	3,713	4,441
Pakistan	15,135	17,749
Singapore	3,597	3,973
Sri Lanka	6,463	9,606
Taiwan	1,155	1,881
Thailand	6,503	10,497
Vietnam	13,801	16,865
Other Asia	27,583	39,168
Total Asia	603,031	690,517

Source: Statistics Netherlands (CBS, Statline).

Note
Includes first and second-generation immigrants; 'Other Asia' includes the Gulf States, Iraq, Israel, Jordan, Syria, Lebanon, Yemen, Bangladesh, Bhutan, Nepal, the Maldives, Cambodia, Laos, Myanmar, Macau, Brunei, Mongolia, Japan, North Korea, South Korea. Among these, migrants of Iraq (mainly Kurds) are dominant with a population of 22,295 and 41,959 in 1998 and 2003 respectively.

are concentrated in the west of the country and in particular in the four largest cities, Amsterdam, Rotterdam, The Hague and Utrecht. This is also true of the South Asian population. The Chinese and Hong Kong communities are concentrated in the large cities but with sizeable populations in the eastern urban areas. The Vietnamese differ from this pattern in that they are most numerous in cities with up to 100,000 inhabitants in the southern provinces (Roelandt *et al.*, 1991; Tillaart *et al.*, 2000). The sizeable Indonesian group is spread out over the Dutch provinces but there are concentrations in the provinces of South Holland (26 per cent mainly in The Hague), North Holland (20 per cent; Amsterdam), North Brabant (13 per cent) in the south and Gelderland (12 per cent) in the east (Prins, 1997: 9). The Moluccans are also dispersed across the Netherlands, with concentrations in the eastern and north-eastern parts of the country.

Employment

A fairly consistent given is that the unemployment rates of the allochthonous population in the Netherlands has remained higher than that of the autochthonous population. During the 1980s, when the economy was stagnating, the unemployment rate of the non-Western allochthonous[7] populations was up to three times higher than that of the autochthonous population. During the 1990s this trend persisted, although the unemployment of the non-Western allochthonous population declined from 26 per cent in 1995 to 11 per cent in 2000. Until 2002, due to a booming economy and the ageing of the population, there were labour shortages in the Dutch labour market. Despite this fact, and although the labour market participation of the allochthonous population has improved during the 1990s, in comparison with the autochthonous population, immigrants still lag behind (WRR, 2000).

Immigrants have a significantly lower educational level and are more often unemployed. The unemployment rate among the allochthonous population was four times that of the native population in 1995 and 2000 (CBS, 2001b: 10). In 2000 this was 3 per cent for the autochthonous population against 11 per cent of the allochthonous population of non-Western origin (CBS, 2003: 119). The net labour participation level[8] increased from 37 per cent in 1995 to 48 per cent in 2000 for the non-western immigrant population; for the autochtonous population this increased from 60 per cent to 67 per cent (CBS, 2003: 119). Excluding the large immigrant groups (Turks, Moroccans, Surinamese, Antillians), the figures for the non-Western immigrant population are 33 per cent (1995) and 45 per cent (2000) (CBS, Statline). As for the unemployment rates of Asian minorities in the Netherlands the data are scarce. For Afghans and Iranians the unemployment rates were estimated at 38 per cent and 46 per cent in 2000 (ITS, 2000).

The vulnerable position of the immigrant populations also shows in the statistics of those dependent on state social benefits. The percentage of non-Western immigrants having social benefits[9] was 24 per cent, which is twice as high as that for the native population. The high dependence on such benefits was particularly true in the case of newly arrived groups such as Afghans (32 per cent) and Iranians (28 per cent) but less so for the Indonesian group (12 per cent) or Chinese/Tibetans (15 per cent) (CBS, 2003: 136).

A survey among 200,000 asylum seekers coming to the Netherlands between 1995 and 2000 showed that the majority had relatively low educational qualifications. Among the Chinese, for instance, only 17 per cent had a secondary or tertiary education; the figures for Sri Lankans, Pakistanis and Afghans was more favourable, however, namely 54 per cent, 63 per cent and 57 per cent respectively (Warmerdam and Tillaart, 2002: 184). Consequently, the range of professions of the asylum seekers in their country of origin is quite broad. Be that as it may, it turns out that asylum seekers have difficulty finding suitable employment and in putting their skills and qualifications effectively to use after their arrival in the Netherlands.

The Dutch government has undertaken several initiatives to curb unemployment among minority groups, not all of them successful. A report of the Minorities Research Advisory Committee Netherlands (ACOM: Adviescommissie Onderzoek Minderheden) in 1986 called for affirmative action, but met with resistance from employers and employees, due to the unpopularity of implementing a statutory requirement for firms and ethnicity registration (Kruyt and Niessen, 1997: 49). More successful initiatives were the 1000 Jobs Scheme for Moluccans,[10] and the 1994 Fair Employment of Ethnic Minorities Bill (WBEAA),[11] requiring medium and large firms to report on the number of ethnic minority employees to the chambers of commerce. The more recent Integration Policy for Newcomers aims at increasing the chances of new immigrants on the Dutch labour market by way of oblig-atory language courses and vocational training (*ibid.*).

Another way in which the Dutch government attempted to diminish the un(der)employment of the population of foreign origin is by way of giving incentives to the medium and small business sector (Memorandum Minister of Social Affairs and Employment, 2000). The expansion of this sector is seen as imperative in the fight against unemployment, but also as a means to revitalize deprived areas in the cities (OECD, 1998). However, stimulating entrepreneurship is no guarantee for improving immigrants' socio-economic position, as the same factors constraining immigrants' access to the regular labour market, such as low educational attainment and lack of social capital, also play a part in developing business (Kloosterman and Rath, 1999: 3). This translates into a process of market segmentation, whereby immigrant entrepreneurs mainly operate in the lower segments of the market, requiring less skills, capital and linkages with autochthonous business networks (*ibid.*). It has been recognized that immigrants starting up in business face all kinds of constraints, such as a myriad of laws and procedures, lack of information on financial markets and business networks and organizations, low educa-tional attainment and/or lack of business skills (Tillaart and Poutsma, 1998; Steensel *et al.*, 1999; Smeets *et al.*, 1999; Kloosterman and Rath, 1999, 2001).

Central and local government have aimed at diminishing all kinds of bar-riers to small-scale entrepreneurship, through fiscal measures, the reduction of red tape, easing of regulations and special programmes aimed at stimulat-ing ethnic entrepreneurship.[12] Helping minority groups into ethnic entre-preneurship is also perceived as a strategy to foster the process of integration of immigrant groups. There have been initiatives to promote the labour force participation rates of immigrants, and policies were formulated aimed at removing barriers to the proliferation of ethnic entrepreneurship in general (Ministry of Economic Affairs, 1998), but these were not *specifically* aimed at stimulating entrepreneurship among Asian immigrant groups.

In general, there was an increase in immigrant entrepreneurship in the Netherlands during the 1990s. The total number of immigrant businesses increased from 25,939 in 1994 to 36,641 five years later, an increase of 41 per cent (Tillaart, 2001: 43).

Similarly, the number of Asian entrepreneurs (excluding the Indo-Euro-pean and Indonesian group) showed an increase during 1990–2000 from

4,786 to 9,091, in particular in the large urban areas. It should be noted that, until 2000, only first-generation immigrant entrepreneurs were counted. Data from 2004 show a fairly large increase for most Asian groups, partially due to the inclusion of the second generation (Tillaart, 2001). The most important Asian entrepreneurs in the Netherlands in terms of numbers are the Chinese (including Hong Kong), Indians, Pakistanis, Iranians and Vietnamese. New groups include Filipinos and Malaysians (Table 13.2). In 2000 enterprises set up by Indo-Europeans (for example, people from the former Dutch East Indies, including Dutch New Guinea, numbered 2,456 businesses. Indonesian businesses in the Netherlands were also numerous in 2000, namely 2,189. The most recent data include also second-generation businesses. For most Asian entrepreneurs it holds true that only a small share (less than 10 per cent) belong to the second-generation. The former Dutch East Indies are an exception. For this category the share of second generation entrepreneurs is 85 per cent. However, considering the large population of Indo-Europeans (around 400,000 people), the number of entrepreneurs is still rather low (around 3.7 per cent).

In many countries, ethnic minorities tend to concentrate in certain urban agglomerations functioning as industrial and service centres (Sassen, 1991). Similarly, the geographical spread of ethnic minority businesses is often an urban phenomenon. In the Netherlands, the largest concentrations of non-

Table 13.2 Number of Asian immigrant businesses in the Netherlands, 1990–2000

Provenance	1990[a]	1995[a]	2000[a]	2004[b]
China	1,519	2,019	2,296	3,221
Hong Kong	877	1,171	1,117	1,231
Iran	197	434	624	1,054
Pakistan	585	1,006	1,033	1,332
India	478	931	955	1,134
Vietnam	450	627	619	744
Malaysia	143	177	168	170
Thailand	64	131	177	286
Philippines	65	140	154	194
Other Asia	408	914	1,948	2,363
Former Dutch East Indies	n.a.	n.a.	2,456	14,777[c]
Indonesia	n.a.	n.a.	2,189	2,913
Total	4,786	7,550	13,736	29,419

Source: Tillaart and Poutsma (1998); Tillaart (2003); datasheets, ITS.

Notes
'Other Asia' includes Lebanon, Syria, Iraq and the remaining Asian nations, excluding Turkey. Netherlands-Indies includes former Dutch New Guinea (292 enterprises in 2000).
a These figures are restricted to first-generation enterprises only.
b In 2004 the figures include second-generation enterprises.
c The share of second-generation business is 85%.

native populations can be found in the four largest cities, for example, Amsterdam, Rotterdam, The Hague and Utrecht (in the so-called Randstad) (Heering *et al.*, 2002: 276). Similarly, we find concentrations of businesses of the largest immigrants groups (Turks, Moroccans) in these cities (45 per cent and 50 per cent respectively). The situation of Asian businesses deviates from this pattern, however. Figure 13.1 shows that the large majority of Asian businesses are found outside the four largest cities, with the exception of the Pakistani and Indian groups, which are concentrated in the Randstad.

Examining the location of Asian *business* by province (not shown), reveals the same pattern, for example South Asian (and Hong Kong) businesses are highly concentrated in the western part of the Netherlands (Randstad). This is less the case for the other groups such as the Chinese, Vietnamese and Malaysian businesses, which are more spread out across the Dutch provinces. The general pattern did not change much since 1997, with the exception of Thai businesses that seemed to concentrate in North and South Holland provinces at the expense of North Brabant and Limburg.

As Table 13.3 shows, most Asian entrepreneurs in 2000 were active in three branches, namely wholesale trade, retail trade and the restaurant/ catering business. Chinese immigrants (including Hong Kong) especially

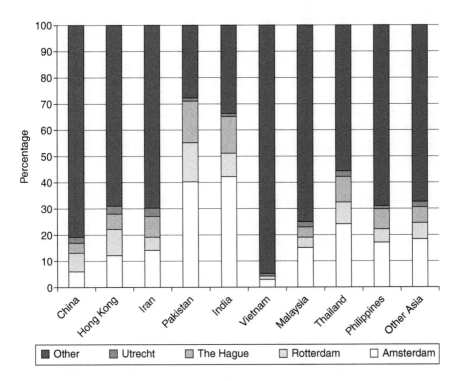

Figure 13.1 Asian businesses in the Netherlands by location, 2000 (source: Tillaart (2001)).

Table 13.3 Asian businesses, by sector, 2000 (%)

Origin	1	2	3	4	5	6	7	8	9	10	No.
China	0	0	0	10	4	80	0	0	4	2	2,297
Hongkong	1	0	0	8	6	75	1	1	5	3	1,113
Iran	3	2	2	26	24	12	2	0	19	10	620
Pakistan	1	1	1	25	30	16	4	1	15	6	1,038
India	2	1	1	25	44	12	1	1	9	5	954
Vietnam	16	1	0	7	11	60	0	0	4	1	619
Malaysia	2	1	1	10	6	61	1	1	11	7	170
Thailand	5	1	0	8	7	37	1	0	9	32	177
Philippines	3	3	3	27	13	14	5	1	20	11	153
Other Asia	2	2	2	34	14	14	2	1	1	28	688

Source: Van den Tillaart (2001).

Notes
1 industry; 2 construction; 3 vehicle trade/repair; 4 wholesale; 5 retail; 6 catering/restaurant; 7 transport; 8 finance; 9 other commercial services; 10 other. 'Other Asia' excludes Syria, Lebanon, Iraq and other Asian (excludes former Netherlands Indies).

are overwhelmingly employed in the catering and restaurant sector. Commercial services were also a significant activity, but less so for the Chinese (including Hong Kong) and Vietnamese. The latter group was also involved in industry, next to the three main sectors mentioned.

Through the years, little has changed in the sectoral spread of immigrants' businesses in the Netherlands. Between 1997 and 2000 the major changes were an increase in the category 'other', in particular for Thai, Filipino, Malaysian and Iranian business, and to a lesser extent for the Pakistani and Indians. Among the Thai there was notable decline of wholesale trade and other commercial services; the latter category also declined for the Filipino and Malaysian groups.

Case study: the Moluccans in the Netherlands

A specific group with roots in the former Dutch East Indies is the Moluccans. Currently, the Moluccan population in the Netherlands consists of the first generation, who came in 1951, after having been demobilized as military of the former Dutch East Indies armed forces (KNIL), and the second and third generations. The first group, which was initially considered temporary immigrants, as they (were) expected to return to the Moluccas after the planned establishment of an independent Moluccan Republic (Republik Maluku Selatan). Therefore, the Dutch government provided them with accommodation and income, and they could not participate in the local labour market. After 1953, however, when it became clearer that the Moluccans would remain in the Netherlands, government support was gradually diminished and the Moluccans became available for the labour market. The

Dutch government formulated specific policies to improve the socio-economic position of the Moluccans, including their housing (Smeets and Veenman, 1994, 2000; Choenni, 2000). Because the Dutch registration system does not differentiate population according to ethnicity, the number of Moluccans in the Netherlands cannot be determined exactly. The Moluccans were either stateless or had Dutch or Indonesian nationality, so would fall within these categories (Smeets and Veenman, 2000: 29). However, their number has been estimated at about 38,000 people in 1997 (SCP, 1998) and about 40,000 in 2000 (Veenman, 2001: 42 n. 2).

Due to a number factors, among them their background as military personnel, the Moluccans have experienced problems in finding suitable work and the majority did not enjoy a stable position on the labour market or remained unemployed for long periods. The second generation of Moluccans entered the labour market during the 1970s and early 1980s, a period of economic recession. Thus, their position on the labour market was problematic too, due to the recession, cultural factors, low educational attainment, inadequate integration and discrimination (Veenman, 1985, 1990: 46–66). By the end of the 1980s their position on the labour market has progressed, as indicated by a significant decrease in unemployment, particularly among Moluccan men of the second generation (Smeets and Veenman, 2000: 53).

Surveys among Moluccans in paid employment in the 1990s showed that Moluccans had a higher labour force participation than Mediterranean minorities, but compared with the native Dutch labour force the unemployment rate of Moluccan men and women remained far higher (Dagevos *et al.*, 1992: 35; Bruggink *et al.*, 1997). This situation persisted to 2000, when the Moluccan labour force participation rate was considerably higher (67 per cent) than that of Turks (47 per cent) and Moroccans (44 per cent). This was still slightly lower than that of the native population (69 per cent). Moluccan men had a higher labour force participation rate than women in that year (75 per cent versus 58 per cent). Most Moluccan men in employment worked in management, administrative and commercial positions, while women were in administrative positions (Dagevos *et al.*, 1992: 39).

According to 1998 data (SPVA, 1998, Veenman, 2001: 82) the economic sectors where most Moluccans worked were industry, repair shops, trade, restaurants and catering, commercial services, education and health services. This is fairly similar to the pattern observed for Turks and Moroccans, although the latter groups are more represented in the primary sector and Moluccans relatively more in industry and health services (Table 13.4). Veenman (2001a) indicates that the position of third-generation Moluccans is not favourable. Although the unemployment rate among Moluccans is a mere 4 per cent, the third generation holds a structurally weak position in the labour market, due to low educational attainment and skill levels. In fact, the third generation has a lower educational level than the second generation of Moluccans. Many still hold insecure and casual jobs. An

Table 13.4 Employment by sector, ethnic minorities in the Netherlands, 2000 (%)

Sector	Moluccan	Turks	Moroccan	Surinam	A/A	Dutch
Primary	0	12	10	2	3	2
Industry	29	31	22	16	15	13
Construct	5	4	5	3	3	3
Trade/restaurant and catering	14	15	20	11	12	15
Transport/communications	7	8	5	8	9	7
Commercial services	17	12	17	19	19	18
Education/government	10	8	9	20	15	23
Health	12	6	7	18	20	14
Culture, services	5	4	5	4	5	5
No.	741	1,084	800	1,587	736	1,396

Source: ISEO (2000); Veenman (2001a: 82).

Note
A/A signifies Arubans and Antilleans.

encouraging sign, however, is that more young Moluccans manage to move out of the low-paid industrial sector and into the service sector.

Although the range of job positions held by Moluccans has broadened during recent decades from mainly factory workers to artists, civil servants, administrative positions and the professions (Moluks Historisch Museum, 2000), the question is whether entrepreneurship also figures among this group? According to Smeets and Veenman (2000), Moluccans turn to entrepreneurship to a limited extent. However, the authors did not qualify this statement further. In a subsequent publication (Veenman, 2001), data from three surveys conducted in the early 1980s, in 1990 and in 2000, were brought together, which show that entrepreneurship is indeed still limited among Moluccans (Table 13.5). This is explained by referring to the military background of first-generation Moluccans in the Netherlands and the absence of a 'tradition' of entrepreneurship (Veenman, 2001: 87).[13] The second generation was not familiar with entrepreneurship either and was hardly stimulated to move in that direction. The lack of a middleman

Table 13.5 Entrepreneurship among the Moluccan population, 1983, 1990 and 2000

Variable	1983	1990	2000
Surveyed population	5,140	7,101	2,063
Labor force (15–65 years)	2,057	2,394	795
Entrepreneurs	8	16	17
% entrepreneurs/labour force	0.39	0.67	2.14

Source: Based on Veenman (2001a: 87).

tradition among Moluccans was corroborated by statements by some Moluccan entrepreneurs themselves. As one Moluccan put it: 'this is because Moluccans are traditionally no merchants. That is not in their blood, also because we are not stimulated to go into business' (interview, in Worung, 1999). Other informants noted that another constraint is the lack of business skills and business networks among Moluccan youth (personal communication, BP).[14] Nevertheless, a slight increase in setting up business ventures has been noted. Not only have Moluccans (temporarily) set up shops selling ethnic foodstuffs and commodities during night fairs (*pasar malams*), they have diversified their economic activities to other sectors as well.[15] As became evident from our interviews and desk research, Moluccan entrepreneurs are currently active in building contracting, tourism, retail trade, catering, driving schools, publishing and commercial services.

However, among most of the large immigrant populations in the Netherlands the percentage of the labour force that are entrepreneurs is limited too. It was less that 5 per cent among Moroccans, Surinamese and Antilleans. Only among the Turks is the percentage as high as in the native Dutch labour force, namely 10 per cent. It can be doubted whether entrepreneurship is a viable avenue for improving the socio-economic position of immigrants, considering the fact that many immigrant entrepreneurs fail to generate satisfactory business results and sufficient income (Dagevos *et al.*, 1992: 45; Bakker and Tap, 1987). More recent surveys indicate, however, that the performance of immigrant entrepreneurs is improving. The rate of failure during the first years of operation among ethnic businesses improved from 28 per cent in 1993 to 21 per cent in 1998 (Memorandum Minister of Social Affairs and Employment, 2000).

The economic boom years translated into more paid employment for the Moluccan minority but not into higher average educational attainment, employment level or entrepreneurship (Veenman, 2001: 152). Although some preconditions of the development of ethnic business have been met (e.g. an ethnic market with specific demands for ethnic products, like foodstuffs, herbal medicines, magazines, etc.), part of this market is filled by businesses catering to the larger Indonesian/Indo-European population, with similar demand. However, it was noted that some Moluccans do market distinctive products at night fairs (*pasar malam Maluku*) but these ventures were often merely occasional and a sideline to their regular activities.

Cultural factors have a bearing on the slow development of entrepreneurship among the Moluccans. Moluccans tended to emphasize their ethnic identity, latterly reinforced as a result of the volatile situation in the Moluccas, characterized by religious conflict and communal violence. Particularly the younger generation are engrossed with their Moluccan identity, and emphasize Moluccan values and the Malay language (Bruggink *et al.*, 1997; Knibbe *et al.*, 1998; Verkuyten *et al.*, 1999). They tend to strive to achieve harmonious relations among friends and relatives rather than commercial success (Veenman, 2001). As one key informant noted, inter-group relations are based on mutual respect and reciprocity, in which there is no room for commercial transactions.

In addition, rather than values emphasizing individual achievement, independence and accumulating capital, as in business, more communal values such as reciprocity, unconditional help to co-ethnics, reliability and group (*pela*) solidarity are important in Moluccan culture (Wittermans, 1991: 136; Knibbe *et al.*, 1998). There are indications that, among the Moluccan group, striving toward higher education and a successful career is less valued than playing a role in the Moluccan community or Moluccan organizations (Dagevos *et al.*, 1992). Unlike the Chinese and Vietnamese, where having one's own business is the ideal and adds to one's prestige (Tillaart *et al.*, 2000; Ma Mung, Chapter 3 of this volume), status in the Moluccan community was derived more from one's role in the community, church or former position in the army (Dagevos *et al.*, 1992: 65). Even if such (unpaid) activities impeded one's career or economic activities, they were highly valued (*ibid.*: 130). As was observed among Iranians in Sweden (Khosravi, 1999), entrepreneurship is less valued among Moluccans.

Another factor reinforcing these cultural factors is the fact that many Moluccans resided in specific ethnically homogeneous neighbourhoods. These areas were not conducive to the development of entrepreneurship as they were rather self-contained and the build-up of (business) networks outside these communities was inhibited.

Case study: Filipinos in the Netherlands

The history of emigrants from the Moluccas is very specific and, in consequence, differs in many aspects from the Philippine one. People from the Philippines belong to one of those countries which have already experienced international migration for many decades. Philippine households are accustomed to international migration. Around 10 per cent of the population live abroad, ashore or as seamen. The number of Philippine emigrants all over the world is estimated at 7.2 million, of whom three million are overseas Filipino workers (OFWs), some 2.3 million are permanent residents abroad and 1.9 million are undocumented (illegal, informal) Filipinos abroad. Due to colonial history, the majority of the permanent residents are found in the United States. Most of the OFWs are employed in the Middle East and in the richer (newly industrialized) countries of East Asia, but there are also large numbers of OFWs in the Roman Catholic countries of southern Europe: Spain, Italy and Greece. While male workers dominate the OFWs in the Middle East, in southern Europe most of them are female and employed as domestic workers.

The number of Philippine immigrants in the Netherlands is very modest. According to the Dutch Central Office of Statistics there are about 9,400 'allochthones' of Philippine descent, which includes some 3,000 legal immigrants from the Philippines and 6,400 of their children. To these we should add an estimated 3,000 undocumented immigrants, which gives a total of around 12,400 first and second-generation Filipinos in the Netherlands. This is indeed a very small group compared with other immigrant groups in the Netherlands, but as an immigrant group with specific characteristics it is interesting to see whether entrepreneurship has been developed, what kind of businesses and for what reasons. Unlike the Moluccan group, so far not much research has taken place on the Philippine migrant group and there is a lack of secondary sources.[16]

The first Philippine immigrants arrived in the early 1960s and were female OFWs. Their total number was around 1,000 of whom around one-third were nurses, while two-thirds were employed by a clothing factory with its major establishment in the east of the Netherlands. As for the latter, during eight years (1966–74) they arrived in twelve batches of around sixty girls each. The women got a contract for three years, after which it could be renewed. Intermediary for the nurses was a Dutch nun, while in the case of the factory workers it was a Dutch missionary. After the contract period, most of these OFWs went to Canada, about 100 stayed in the Netherlands and only a few went back to the Philippines.

A second 'wave' of immigrants occurred in the late 1970s and early 1980s, when political refugees from the National Democratic Front (NDF) sought and found asylum in the Netherlands. This group comprises about 100 families, living in or near the city of Utrecht, where the office in exile of the National Democratic Front is located. Besides, a number of around fifty Philippine women, married to former missionaries or members of the Dutch volunteers (SNV), went with their husbands to the Netherlands. Although small in number, for obvious reasons these groups are the most vocal ones within the Philippine community. From the second half of the 1980s onward, new immigrants came as 'postal brides', as workers in the entertainment industry or as au pairs and domestic workers. Not so much is known about these immigrants but one can assume that many of them are undocumented (sometimes semi-legal) workers or entered the country as such. In any case, women are predominant among the Philippine immigrants in the Netherlands. In the period 1988–98 about 75 per cent of the Philippine migrants were women (Muijzenberg, 2000: 96–7).

Although very different, the profile of the Philippine immigrants is as clear as that of the Moluccan migrant group. In the case of the Philippine migrants, the majority concerns women who did not come to the Netherlands with the aim to start a legal business nor through economic circumstances were forced to entrepreneurship. Thus it is hardly surprising that there are not many Philippine entrepreneurs in the Netherlands. Data of the Dutch Association of Chambers of Commerce mentions 194 Philippine-registered enterprises in the Netherlands (Table 13.2) and 238 Philippine entrepreneurs in 2004. The latter implies that only 0.8 per cent of the Philippine legal migrants are involved in private entrepreneurship.

The majority of the businesses of the registered Philippine entrepreneurs are small. Excluding eleven entrepreneurs on whom no data on employment are available, 111 out of 227 (49.9 per cent) run one-person ventures, and twenty-four of these entrepreneurs are part-timers. Only fourteen entrepreneurs (6.2 per cent) have five or more full-time working places. The largest business, a trade promotion business, provides employment to twenty-five persons. More than half the entrepreneurs (126, 53 per cent) live in Randstad Holland and of these seventy-five live in one of the three largest cities of the Netherlands, respectively thirty-nine in Amsterdam, nineteen in The Hague and seventeen in Rotterdam. The other 47 per cent of Philippine entrepreneurs are spread over the southern and eastern provinces of the Netherlands. Philippine business is notably negligible in the northern provinces. This pattern differs from other South East Asian immigrant groups such as Vietnamese and Malaysians, who, as we have seen earlier, have their business mainly outside the Randstad. Similar to other immigrant groups, Philippine business is only marginally represented in agriculture and manufacturing industry, and mainly concerns wholesale and retail trade, restaurants and catering services and business services. As Table 13.6 shows, the latter is the major sector for Philippine entrepreneurs.

The majority of the enterprises are run by women. Many of them have had high-school education, as have most Philippine migrant women in Europe. Finance is often obtained from the family network, which here implies financial support from the Dutch spouse. It is sometimes difficult to draw a line between the activities of the woman and her spouse, in the sense that the husband shows an active interest although usually he has his own job. Significantly, this job is usually not that of a self employed entrepreneur, because otherwise she might be employed in his business. These and other specific features of the Philippine women business in the Netherlands are the subject of research by Marisha Maas.[17] However, although an overall majority of the Philippine entrepreneurs are women, Table 13.7 shows that a gradual change in the sex composition may occur since Philippine males are relatively stronger represented in the younger age groups.

Although most Philippine business aims to attract a mixed public of both the migrant community and mainstream Dutch society, they are often char-

Table 13.6 Philippine entrepreneurs by sector and sex, 2004 (%)

Sector	Male	Female	Unknown	Total
Wholesale trade	20.4	18.7	23.5	19.7
Retail trade	7.4	16.0	20.6	14.7
Restaurant/catering	5.6	10.7	0.0	8.0
Business services	31.5	24.7	8.8	23.9
Other	35.2	30.0	47.1	33.6
No.	54	150	34	238

Source: Association of Chambers of Commerce, May 2004.

Table 13.7 Philippine entrepreneurs by age and sex (excluding unknown), 2004 (%)

Age	Male	Female	Total
20–29	13.0	4.7	6.9
30–39	35.2	24.7	27.5
40–49	25.9	46.0	40.7
50–59	20.4	20.7	20.6
60-above	5.6	4.0	4.4
No.	54	150	204

Source: Association of Chambers of Commerce, May 2004.

acteristically ethnic-oriented. A typical example is a Philippine food store situated near the Albert Cuypstraat, a well known multicultural market street in Amsterdam. The shop sells South East Asian food such as rice, *mie* (noodles) and *kecap* (soy sauce), but is distinctly Philippine in selling specialities such as Tanduay rum, San Miguel beer, *balut* (fertilized duck eggs) and *danguan* (dried pig blood). It is also a meeting point for Philippine people and on Saturday: 'Philippine customers come from all over the country to buy the specialities' (interview in the journal *Tambuli*, October 1999). Nevertheless, it is questionable whether the shop would survive without selling to customers from the neighbourhood.

Another example is the publisher of a bi-monthly journal, *Philippine News*. Its stated objective is to inform the Dutch public about the Philippines. It comprises articles on Philippine policies and contributions on Philippine culture and the Philippines as a tourist destination. On the other hand, it also contains a series of photographs of the inaugural fiesta of Grace, who '. . . reached the age of 18 in July'. Also, advertisements about cheap ways to phone and to travel to the home country and sentences in Tagalog, the national language of the Philippines, indicate that many of its readers will be found within the Philippine community in the Netherlands. The magazine is run by a Philippine-Dutch couple, and can be considered as a mixed endeavour. Investments in the printing machine and related computers were paid by way of profits made by an earlier business of the Dutch husband but the Philippine woman determines the content of the journal. Moreover, the husband has worked for more than twenty years in the Philippines and speaks the Philippine language, which justifies classifying this enterprise as Philippine.

Characteristic of first-generation migrant entrepreneurs is their reskilling, since they have to find their way in a new environment. Such is the case with a Philippine restaurant owner who migrated to the Netherlands about fifteen years ago. He was skilled at a school of dress design and together with his sister started a clothing shop in his home country. 'That is what I wanted to do, to start an own business. It went well. My sister was responsible for women's clothes and I for men's clothes. We got a lot of orders, in particular for weddings. We did everything ourselves, including designing and sewing.' After two years, 'to become more independent from my parents', he decided to migrate and – on the advice of an uncle – travelled to Amsterdam. During the first years he earned his money by cleaning the houses of rich families. However, after a while his uncle gave him the chance to start with the restaurant and he joined his uncle as assistant and a few years later became the owner. After thirteen years in the catering trade, he wants to change again and to start a bed-and-breakfast hotel or a clothing shop (*Tambuli*, May 2000).

It is interesting to see how some entrepreneurs make use of resources from both the country of origin and the host country. These are *transnational entrepreneurs* who connect two geographically and culturally distant areas. A characteristic case was a trade service centre, specializing 'in offering agricultural raw materials that are part of the daily lives of Filipinos and their communities. These are products found and grown in the Philippines with high demand and value in the international market. They include coconuts, rattan and bamboo, natural rubber, herbs and spices, nuts and tropical fruits' (leaflet). The company was set up to serve as middleman between Dutch buyers and Philippine producers. It intended to offer information and research with regard to producers in the Philippines, Dutch and Philippine trade relations and market opportunities, as well as a brokering service and market matching. It was, however, a short-lived enterprise. Another, more successful, transnational endeavour is a door-to-door service, set up in 1983. Today the company comprises several enterprises, both in the Netherlands and in the Philippines. It started with delivering boxes with goods from migrants in the Netherlands to family members in the Philippines, and the other way round. In due course the company also started to transport food products and second-hand goods, such as furniture, from the Netherlands that is sold by two shops in the Philippines. A small factory producing coconut products was later added (Maas, 2003).

Philippine business in the Netherlands shows all the characteristics of *incipient immigrant entrepreneurship*: engagement in service sectors rather than in production, related to the own ethnic community, small-scale, low capital input, low turnover, and low profits, lack of official support from the municipality and banks, and instability. Besides these general specific features, however, Philippine business in the Netherlands has its own specific features compared with other immigrant groups. In the first place only a very small number of immigrants are engaged in business. Second, a remarkable majority of the enterprises are run by women.

Conclusion

Although it can be debated whether entrepreneurship is a viable avenue for the improvement of the socio-economic position of immigrants, given that immigrant business ventures often fail to generate satisfactory business results and sufficient income (Dagevos *et al.*, 1992: 45; Bakker and Tap, 1987), the decrease in failure rates at the end of the 1990s is an encouraging sign.

Among the large immigrant populations in the Netherlands the percentage of the labour force that are entrepreneurs is relatively limited, but increasing, especially for Turks. The estimated percentage of first-generation entrepreneurs in the total labour force in 2000 (according to chamber of commerce data used by Tillaart, 2001) was for Moroccans (4.6 per cent; up from 3.3 per cent in 1986), Surinamese (4.8 per cent; was 2 per cent) and Antillians (3.8 per cent; was 2.9 per cent). The rate of Turkish entrepreneurship (10.1 per cent in 2000; 4.4 per cent in 1986) is now similar to the native Dutch (10.2 per cent and 8.0 per cent respectively) (Tillaart, 2001: 27; Jansen *et al.*, 2003: 11). For these immigrant groups the second generation tends to have more entrepreneurs (Jansen *et al.*, 2003). Among Moluccans a similar progressive trend can be observed, although entrepreneurship remains relatively limited.

As for Asian entrepreneurship in general, the number of businesses has shown a progressive trend, in particular among the more long-standing Asian immigrant groups (Chinese, Pakistanis and Indo-Europeans). However, despite diversification strategies, failure rates among Asian businesses remain relatively high as compared with Dutch enterprises, pointing to problems of sustainability and market saturation. Chinese and Filipino businesses seem to do better in this respect.

A number of factors have been identified which could push minorities into immigrant entrepreneurship, *viz.* external factors such as high unemployment, inferior education, language barriers and discrimination. Internal group-specific factors which are conducive to the development of ethnic economies and entrepreneurship are, for instance, group solidarity, ethnic networks, mutual trust, a common culture and language and a strong orientation on the homeland. Barrett *et al.* (1996: 791), for example, have observed that immigrant groups that are internally oriented, emphasizing their own ethnic identity, and lacking a strong drive to assimilate to the host society, seem to have a lead in business development. These factors are all relevant in the case of the Moluccan community in the Netherlands, since this group is characterized by strong ethnic consciousness and solidarity, with a spirit of mutual trust, co-operation and collective self-help. Despite this, immigrant entrepreneurship is not well developed within this minority group, as it is among other groups in similar circumstances, e.g. Iranians in Sweden (Khosravi, 1999). In the case of the Filipinos in the Netherlands, entrepreneurship is also limited and concentrated in the service sector. A comparison of the still underdeveloped entrepreneurship among both groups can add to an explanation to the working of structural factors mentioned above. It is possible to formulate at least three hypotheses for further research.

In the first place, the cultural background of the two groups clearly plays a role in the development of entrepreneurial activities. In the case of the Moluccan group, available studies point to cultural values and communal factors, which was corroborated by our respondents. Having one's own business is not so much an ideal in the Moluccan community as it is for instance in Chinese or Vietnamese culture. As one person put it, 'Moluccans are not commercially minded'. In addition, in certain sectors (e.g. catering) there are market factors which could constrain the proliferation of businesses, e.g. competition by Indonesian/Indo-European business ventures. The same lack of tradition – although for other reasons – can be found among the Philippine immigrants. However, as for the Philippine female migrants, the position of women is different compared with the Moluccan one: it is not only an accepted value but also thought to be desirable that women will be economically independent. (This value may partly be rooted in the US colonial tradition.)

A second hypothesis relates to embeddedness in the Netherlands. We argue here that the Moluccan minority initially considered themselves and were seen as transients; this translated into social isolation, lack of acculturation and problematical access to the Dutch labour market. They already occupied a specific position in Dutch East Indies society and for several decades this was maintained in the Netherlands, as symbolized in the neighbourhoods they were and – in the majority of cases – are still living in. These ethnically homogeneous Moluccan neighbourhoods are not conducive to the development of entrepreneurship, as they are rather self-contained and the build-up of (business) networks outside these communities is inhibited. Thus, despite their close-knit community structure, inter-group solidarity and ethnic resources, this has not translated into the development of immigrant entrepreneurship. In the case of the Philippine immigrants, for many of them integration into the Dutch society has been facilitated by marriage. It may be that in future this will facilitate entrepreneurship.

Finally, as a third hypothesis, the policy consequences for the two groups are different. The Moluccan group is formally recognized as a minority group in the Netherlands, which means that specific policies have been implemented for them. However, although policies have been devised aimed at improving the labour market position of Moluccans, these have not included measures specifically geared toward the fostering of entrepreneurship. As for the Philippine group, they have to rely on general policies towards immigrants.

Notes

1 In this chapter 'Asia' excludes Turkey.
2 The Dutch Central Statistical Office (CBS) uses a broad and a narrow definition for foreigners: according to the wide definition anyone who is born outside the Netherlands, or, when born in the Netherlands, with at least one foreign-born parent, is considered a non-native (allochthonous). Using the narrow definition implies that only those foreign-born with at least one parent born abroad, or, if

born in the Netherlands, both parents are foreign-born, are considered allochthonous. Since 1999 a non-native has been defined as all persons with at least one parent born abroad.

3 The first wave of immigrants arrived between 1945 and 1949 and consisted of 44,000 people who were evacuees or who came to recuperate in the Netherlands after the war. The second group (1950–51) of 68,000 people (including the Moluccans) came after the declaration of independence and the subsequent dismantling of the Dutch colonial administration. The increasing political conflict between Indonesia and the Netherlands over western New Guinea, the nationalization of Dutch firms and the forced expulsion of Dutch in 1957 prompted a third wave of about 71,000 between 1958 and 1963 (Ellemers and Vaillant, 1985: 40–2).

4 Including those from the former Netherlands East Indies (e.g. Moluccans, Papuans, repatriates) and Indonesians.

5 This includes both the first and the second generation. According to Statistics Netherlands, 'first-generation' implies that the person was born outside the Netherlands and at least one parent was born abroad. Second is similar except that the person concerned was born in the Netherlands.

6 This is the most recent estimate for the year 2001 (Beets *et al.*, 2002a, b: 62).

7 The category non-Western includes foreigners from Turkey, Africa, Latin America and Asia except for Indonesia and Japan (categorized as Western).

8 Net labour market participation signifies the number of employed persons (sixteen to sixty-four years; over twelve hours/week) as a percentage of the labour force (WRR, 2000)

9 I.e. unemployment, disability and social welfare benefits.

10 For example, the so-called Thousand Jobs Plan (1986–89) aimed at creating jobs in the public sector at national, provincial and municipal level for Moluccans, who showed high unemployment rates (Algemene Rekenkamer, 1991).

11 *Wet Bevordering Evenredige Arbeidskansen Allochtonen.*

12 For example, MOTOR (Migranten Ondernemer: Talent, Opleiding, Resultaat), migrant entrepreneur, talent, training and result. This project (from 1997 to 2000) is aimed at stimulating entrepreneurship through (language) training and improving access to financial markets.

13 It should be noted, however, that the surveys on which these figures are based are somewhat biased in that the surveyed Moluccan population was mainly selected from so-called 'Moluccan communities'. These are municipalities where the Moluccan population is concentrated in separate neighbourhoods. Moluccans living outside the Moluccan neighbourhoods, of whom it is said that they possibly have a higher educational attainment and more favourable labour market position, are underrepresented (Veenman, 2001: 8–9).

14 A few years ago a network of Moluccan businesses had been set up, to foster entrepreneurship among Moluccans. Over sixty entrepreneurs working in construction, graphic design, publishing, advertising, legal counselling and commercial services initially participated. However, this initiative was discontinued after a few meetings (personal communication, BP220501).

15 Current research on Asian migrant entrepreneurs in the eastern cities of Arnhem and Nijmegen (funded by the EQUAL programme of the European Social Fund, 2002–04) shows a general trend to diversification. This is linked with the period of stay in the Netherlands and the degree of integration as indicated by intermarriage and knowledge of the Dutch language. It also shows that second-generation migrant entrepreneurs rely less on informal social networks and mostly relate to support services of the municipalities and the regional chamber of commerce. Contrary to the situation in the large cities of the Randstad, where

more than 40 per cent of the population consists of immigrants and their descendants, there were no indications of 'ethnic clustering' although it is clear that immigrant business in Arnhem and Nijmegen is often found in specific streets.

16 This section is mainly based on articles in the Dutch–Philippine magazine *Tambuli*, interviews with people from the Philippine community in the Netherlands and information from Marisha Maas, who is doing research on Philippine transnational entrepreneurship.

17 See the list of references. The research of Maas is a Ph.D. project of the Department of Geography and the Centre for International Development Issues of the Radboud University Nijmegen.

References

Association of Chambers of Commerce (2004) data sheets on foreign entrepreneurs, May.

Alders, M. (2003) 'Allochtonenprognose 2002–2050: bijna twee miljoen niet-westerse allochtonen in 2010', *Bevolkingstrends*, 51 (1): 34–41. Voorburg: CBS.

Algemene Rekenkamer (1991) *1000-Banenplan Molukkers*. The Hague: Algemene Rekenkamer.

Bakker, E. and Tap, L. (1987) *Etnische ondernemers in Utrecht en Rotterdam*. The Hague: Hoofdbedrijfschap Ambachten.

Barrett, G., Jones, T. and McEvoy, D. (1996) 'Ethnic minority business: theoretical discourse in Britain and North America', *Urban Studies*, 33 (4/5): 783–809.

Bates, T. (1994) 'Social resources generated by group support may not be beneficial to Asian immigrant-owned small businesses', *Social Forces*, 72: 671–89.

Beets, G. and Koesoebjono, S. (1991) 'Indische Nederlanders: een vergeten groep', *Demos (NIDI)*, 7 (8): 60–4.

Beets, G., Walhout, E. and Koesoebjono, S. (2002a) 'Demografische ontwikkeling van de Molukse bevolkingsgroep in Nederland', *Maandstatistiek van de Bevolking*, 6: 13–17.

Beets, G., Huisman, C., van Imhoff, E., Koesoebjono, S. and Walhout, E. (2002b) *De demografische geschiedenis van de Indische Nederlanders*. Report 64. The Hague: NIDI.

Bonacich, E. (1973) 'A theory of Middleman Minorities', *American Scociological Review*, 38 (5): 583–94.

Bruggink, G., Haasken, R. and Keen, R. (1997) *Allochtonen in Groningen. Een inventariserend onderzoek naar de maatschappelijke positie van Antillianen/Arubanen, Marokkanen, Molukkers/Indonesiers, Surinamers, Turken en nieuwkomers iun de provincie Groningen*. Groningen: Maatstaf, Bureau voor Sociaal-wetenschappelijk Onderzoek.

CBS (2001a) *Statistisch Jaarboek* (Statistical Yearbook) *2001*. Voorburg: Central Statistical Bureau.

CBS (2001b) *Allochtonen in Nederland 2001*. Voorburg: Central Statistical Bureau.

CBS (2003) *Allochtonen in Nederland 2002*. Voorburg: Central Statistical Bureau.

Choenni, C. (2000) 'Ontwikkeling van het rijksoverheidsbeleid voor etnische minderheden', in Nico van Nimwegen and Gijs Beets (eds) *Bevolkingsvraagstukken in Nederland anno 2000*. Report 58. The Hague: NIDI, 131–44.

Dagevos, J., Veenman, J. and Tunjanan, T. (1992) *Succesvolle allochtonen*. Meppel: Boom.

Ellemers, J. and Vaillant, R. (1985) *Indische Nederlanders en gerepatrieerden.* Muiderberg: Coutinho.

Folkerts, H. (1999) 'Allochtonen in Nederland: vijf grote groepen' (Non-natives in the Netherlands: five large groups), *Maandstatistiek van de Bevolking* (CBS), April, 9–19.

Heering, L., de Valk, H., Spaan, E., Huisman, C. and van de Erf, R. (2002) 'The demographic characteristics of immigrant populations in the Netherlands', in P. Compton and Y. Courbage (eds) *The Demographic Characteristics of Immigrant Populations.* Population Studies no. 38. Strasbourg: Council of Europe, 245–98.

ISEO (2000) *Minderhedenmonitor 2000.* Rotterdam: Instituut voor Sociaal Economisch Onderzoek (ISEO).

ITS (2000) *Nieuwe etnische groepen in Nederland.* Nijmegen: ITS.

Jansen, M., de Kok, J., van Spronsen, J. and Willemsen, S. (2003) *Immigrant Entrepreneurship in the Netherlands. Demographic Determinants of Entrepreneurship of Immigrants from non-Western Countries.* Report H200304. Zoetermeer: EIM.

Khosravi, S. (1999) 'Displacement and entrepreneurship: Iranian small businesses in Stockholm', *Journal of Ethnic and Migration Studies,* 25 (3): 493–508.

Kloosterman, R. and Rath, J. (1999) 'Het ondernemerschap van immigranten: de overheid een zorg?' *Rooilijn,* 32 (3): 108–14.

Kloosterman, R. and Rath, J. (2001) 'Immigrant entrepreneurs in advanced economies: mixed embeddedness further explored', *Journal of Ethnic and Migration Studies,* 27 (2): 189–202. Special issue on immigrant entrepreneurship.

Knibbe, K. *et al.* (1998) *Molukkers langs de maatschappelijke ladder. Een onderzoek naar de sociaal-economische positie van de Molukse gemeenschap in Noord- en Zuid-Holland.* Amsterdam: Wetenschapswinkel Vrije Universiteit.

Kruyt, A. and Niessen, J. (1997) 'Integration', in Hans Vermeulen (ed.) 'Immigrant Policy for a Multicultural Society'. A Comparative Study of Integration, Language and Religious Policy in five Western European Countries. Brussels: Migration Policy Group, 15–51.

Light, I. (1972) *Ethnic Enterprise in America.* Berkeley CA: University of California Press.

Light, I. (1980) 'Asian enterprise in America', in S. Cummings (ed.) *Self-help in Urban America.* New York: Kennikat Press, 33–57.

Light, I. and Rosenstein, C. (1995) *Race, Ethnicity and Entrepreneurs in Urban America.* New York: De Gruyter.

Light, I. and Gold, S. (2000) *Ethnic Economies.* San Diego CA and London: Academic Press.

Maas, M. (2003) 'Transnationale migrantenonderneming in opkomst. Dienstverlener tussen Nederland en vaderland', *Geografie* 12 (6): 38–240.

Maas, M. (2004) 'Filipino Entrepreneurship in the Netherlands. Male and Female Business Activity Compared'. Paper presented at the seventh International Conference on Philippine Studies (ICOPHIL), Leiden (Netherlands), 16–19 June.

Martina, J. (1999) *Etnisch ondernemerschap in Gelderland* II. *Studie naar de knelpunten van etnisch ondernemen in Arnhem en de behoefte aan ondersteuning.* Arnhem: Osmose.

Memorandum Minister of Social Affairs and Employment (2000) *Nota Arbeidsmarktbeleid etnische minderheden 2000–2003.* Kamerbrief 27–223, vergaderjaar 1999–2000. The Hague: SDU.

Ministry of Economic Affairs (1998) *Kabinetstandpunt. Beleid inzake het ondernemerschap van personen uit etnische minderheidsgroepen.* The Hague: Ministry of Economic Affairs.

Moluks Historisch Museum (2000) *Molukkers in Nederland.* Utrecht: MHM.

Muijzenberg, O. van den (2000) *Four Centuries of Dutch–Philippine Economic Relations, 1600–2000.* Manila: Royal Netherlands Embassy.

OECD (1998) *Fostering Entrepreneurship.* Paris: OECD.

Piard, M., van Dijk, J. and Dermijn, S. (1998) *Etnisch ondernemerschap in Gelderland* I. *Inventarisatie van kerncijfers.* Arnhem: Osmose.

Prins, C. (1997) 'Population born in Indonesia or in the former Dutch East Indies' (in Dutch), *Maandstatistiek van de Bevolking* (CBS), 4: 6–10.

Prins, C. and Verhoef, R. (2000) 'Demografische ontwikkelingen in Nederland', in Nico van Nimwegen and Gijs Beets (eds) *Bevolkingsvraagstukken in Nederland anno 2000.* Report 58. The Hague: NIDI, 77–121.

Rath, J. (2000) 'Introduction: immigrant businesses and their economic, politico-institutional and social environment, in Jan Rath (ed.) *Immigrant Businesses. The Economic, Political and Social Environment.* Basingstoke: Macmillan, 1–17.

Roelandt, T., Roijen, J.H.M. and Veenman, J. (1991) *Minderheden in Nederland. Statistische Vademecum 1991.* The Hague: Sdu.

Sassen, S. (1991) *The Global City. New York, London, Tokyo.* Princeton NJ and Oxford: Princeton University Press (2nd edn 2001).

SCP (1998) *Sociaal en Cultureel Rapport 1998.* Rijswijk: Sociaal Cultureel Planbureau.

Smeets, H. and Veenman, J. (1994) 'Steeds meer "thuis" in Nederland: tien jaar ontwikkelingen in de Molukse bevolkingsgroep', in Hans Vermeulen and Rinus Penninx (eds) *Het Democratisch Ongeduld. De emancipatie en integratie van zes doelgroepen van het minderhedenbeleid.* Amsterdam: Spinhuis, 15–44.

Smeets, H. and J. Veenman (2000) 'More and more at home: three generations of Moluccans in the Netherlands', in Hans Vermeulen and Rinus Penninx (eds) *Immigrant Integration. The Dutch Case.* Amsterdam: Spinhuis, 36–63.

Smeets, H., Martens, E. and Veenman, J. (1999) *Jaarboek Minderheden 1999.* Houten: Bohn Stafleu Van Loghum.

SPVA (1998) *Sociale positie en voorzieningengebruik allochtonen (SPVA).* Rotterdam: ISEO.

Steensel, K., el Achkar, M. and de Jong, J. (1999) *Dualiteit in asiel en arbeid. Nieuwe modellen voor integratie en participatie.* The Hague: Stichting Maatschappij en Onderneming (SMO).

Tambuli (magazine on the Philippines), Utrecht and Dordrecht: Filippijnenwerkgroep, FIDOC.

Tillaart, H. van den and Poutsma, E. (1998) *Een factor van betekenis. Zelfstandig ondernemerschap van allochtonen in Nederland.* Nijmegen: ITS.

Tillaart, H. van den, Olde Monnikhof, M. *et al.* (2000) *Nieuwe etnische groepen in Nederland. Een onderzoek onder vluchtelingen en statushouders uit Afghanistan, Ethiopië en Eritrea, Iran, Somalie en Vietnam.* Nijmegen: ITS.

Tillaart, H. van den (2001) *Monitor etnisch ondernemerschap 2000. Zelfstandig ondernemerschap van etnische minderheden in Nederland in de periode 1990–2000.* Nijmegen: ITS.

Veenman, J. (1985) *De arbeidsmarktproblematiek van Molukkers.* Rotterdam: ISEO, Erasmus University Rotterdam.

Veenman, J. (1990) *De arbeidsmarktpositie van allochtonen in Nederland, in het bijzonder van Molukkers.* Groningen: Wolters Noordhoff.

Veenman, J. (2001) *Molukse Jongeren in Nederland. Integratie met de rem erop.* Assen: Van Gorcum.

Verkuyten, M., van de Calseijde, S. and de Leur, W. (1999) 'Third-generation Moluccans in the Netherlands: the nature of their identity', *Journal of Ethnic and Migration Studies*, 25 (1): 63–79.

Vries, M. de (1999) 'Why ethnicity? The ethnicity of Dutch Eurasians raised in the Netherlands', in M. Crul, F. Lindo and Ching Lin Pang (eds) *Culture, Structure and Beyond. Changing Identities and Social Positions of Immigrants and their Children.* Amsterdam: Spinhuis.

Waldinger, R. (1995) 'The "other side" of embeddedness: a case study of the interplay of economy and ethnicity', *Ethnic and Racial Studies*, 18 (3): 555–80.

Waldinger, R., Aldrich, H., Ward, R. and associates (1990) *Ethnic Entrepreneurs. Immigrant Business in Industrial Societies.* London: Sage.

Warmerdam, J. and van den Tillaart, H. (2002) *Arbeidspotentieel en arbeidsmarktloppbanen van vluchtelingen en asielgerechtigden.* OSA publicatie A189 (July). Tilburg: Organisatie voor Strategisch Arbeidsmarktonderzoek.

Wittermans, T. (1991) *Social Organization among Ambonese Refugees in Holland.* Amsterdam: Spinhuis.

Worung, J. (1999) 'Molukkers zijn van oudsher geen handelsmensen', *Marinjo* (independent Moluccan journal), 9: 23.

WRR (2000) *Doorgroei van Arbeidsparticipatie.* Rapport van de Wetenschappelijke Raad voor het Regeringsbeleid (WRR). The Hague: SDU.

14 Chinese immigrant entrepreneurs in Italy

Strengths and weaknesses of an ethnic enclave economy

Daniele Cologna

Though this article's goal is to present and analyse several key characteristics of the social and economic insertion process of Chinese immigrant entrepreneurs in Italy, the fieldwork on which it is based has been carried out in Milan, and much of the data collected also refer to this specific urban context.[1] Since it is in this city that Chinese migration to Italy first manifested itself, focusing on this particular territory offers a unique opportunity to gather information regarding its history. Moreover, since about 18 per cent of Italy's Chinese population live in this city, research findings concerning the Chinese migrants' socio-economic insertion strategies developed since the early 1990s certainly do have a bearing on this immigrant group's situation in the whole country today. Nevertheless, though the picture one may get from this particular vantage point may be considered fairly representative of the Chinese migrant's reality in those Italian cities (Milan, Turin, Bologna, Rome) where the Chinese entrepreneurs' insertion in the service sector plays a significant role, in order to gain a better understanding of the Chinese ethnic business in Italy's industrial clusters (such as the garment manufacturing clusters in the Capri–Modena–Reggio Emilia, Prato–Empoli and Naples–San Giuseppe Vesuviano areas), where Chinese entrepreneurship has evolved along slightly different lines, one may need to complement the research findings presented here with those resulting from the research work carried out in those areas by other authors.[2]

The historical development of migration from China to Italy, as documented by the relevant literature and the fieldwork data, is detailed in the first part of this chapter. The second part deals with the development of the Chinese ethnic enclave economy in Milan during the 1990s, and purports to highlight its double-edged character as a social and economic insertion strategy. Though the Chinese socio-economic enclave may indeed provide both newcomers as well as long-time immigrants with an ethnic opportunity structure that can be accessed through patient networking tactics, it does also expose immigrants to the risks entailed by protracted cultural and social exclusion, thereby eventually increasing their social and economic vulnerability.

Emigration from the People's Republic of China to Italy: historical background and present trends

According to the latest data (31 December 2001) provided by the Milan Province Communal Registry offices, 12,476 Chinese citizens are registered as resident in the province of Milan (10,271 of them, for example, about 80 per cent, live within the city proper) (table 14.1). An approximate 20–30 per cent more are estimated to be irregularly present on the same territory, putting the total of Chinese immigrants living and working in Milan at about 16,000 people. At least three generations of Chinese immigrants have been choosing Milan as a destination of their quest for success and prosperity in the last seventy years, and the Chinese could very well become the largest and most demographically complex immigrant population in Milan within the next decade. (They are presently ranked as the third largest, behind the Philippines and Egypt.) If the demand for outmigration in southern Zhejiang remains strong and Zhejiang migrants continue to find in Milan a favourable environment for the deployment of their migratory careers, the newcomers will keep on coming, joining a population largely made up of families who have a high fertility rate,[3] hereby securing high growth rates for the near future. Between 1984 and 2001 the Chinese population in Milan city proper grew twenty times in size, from 500 residents in 1984 to 10,271 in 2001. In the last ten years, each new legalization process has registered an increase of over 30 per cent in the total of Chinese immigrants.

The majority of these People's Republic of China (PRC) immigrants come from a specific area in southern Zhejiang: the small villages strewn about the mountainous rural areas to the south-west of the city of Wenzhou. The main districts of origin of Zhejiang migrants to Italy are Qingtian, Wencheng, Rui'an and Wenzhou-Ouhai. Most Zhejiang immigrants in Milan are nowadays Wencheng people, mainly rural residents of the villages surrounding the town of Yuhu. Other Zhejiang migrants in Milan are generally people who migrated from the rural areas of the Qingtian and Rui'an districts, while former Wenzhou-Ouhai residents nowadays are very few.

A small proportion (probably less than 10 per cent) of the PRC citizens residing in the Milan area can be ascribed to two other different migration flows: migrants who hail from the Sanming area in central Fujian province and people who migrated from the cities of the Chinese north-east (known in Chinese as Dongbei), mostly from Liaoning province. The migration of Sanming Fujianese appears to be closely linked with Wenzhou migratory chains. People from the Wencheng and Rui'an districts migrated to Sanming during the late 1950s and early 1960s to work as lumberjacks, carpenters and masons for the large State-owned enterprises established there after the Great Leap Forward. Many of them eventually settled down in Sanming and married Fujianese women. As a result, during the late 1980s and early 1990s people of the Sanming area who had family ties with

Table 14.1 Chinese residents in Italy's regions and provinces, 31 December 2000

Region	Males		Females		Both		Provinces with the highest No. of Chinese residents	Total	% of regional total	% of national total
	No.	%	No.	%	No.	%				
Valle d'Aosta	22	59.5	15	40.5	37	0.1	Aosta	37	100.0	0.1
Piedmont	2,145	54.2	1,816	45.8	3,961	6.7	Turin	2,477	62.5	4.2
							Cuneo	488	12.3	0.8
Lombardy	8,490	53.3	7,441	46.7	15,931	27.1	Milan	10,861	68.2	18.5
							Brescia	1,895	11.9	3.2
							Mantua	844	5.3	1.4
							Varese	592	3.7	1.0
							Bergamo	561	3.5	1.0
Liguria	529	51.5	499	48.5	1,028	1.7	Genoa	657	63.9	1.1
Trentino Alto Adige	184	54.0	157	46.0	341	0.6	Trento	173	50.7	0.3
							Bolzano	168	49.3	0.3
Veneto	2,654	53.8	2,283	46.2	4,937	8.4	Treviso	1,178	23.9	2.0
							Verona	1,005	20.4	1.7
							Padua	940	19.0	1.6
							Venice	764	15.5	1.3
							Vicenza	554	11.2	0.9
Friuli Venezia Giulia	439	54.8	362	45.2	801	1.4	Trieste	336	41.9	0.6
Emilia Romagna	3,271	53.7	2,821	46.3	6,092	10.4	Bologna	1,772	29.1	3.0
							Reggio Emilia	1,347	22.1	2.3
							Modena	1,256	20.6	2.1
							Rimini	502	8.2	0.9

Tuscany	6,784	55.1	5,527	44.9	12,311	20.9
Umbria	323	61.1	206	38.9	529	0.9
Marche	512	53.4	446	46.6	958	1.6
Lazio	3,355	53.6	2,904	46.4	6,259	10.6
Abruzzo	402	54.2	340	45.8	742	1.3
Molise	6	66.7	3	33.3	9	0.0
Campania	1,521	57.7	1,114	42.3	2,635	4.5
Puglia	461	57.4	342	42.6	803	1.4
Basilicata	28	65.1	15	34.9	43	0.1
Calabria	127	58.0	92	42.0	219	0.4
Sicily	386	52.8	345	47.2	731	1.2
Sardinia	277	58.1	200	41.9	477	0.8
Total	31,916	54.2	26,928	45.8	58,844	100.0

Florence	6,216	50.5	10.6
Prato	4,814	39.1	8.2
Perugia	462	87.3	0.8
Ancona	350	36.5	0.6
Rome	6,020	96.2	10.2
Teramo	359	48.4	0.6
Pescara	276	37.2	0.5
Campobasso	9	100.0	0.0
Naples	2,224	84.4	3.8
Bari	365	45.5	0.6
Lecce	171	21.3	0.3
Taranto	159	19.8	0.3
Matera	34	79.1	0.1
Catanzaro	84	38.4	0.1
Reggio Cantabria	51	23.3	0.1
Cosenza	54	24.7	0.1
Palermo	250	34.2	0.4
Catania	193	26.4	0.3
Cagliari	295	61.8	0.5
Sassari	150	31.4	0.3

Zhejiangese who had relatives abroad suddenly realized they could access the 'migratory capital' owned by their kin in Wencheng. By 'migratory capital' I mean the whole set of material, social and cultural resources that an individual born in a *qiaoxiang* (lit. 'overseas Chinese village', for example, a village where the majority of its residents have relatives abroad) family can access through his/her social networks in order to leave China (by legal or illegal means) and settle successfully in another country. According to the migrants I interviewed in Milan, the first Sanming Fujianese to reach continental Europe availed themselves of the human trafficking networks set up by Wencheng and Rui'an migrants, and the ones who entered Italy settled down primarily in locales where Wencheng and Rui'an migrants are prominent, like Milan and the Prato–Empoli area in Tuscany. Although they speak a very different dialect (*minnanhua*, also known as *hokkien*) and their social background may differ from that of the average Wencheng or Rui'an immigrant, their social and economic insertion strategies in the Italian context closely follow the pattern set by Zhejiang immigrants.

Immigration from Dongbei, a very recent phenomenon, is an altogether different matter, for the typical north-eastern migrant is generally an urban dweller (many come from the regions main cities, particularly Shenyang, in Liaoning province, and Harbin, in Heilongjiang province), and is usually a laid-off industrial worker, a so-called *xiagang* (lit. 'off-post'). The link with Zhejiang migrants in this case is rather controversial. According to a few key informers, the first batch of Dongbei migrants to reach Italy and Milan in the late 1990s where *xiagang* women who had been recruited by Zhejiang migrants in Liaoning's coal-mining districts (especially Xinbin, in the Fushun area) to provide certain services, and especially to work as *baomu*, 'nannies'. These women were generally over thirty and considered themselves permanently discriminated against on the local labour market, to the point that they were more than willing to accept unskilled jobs abroad, even if the pay and living conditions were not very good. Some of these *Fushun dasao*, or 'Fushun older sisters', as they are known among Italian Dongbei migrants, actually did resort to prostitution while in China, and according to one story, many had originally been hired by Zhejiang women who had been prostitutes themselves to 'take their place' in local Chinese-run secret whorehouses and clandestine hotels and boarding houses that catered only to PRC travellers and businessmen. Though it is difficult to substantiate this claim, it is a fact that many of the first Dongbei migrants who arrived in Milan after 1995 really did hail from the Fushun area, and that they were mostly women who worked as *baomu* for Zhejiang families. And Chinese *street* prostitution (previously unheard of in Milan), for instance, definitely did involve only Dongbei migrants in 1997 and 1998. The street experience has been short-lived, and now most Dongbei prostitutes work behind closed doors, in private apartments or massage parlours.

Yet this process, though it may account for the origins of Fushun chain migration, certainly does not suffice to explain the growing flow of Dongbei migrants hailing from the region's bigger cities. Many of these, both men and women, during the nineties first left China to go working in Siberia, and some pushed westwards to reach Moscow and other Eastern European 'way stations' on the northern illegal migration route from China to Europe, where they may have came in touch with Zhejiang Chinese. The latter may have highlighted for them the opportunities open to Chinese workers in those Western European countries where they had a strong base. Some Dongbei migrants in the late 1990s may have actually entered Italy (or Germany, or France) thanks to the services offered to them by Zhejiang *shetou* (lit. 'leader of snakes' or 'snakehead', for example, a human smuggler, a dealer in *renshe*, or 'human snakes', which is what illegal immigrants are called in the smugglers' jargon, due to their having to 'snake their way' through borders and customs), but most of them have been able to make their journey on their own. As laid-off State workers coming from big industrial cities, it must have been much easier for them to get passports and visas to go abroad than it is for Zhejiang migrants. According to key informers, in the Chinese north-east the requirements and procedures for issuing passports to common citizens have eased considerably since the late 1990s, and the only difficult part for a would-be migrant nowadays seems to be obtaining a visa for a European country. This 'service' is now provided legally by operating agencies in Shenyang and other cities across Dongbei.[4] Italian visas have been comparatively hard to get (this may change now, though, as Italy is getting ready to welcome Chinese mass tourism), but Switzerland, Germany and France have been popular destinations for years. That way, only the last leg of the journey is accomplished illegally, and Dongbei migrants spend considerably less to leave the country, and at an average €4,000 the cost of emigration for them is nowhere near the out-landish sums of money that Zhejiang migrants still have to pay the *shetou* who safely carry them from China to Europe without valid papers.

Unlike their Zhejiang counterparts, most north-easterners who reach Milan do not have a clearly set migratory project. Formerly employed in big State-owned factories, they generally are not as driven by an entrepreneurial mind set as most immigrants hailing from the Wenzhou area. These new-comers usually get started either by working as unskilled hired labour for Chinese or Italian employers or by exploiting economic niches in the shadier corners of Italy's underground economy. Yet on the whole Dongbei immi-grants are also better educated and more likely to integrate themselves cul-turally in Italian society than Zhejiang immigrants, because they rarely manage to fit into the ethnic enclave economy set up by the latter. Dongbei immigrants do not seem to be strongly motivated by the aspiration of even-tually becoming a *laoban*, the 'boss' of one's own family business, a Zhejiang migrant's first and foremost priority. Those whom I have managed to inter-view so far usually think that for them a good job in an Italian-run business, with a decent pay cheque, a valid contract and no hassles, would be the most satisfying option. They like to point out that 'Zhejiang immigrants are obsessed with money and their sole purpose in life seems to be work, while we work for a better living, we like to spend money to enjoy life.' It is yet unclear whether Dongbei immigration to Europe – which appears to be still a lot less structured than Zhejiang or Fujian immigration – will eventually trigger widespread family-based chain migration. Most Dongbei immigrants at this stage are single and do not voice an overt desire to settle permanently in Europe, but couples are starting to consider having their children born and/or and educated in Italy, and the number of family reunion permits issued to them is likely to grow.

On the other hand, Zhejiang immigration has by now a deep-rooted tradition in Europe (Thunø, 1996; Benton and Pieke, 1998). In Italy, the first Chinese immigrants to settle more or less permanently were immigrants from the Qingtian district, who moved to Milan from France during the late 1920s. Milan being conveniently located near Europe's most important silk-producing and processing area at that time (the town of Como and its surroundings), those migrants who had started out as peddlers of necklaces, nicknacks and other petty wares imported from France and Central Europe soon seized the opportunity to step out of a trade that was still too closely tied to the supply of necklaces and other petty wares from abroad. They therefore began to peddle merchandise that they could manufacture and sell locally: silk ties. Some immigrants opened small shops and employed local workers, generally women who had left the northern Italian countryside to come to work in the city's garment factories, to help in cutting and sewing the ties, while they relied on fellow Chinese to distribute their products in Milan and in other northern Italian cities. Before and during the Second World War the number of small Chinese entrepreneurs increased, many married Italian seamstresses and were joined by their brothers and relatives, who left France, the Netherlands or even China knowing they could count on the support of their kin to find a job and quickly make themselves at home in their new surroundings. After the war, they gradually abandoned the silk tie business and restructured their small shops in order to produce leather belts and bags, as well as satchels and duffel bags made of cloth.[5] This first generation of Chinese immigrants was very successful in creating an ethnic opportunity structure that enabled most newcomers to rapidly move a big step up the social ladder and open their own shop, after having worked for or with a relative or a *tongxiang* (lit. 'same village', for example, a person from their own village). They built what back then could rightly be considered a closely knit ethnic community, held together by bonds of trust and by a network of mutual loans that offered everyone a chance to start a business of their own, generally at no interest.[6] Yet this pattern of ethnic solidarity lasted only until the new wave of Zhejiang immigration reached Italy during the 1980s, and, as will be described in the following pages, definitely did not survive the 1990s.

During the early phase of Chinese immigration to Italy (1929–79), each little shop was run as a family business. The largest ones had Italian employees (mostly women), and during the 1950s and 1960s the number and size of Chinese-run shops increased, eventually attracting a growing number of small collateral shops, run by Italians, to what began to be known as Milan's 'Chinese district' (the via Canonica–via Paolo Sarpi area). These shops supplied the Chinese manufacturers with small metal parts, zips, rivets as well as sewing machines. The Chinese, who were still a very small community of no more than 100 men, were well liked and reputed to be an industrious lot. Their businesses had contributed a good deal to the development of the economic structure of the city district in which they lived, their Italian wives provided important links with everyday Milanese life, while their children were taught to assimilate as quickly and thoroughly as possible, in order to help carry on the family business without the cultural and linguistic handicaps that still hampered their fathers. Consequently, these second-generation Chinese-Italians grew up as accomplished Milanese, and most of them slowly lost whatever links they might have had with the language and culture of their fathers, eventually blending into the mainstream society with little more than their alien facial traits and peculiar food habits to differentiate them from the average Milanese.[7]

During the 1960s and 1970s many Milan Chinese left Lombardy, spreading out towards other parts of the country, such as Rome, Bologna and Florence. But their numbers remained small and their presence in the local economy went largely unnoticed until the People's Republic of China ended its thirty-year-long isolation and the leadership launched its policy of 'reform and openness' after 1979. The 1980s were a decade of great economic and social development in the People's Republic, and a more relaxed policy towards emigrants and their families paved the way for the reprise of migratory flows from southern China to the West. Old migratory chains that linked tiny villages in the Wenzhou area in southern Zhejiang province with cities in continental Europe were revived. In Italy, the main target areas were Milan, Rome and Florence.

Wenzhou is famous in China for its entrepreneurial ebullience. During the Maoist era it was frequently denounced for its 'creeping capitalism' (Boutellier, 1997: 79–82), while in the reform era the area became a beacon of free enterprise. In Wenzhou, economic development was originally based on farm family units in a rural environment of small towns and villages where local entrepreneurs played a crucial role. As Koo and Yeh point out in their 1995 report:

> Its major comparative advantage, the Wenzhou people's craftsmanship and marketing skills, stems from a long tradition of its handicraftsmen and merchants emigrating to other parts of the country. The emigration originated in the peasants' attempt to supplement their meagre income by exporting labour. The economic consequences, however, were profound. These wandering craftsmen and peddlers brought back new concepts, ideas, business experience and accumulated savings. Many

have become enterprising traders, roaming all over China and forming a nationwide network supplying market information, goods, and technology.

(Koo and Yeh, 1995: 325)

When the reform years provided new opportunities for economic growth, Wenzhou businessmen could already count on a fairly widespread entrepreneurial culture, and rapidly developed a need for financial services. The first private banks in China were born in Wenzhou, as well as credit associations and pawnshops, ready to offer credit at interest rates that reflected supply and demand in the local money market. As local cadres decided to support the newborn market economy, Wenzhou became a portentous boom town, and all through the 1990s the so called *Wenzhou moshi* (lit. 'Wenzhou model') was all the rage in China. To villagers inhabiting the periphery of the economic boom, this lightning-quick development also entailed an increase in the cost of living, and they rapidly felt the need to keep up with the times. Those who did not have any real chance to join in the small-business bonanza, but did possess 'migratory capital', decided that going abroad could ensure their families just as quick a progress up the social and economic ladder. According to Li Minghuan (Li, 1999), relative deprivation,[8] for example, the gap in household incomes between the more developed core of the economic boom and its backward rural periphery, and chain migration appear to be the main factors influencing outmigration from the Wenzhou area. Since the 1980s, *chuguo*, or 'leaving the country', has been the most popular option for all those who happened to be born in a *qiaoxiang*, a village with a tradition of people working overseas.

Chinese immigrant entrepreneurs in Milan: the evolution of an ethnic enclave economy, 1990–2003

The ideal migratory career of a Wenzhou migrant follows a set pattern that, to some extent, outlines what was the most common strategy of socio-economic insertion during the 1980s and early 1990s. Upon the migrants' arrival in the country of their destination, a rapid process of disenchantment ensues. The promised land soon reveals itself to be a place of hard work and little reward, yet the huge amount of money the newcomers have come to owe their family and friends, as well as the great expectations of success and wealth cherished by their loved ones at home, are deterrents strong enough to deny them a sudden return. To go back empty-handed would entail a great loss of face, and no Chinese immigrant would accept defeat so soon. It is part of the migrants' mind set to see themselves – and the overseas Chinese in general – as able and willing to *chiku nailao* (lit. 'to endure bitterness and work patiently') until success comes their way.

Whether an illegal alien or a spouse/child who obtained a permit to stay on grounds of family unity, for the newcomer the general rule in the recent past would have been to be taken under the wing of the relative or friend – generally the owner of a small business, a restaurant or a sweatshop – who paid for a safe passage from China to Italy. The same person would provide food, shelter and a job. In order to repay the debt, newcomers would work basically for nothing for a given period of time, generally two or three years. Yet nowadays newcomers may be forced to work for other employers, as controls on illegal immigrants' work in Chinese-run business have been sharpened and job opportunities in traditional sectors such as restaurants or leather/garment sweatshops have become harder to come by.

A certain saturation of these two traditional sectors of economic insertion and the risks involved in hiring illegal migrants in an ethnic labour market that can now count on a relatively large legalized ethnic work force often only leaves extremely precarious menial jobs open to those who cannot muster a permit to stay. These jobs are generally just short-term occupations, in businesses with a high work force turnover. The garment industry still offers a certain number of job opportunities, and so do some Italian-run restaurants, bars and pizza places. Most migrants try to cope with these unfavourable conditions anyway. No matter how slow and demeaning the start of a migrant's career (as it often entails working long hours as a dishwasher or at the sewing machine), the newcomers hope to use this difficult first stage to learn the ropes, and in the course of time to be able to move up the internal hierarchy of the restaurant's kitchen or the sweatshop. The goal is to eventually be able to pay back most of their debts and to strengthen their local network of *guanxi* ('connections', privileged relationships and particularistic ties[9]). Expenses are kept to a bare minimum: food and shelter are generally provided by the employer, clothes have been brought from China or bought at bargain prices in local markets, and everyday life focuses on work. Every cent that is not used to repay debts – as *huanqian* (lit. 'paying back money') soon becomes a migrant's first and foremost priority – is generally saved, sent back home or lent to relatives, friends or *tongxiang* who are just about to start their own business and need a helping hand. These loans are generally offered interest free, the purpose being the gradual expansion of one's own support networks. Those *guanxi* will be precious in the future, when the time will come for the seasoned immigrant to open up his or her own little sweatshop or fast-food take-away. According to several key informers, at times an entire month's salary is lent out. Within a couple of years a Chinese immigrant could very well have 'invested' tens of thousands of euros in his or her own support network.

This long apprenticeship also provides newcomers with an indispensable minimum of language skills (especially if they happen to be employed in restaurants), and with a basic knowledge of the laws that regulate private businesses in Italy. Eventually, they will get a chance to regularize their position thanks to a *sanatoria*, a general regularization of illegal aliens carried out on the basis of new immigration laws (such was the case in 1986, 1990, 1998 and 2002) or executive orders (1995). Once having obtained a permit to stay, and having strengthened their own support network, if a good opportunity to start a business is at hand, a Chinese immigrant may decide that the time has come to collect the money lent out within their own network of friends and perhaps to borrow some more on their own account, This ethnic credit system, based on trust, enables even the humblest kitchen hand to put together a substantial amount of cash (sometimes up to €30,000) within a very short time. In a cash-starved real estate market, a lump sum of several tens of thousands of euros and a number of IOUs will be more than welcome to settle a deal, making it possible to buy out a former small *trattoria* or pizza take-away. An immigrant's first business will generally be a small and low-cost operation, where labour costs can be kept down to a minimum thanks to the employment of one's own family members and to the strict limitation of all collateral expenses. A small garment sweatshop, with three or four sewing machines, has been the most common way to take the first step as entrepreneur throughout the 1990s. Among young couples it is generally the men who leave first to work as wage earners until they are able to regularize their position and ask for a family reunion permit. The garment sweatshop – which usually serves as a home, too – may enable the husband to keep his job while providing the wife with a job that allows her to tend the children as well. With a small investment, solid connections, sufficient skill and a bit of good luck the family may thus double its income and move further towards independence. Becoming *laoban*, being 'one's own boss', still is the ultimate goal.

The subsequent steps may be taken on various different paths, as it is common to switch from a garment shop to a Chinese take-away, a small restaurant or even a tiny import–export company, though generally the main goal is to end up as owners of a good medium-sized restaurant. A restaurant is still considered the operation that 'has the most face', second only perhaps to a large import–export company or a chain of supermarkets. But even if that stage is attained, the ideal career of the Wenzhou migrant doesn't stop there. The imperative is *fazhan*, 'to develop and to flourish': opening another restaurant, if it is considered viable, or branching out into real estate, buying apartments that will be rented to one's own restaurant workers or to other *tongxiang* looking for a place to stay and work. The steadfast entrepreneurial drive of the Wenzhou migrant may be considered a survival strategy during the early stages of a migrant's career, when the lack of language skills, as well as skills mismatch phenomena combined with the illegal status of most newcomers, makes their economic insertion as wage earners almost impossible. But it can also be seen as a vital component of this immigrant group's cultural capital, one providing the strongest motivating factors for all those who actually struggle for years to find a way out of the sweatshop.

It is still very rare for the average Chinese migrant to remain unemployed for longer than a couple of months, while most hardly ever had any trouble in finding at least some precarious occupation of sorts. Those who are working without a permit to stay see their time as illegal aliens as a time of utter precariousness and extreme vulnerability, but are usually able to cope with the insecurity and apparent hopelessness of their situation thanks to the resources offered them by their ethnic networks and by the actual need of low-cost labour within certain niches of the domestic economy – most notably the garment industry. The ethnic opportunity structure[10] that enables Chinese immigrants from the Wenzhou area to promote their economic insertion in the local economy and that supports their entrepreneurial drive centres on the family, for example, on a network of blood relations that has the single nuclear family at its core and encompasses even very distant relatives, with an essential periphery of 'close friends' and *tongxiang* to provide links with other clans or networks. From the very start of a Wenzhou migrant's career, the size and extension of one's family network is of crucial importance. It is within the family clan that the capital needed to leave the country is raised, and it is thanks to its good connections with the people who provide the necessary papers and organize the passage – or at least part of it – from China to Italy that emigration becomes a feasible option. Those who cannot rely on family support networks, with no relatives abroad to guarantee their solvency and but very weak ties with the *shetou*, have only an extremely low chance of succeeding in their migrant venture.

The Chinese in Milan have managed to adapt themselves and often prosper in a new and alien social and economic environment by building a socio-economic enclave.[11] In a segmented labour market, split into a core of stable, regular jobs, a periphery dominated by the informal economy (part-time or temporary jobs, less regulated and more precarious) and an outer, 'dark' segment where the underground economy thrives, a socio-economic enclave can be roughly outlined as a further segment, which overlaps all the others. Within this particular segment of the labour market, the boundaries of which are ethnically defined, Chinese immigrants can deploy their informal strategies of resource allocation and redistribution within an ethnic economy, for example, moving along the lines of the networks that bind them to other Chinese. During the 1980s, the Chinese socio-economic enclave had been able to absorb basically all the extant Chinese labour force, promoting the expansion of Chinese ethnic business in Milan and throughout Italy as well. But in the 1990s, this segment was subjected to strong change. It got 'leaner', as a growing number of workers moved within the enclave from its large informal segment to both the core *and* the underground segments, while an even larger number of newcomers opted to forsake the dwindling opportunities offered by the ethnic socio-economic enclave and tried to find jobs working for Italian employers in all three segments of the domestic labour market. Like most other immigrants, they generally were absorbed by the Italian informal economy and, to an even greater extent, by the Italian underground economy.

Thus, those who gambled their migratory capital successfully during the 1980s, when the ethnic opportunity structure that made the socio-economic

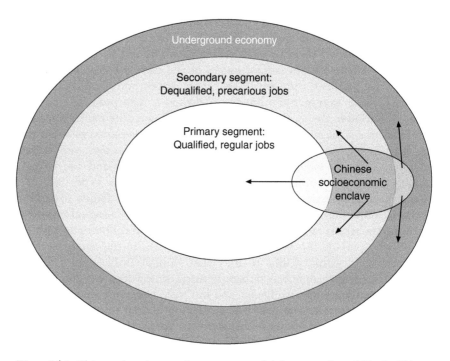

Figure 14.1 Chinese immigrants in a segmented labour market: Milan's Chinese socio-economic enclave in the 1990s.

enclave possible was indeed providing many immigrants with a fast track career, by and large really did make it. Those were the glory days of Chinese ethnic business in Milan, when restaurants and family workshops were thriving and many newcomers managed to be able to run their own small business in the space of two or three years. The informal credit system that supports enterprise creation among Chinese immigrants is tightly bound to the overall economic expansion of the immigrant population. The more successful entrepreneurs keep prospering the more resources flow through *guanxi* networks to boost other immigrants' entrepreneurial careers. But if stagnation sets in – as eventually became the case in the restaurant/catering sector as well as in the garment/leather or textile goods sector – the pool of resources available readily decreases. With less resources to fuel their rapid progression from the sweatshop to the restaurant or grocery chain, in the early 1990s a growing number of immigrants remained stuck in the very early stages of their migratory career. People who thought they would be running a sweatshop for no more than two years actually ended up keeping up that business for a whole decade, and with diminishing profits.

Though it is certainly true that a minority of the Chinese immigrant population in Milan – the elite – has been able to adapt quickly to change, diversifying its activities and branching out into the ethnic services sector (supermarkets, video rental shops, restaurants and bars catering to a Chinese clientele, etc.), most Chinese immigrants nowadays either work in subordinate positions in restaurants and sweatshops run by Chinese or even by Italian employers, or they run small subsistence enterprises with little hope of ever being able to branch into more lucrative businesses. This absolutely does not match their expectations. To a typical Wenzhou migrant, being an employee or subordinate worker is just a momentary solution, a tool used to start fuelling his or her own networks, loaning part of their pay cheques to boost their relatives', friends' or fellow villagers' business endeavours, so as to guarantee themselves ample financial support when their own situation is ripe for a change. Being a wage worker for years is not part of the plan. Yet since for those who emigrated to Italy after 1991 prospects have looked rather dreary, to them it became very difficult to grant their own spouses or relatives a sufficiently sound economic environment to start their own careers once they joined them in Italy. As many immigrants comment, 'How can I provide work, food and shelter to a niece or a cousin when I myself can just barely make it through each month?'.

Milan is an expensive place to live, lodging is hard to come by and very expensive even for the locals, and the Chinese are still heavily discriminated against when it comes to finding low-rent housing. Those who struggle to pay back their debts while working in a relative's sweatshop may consider indebting themselves even more to get their spouses and children to join them, for a working wife or husband and children who can both help with the family business and help their parents get a better grasp of Italian society seems to be a better strategy than just toiling away for years on end by themselves.

Even though the socio-economic enclave has lost much of its former appeal, and functions ever less as a booster of successful economic integration, most jobs are still to be found within it. To some extent, this is also a consequence of the high expectations of most newcomers, who just won't settle for dead-end jobs as menial workers in the Italian labour market. This attitude is changing as disenchantment grows, but to many immigrants staying within the enclave still remains the best way to access resources available only to those who regularly play the *guanxi* game. The enclave economy survives because it is tightly linked to the domestic economy's need of cheap labour on the one hand, and to the growing demand for ethnic services on the other. The degree to which Chinese ethnic businesses can be viewed as embedded in the Italian economic and social context varies greatly. If a video rental shop is a typical example of an ethnic business where everyone related to it – the shop's owner, the company's boss, suppliers, customers as well as the workers employed in it – is Chinese, most ethnic businesses generally imply a much higher level of integration. Restaurants are bought from Italian owners, they are run purchasing at least one-third of supplies from Italian suppliers, and they still cater mostly to

Italian customers. Workshops are set up in cellars, garages or shops rented to the Chinese by Italian owners, and it is Italian garment and textile factories that place orders and, of course, fix prices and deadlines.

The main sectors of economic insertion of Chinese-run businesses from the early 1980s to the late 1990s have been restaurants and subcontracting garment or leather factories (family workshops/sweatshops). These are niche economies where the development of family enterprises was pioneered by Chinese ethnic business during the preceding decades, and that subsequent waves of immigration have come to saturate in the last ten years. The expansion of Chinese restaurants has played a crucial role in revitalizing the restaurant/catering sector in Milan, which was stagnant at the beginning of the 1980s. Restaurants are still fantasized as the ideal point of arrival of the average Chinese immigrant's migratory career: it is an enterprise that 'has a lot of face' and also potentially offers ample profit. During the 1980s, Chinese entrepreneurs who succeeded in opening up restaurants found it relatively easy to cut costs by employing relatives and getting most of their supplies from Chinese food importers. At that time Chinese restaurants were all the rage in Italy, as they perfectly matched a growing domestic demand for exotic food and quick, cheap meals. The 1980s were also a time in which most Italians rediscovered the pleasure of having money to spend eating out, after the long economic austerity of the 1970s. The prominence of the two traditional insertion sectors of Chinese-run businesses in Italy stand out clearly in Table 14.2.

Being employed by a Chinese employer – nowadays no longer necessarily a relative – still represents the most viable option for the Chinese immigrant who has just entered the Italian labour market. It is usually the only option open to them: most newcomers are still people who enter the country illegally, do not know Italian nor have any notion regarding the local cultural, social and economic context, and who need to start working on their support networks in order to grant themselves better prospects for the immediate future. They are also exposed to a high degree of social vulnerability and marginalization, as they do not have a home or a job and start out their migratory career literally crushed by debt. Their relatives, friends or fellow villagers are obviously the first people they can turn to for immediate help. The fact that resorting to subsistence enterprises within the traditional niche economies created by the Chinese in Italy is still the first impulse of every newcomer highlights a crucial element of weakness in the Chinese socio-economic enclave. As there is a growing number of restaurants and workshops verging on the brink of economic collapse, the fast decline of which is largely due to widespread ignorance of the Italian market and to the way most businesses are basically copycat versions of already established enterprises, a series of naive mistakes and miscalculations are inevitably perpetuated by every new start-up of businesses of this kind. This practice eventually proves itself fatal to business development in these sectors.

All collateral business around the restaurant – suppliers, interior decorators, business consultants, etc. – is largely provided by other Chinese or by

Table 14.2 Chinese-run businesses registered at the Milan Chamber of Commerce, 1995, 1999 and 2002

Sector	No.	%	Cum. %	No.	%	Cum. %	No.	%	Cum. %
Garment factories (family workshops/sweatshops)	166	28.0	28.0	347	32.7	32.7	769	26.3	26.3
Restaurants, take-aways, pizzerias, bars	143	24.2	52.2	278	26.2	58.9	777	26.6	52.9
Retail shops	27	4.6	56.7	60	5.7	64.6	601	20.5	73.4
Leather or textile bag factories (family workshops/sweatshops)	162	27.4	84.1	211	19.9	84.5	238	8.1	81.5
Wholesale shops	26	4.4	88.5	90	8.5	93.0	266	9.1	90.6
Other businesses	68	11.5	100.0	74	7.0	100.0	275	9.4	100.0
among them:									
real estate agencies	n.a.			n.a.			23	0.8	
construction firms	n.a.			n.a.			15	0.5	
delivery services	n.a.			n.a.			14	0.5	
internet points and phone shops	n.a.			n.a.			10	0.3	
publishing and typesetting services	n.a.			n.a.			6	0.2	
Total	592	100.0		1,060	100.0		2,926	100.0	

Sources: *1995 data:* Chinese-run companies registered as 'active' at the Milan Chamber of Commerce on 31 December 1995. *1999 data:* Chinese-run companies registered as 'active' at the Milan Chamber of Commerce on 31 March 1999. *2002 data:* data processed by the Formaper Research Institute, relating to all Chinese entrepreneurs who are registered as owners or partners of companies registered at the Milan Chamber of Commerce on 31 December 2002. This data set is therefore different from the previous two in that it counts the people, not the firms, and in that it also considers companies who may have ceased to operate but are still registered at the Chamber of Commerce. Comparability is thus limited to the percentages referring to the importance of the different business sectors (Cologna, 2003). *n.a.* data not available.

Italian firms which work primarily for Chinese-run businesses. Most restaurant owners rarely bother to look for alternatives. They will employ a team of interior decorators because it will have been recommended to them by a relative, it will trust an Italian business consultant simply because he/she is known for his/her wide Chinese clientele, etc., regardless of the actual quality of the services provided. In most restaurants' kitchens the cooks are often relatives or friends who improvise, never having had any real work experience in this field in China. The same applies to waiters and maitre d's: they are recruited among family members and hardly ever undergo any kind of training other than 'learning by doing'. As most Chinese restaurant owners rarely set foot in an Italian restaurant, they can only guess what their Italian customers may consider good-quality service. Interior decoration is provided by Chinese firms that import furniture and decorations directly from China, without giving any thought to what their customers may actually like best. To many immigrant families, the restaurant is just a means to grant them survival in a foreign market they do not really understand and in which they blindly feel their way onwards. Considering the very low level of social interaction with Italians, one cannot but admire the courage and determination that so many of these small family businesses are run with. It is usually sons and daughters educated in Italy that make up for their parents' shortcomings, and who may eventually impart a new twist to restaurants otherwise doomed to shut down at short notice. Only a small minority of restaurants actually muster an efficient, knowledgeable management and real chefs, and those are the ones that have become exotic dining mainstays in Milan.

Similar problems also plague many other kinds of Chinese-run business, and most notably subcontracting garment and textile workshops (manufacturers of leather goods are on their way out, 'killed' by cheap imports). What ensures success to such operations is the manager's ability in contacting Italian contractors and in having them ensure the family workshop a steady flow of orders. Most such workshops/sweatshops can barely cover their costs, because they are unable to widen their circle of contractors. They have no clue how to reach the best firms, how to let them know they are there or the quality of work they can ensure. Knowledge of the Italian textile industry is surprisingly limited, even among businessmen who have been operating their businesses for more than a decade. Many Chinese businessmen do not know how to avail themselves of the services provided by the Milan Chamber of Commerce, nor do they understand much of the way Italian banks can support them. Hardly any sweatshop owner interviewed knew how to use a credit card, or how to ask for a loan. Most declared not to understand anything of accounting and fiscal management, which they generally entrust to the hands of Italian business consultants who may blatantly take advantage of the situation. Cut-throat competition among Chinese-run garment factories hinders any passing on of even the most basic business information. This high degree of social and cultural exclusion exposes most Chinese entrepreneurs to scams of every sort. Apartments and workshops are rented to them at prices which are sometimes three or four times the actual market value, contractors often refuse to pay for completed orders, or pay after one or two years. Against such events, Chinese entrepreneurs hardly ever know how to defend themselves. Citizen or workers' rights are basically unknown to most of them, and this partly explains why many entrepreneurs and workers are reluctant to denounce fraud, unless they are facing a situation of the utmost gravity. It is all the more awe-inspiring to see how many Chinese businesses manage to survive and even to succeed in such hostile business conditions.

All the above-mentioned problems play a major role in the experience of most Chinese entrepreneurs and workers, but they hardly affect the elite of the Chinese immigrant population in Italy. Some of the young and upwardly mobile offspring of first and second-generation immigrants have indeed shown remarkable business skills, accessing entirely different economic sectors. The grocery store and restaurant chains, the first financial services, the travel and real estate agencies, the boldest import–export firms have all sprung out of this particular milieu. The commercial elite can count on extensive transnational business networks, often channelled through their own family ties. They are linked with food industry operators across the world, in France, Holland and South East Asia. They have taken over the whole 'ethnic food' sector, providing Chinese as well as other Asians, Africans and Latin Americans with staples imported from their own countries and distributed by commercial networks eminently operated by other overseas Chinese. It is this particular portion of immigrant entrepreneurs that has sponsored the transformation of Milan's traditional Chinese district (the via Canonica–via Paolo Sarpi area) into a centre of ethnic services densely packed with Chinese-run shops of every kind.

Conclusion

The development of Chinese entrepreneurship in Italy – more specifically in Milan – during the 1990s has gone through two distinct phases, whose most significant characteristics are briefly summarized below.

The first phase was that of 1990–96, when subcontracting of garment factories expanded, street peddling returned and ethnic services boomed.

1   *The ethnic services boom.* Supermarkets and grocery stores, video rental shops, barbers and coiffeurs (mostly irregular, due to the difficulty of obtaining a licence to open such shops in Italy), photography studios (mainly catering to Chinese weddings), bookshops, real estate agencies, travel agencies, financial firms, phone shops, bars, clubs and restaurants that mainly cater to Chinese clients.
2   *The return of Chinese street peddling,* almost sixty years after its demise. This form of subsistence enterprise has been a popular insertion strategy for newcomers since the early 1990s.

3   *The wholesale stores boom.* These operations are basically import firms born to supply street peddlers and open-air markets with cheap 'Made in China' products. They were initially meant to cater mainly to a Chinese clientele, but quickly evolved in a flourishing market open to street vendors of every nationality and denomination (including many Italians).

4   *The expansion of Chinese subcontracting factories and sweatshops in the garment and textile sector.* Though these businesses are facing a severe crisis in Milan, in Italy's industrial areas they have flourished and prospered, opening new locales to Chinese immigration (a case in point is the garment manufacturing district in San Giuseppe Vesuviano, near Naples).

The proliferation of garment sweatshops in the early 1990s rapidly saturated an already dwindling market. As Italian firms gradually shifted their production facilities in Eastern Europe and the Far East, the prices they could offer the Chinese subcontracting firms steadily plummeted, prompting many ethnic entrepreneurs to branch out into other business sectors, the most important one being wholesale and retail shops as well as import–export firms and new service companies. The increase in the Chinese population after the 1996 and 1998 regularizations put the socio-economic enclave labour market under renewed pressure, causing even more Chinese would-be entrepreneurs to accept and even actively seek subordinate work from Italian employers. The main new developments of Chinese entrepreneurships during this second phase (1997–2003) can be summarized as follows:

1   The growing crisis of garment manufacturing workshops and sweatshops.

2   The growing saturation of wholesale and retail stores that cater to street vendors and street market salesmen.

3   The ongoing ethnic services boom, especially in sectors which appear to have an ever growing potential for expansion such as the distribution of foodstuffs and the offer of information and communication services (publishers, typesetters, translators, legal and financing services, real estate brokers, Internet points and phone shops).

4   The growth of restaurants and bars catering to an ethnic or multi-ethnic clientele (for example, Chinese restaurants specializing in *chifa*, or Chinese-Andinian food) and the expansion of 'mimetic' enterprises, for example, Chinese-run businesses employing young Italian-speaking, Italian-educated Chinese that offer services (newspaper stands, laundries, butcheries, bars) to a mainstream clientele. These firms are operated pretty much in the same way an Italian owner might do, and though they mostly employ a Chinese work force, they do not have much in common with the typical Chinese ethnic business. In that they tend to merge the incorporation of economic action in social practices that are both Chinese (*guanxi* networks to provide financing, workers and sup-

pliers) as well as Italian (good command of the language and honed social skills to attract and maintain a mainstream clientele) they may be considered budding examples of *mixed embeddedness.*[12]

Of all the above-mentioned developments, the most promising for a more positive integration of Chinese immigrants in Italy's economy and society, and the one that seems well poised to have a very important impact on Italian urban culture as well, is undoubtedly the ethnic and mainstream services boom. These businesses actively promote a wider understanding of the local context, providing access to services which are deeply needed by Chinese immigrants in Italy and at the same time giving a new high-profile visibility to the Chinese presence in Italy. This particular field so far has been little explored by social research in Italy, but there is little doubt that a thorough analysis of Chinese immigrant entrepreneurship will have to focus heavily on these businesses in the near future. Also needed is an accurate exploration of the significance of Chinese subcontracting firms for the domestic textile sector. Although stereotypes and prejudice abound, most notably in the Italian media, much more needs to be known in detail about the actual inner workings that link Chinese sweatshops to Italian firms.

Notes

1 This chapter is mainly based on qualitative fieldwork by the author, who since 1996 has carried out a number of ethnographic research projects on Chinese immigration in Milan (Cologna, 1997a, b, 1998a, b, 1999, 2000, 2002a, b; Cologna and Breveglieri, 2000, 2003; Cologna and Mancini, 2000; Farina *et al.*, 1997). The bulk of the qualitative data I used have been drawn from about sixty in-depth interviews with Chinese workers and entrepreneurs in Milan, repeated contacts with about twenty key informants over six years and participant observation carried out while visiting Chinese families at their homes and in their sweatshops. To get an idea of the scale of Chinese entrepreneurship in the Milan area, I used quantitative data drawn from the Milan Chamber of Commerce database, while the Milan Province Communal Registry offices have been my main source of demographic data.

2 Campani *et al.* (1997); Carchedi and Ferri (1998); Ceccagno (1997, 1998, 1999, 2003); Ceccagno and Rastrelli (1999); Ceccagno and Omodeo (1995); Colombo *et al.* (1995); Colombi *et al.* (2002); Marsden (1994); Miranda (2002).

3 Since many Chinese immigrant women's children were born in China or other countries prior to their arrival in Italy, the births registered by the Communal Residence Registry office cannot be used to calculate a total fertility rate. Fertility rates obtained by dividing the total of Chinese children born in Italy by the number of women aged fifteen to forty-nine can serve only as a means of comparison with the reproductive behaviour of other immigrant groups. Among Asian immigrant residents in the Lombardy region, Chinese and Indian women show the highest fertility rates (38.9 per thousand and 39.4 per thousand respectively. (Patrizia Farina, personal communication, rates based on Lombardy Region Epidemiologic Observatory data, 2001.)

4 On migration from Dongbei to Europe (France), see Pina-Guerassimoff (2003) and Pina-Guerassimoff and Guerassimoff (2003).

5 Cologna (1997b).
6 On the inner workings of a Chinese socio-economic enclave, Zhou (1992).
7 On the development of Chinese immigration to Milan since the 1920s, Farina *et al.* (1997).
8 A good synthesis of Oded Stark's theory of relative deprivation and of its significance for migration theory is provided in Massey *et al.* (1998: 26–8).
9 On *guanxi* see M.H. Bond and K.K. Hwang, 'The social psychology of Chinese people', in Bond (1986); also Yang (1994). For the role of *guanxi* in the dynamics of Chinese emigration also refer to Myers (1997).
10 Seminal work on this concept can be found in Waldinger *et al.* (1990) and Zhou (1992). On the production of social capital that concurs in creating an ethnic opportunity structure, refer to Coleman (1990), Portes (1998) and Waldinger (1997).
11 On the ethnic enclave theory, see Portes (1995), Portes and Manning (1986) and Massey *et al.* (1998).
12 The theoretical framework of embeddedness and mixed embeddedness is too wide and varied to be fully acknowledged in this chapter. My analysis draws on the work of Granovetter (1985), Portes and Sensenbrenner (1993) and Kloostermann and Rath (2001).

References

Benton, G. and Pieke, F.N. (eds) (1998) *The Chinese in Europe*. London: Macmillan.

Bond, M.H. (ed.) (1986) *The Psychology of the Chinese People*. Hong Kong: Oxford University Press.

Boutellier, È. (1997) *Les Nouveaux Empereurs. L'épopée du capitalisme chinois*. Paris: Calmann-Lévy.

Campani, G., Carchedi, F. and Tassinari, A. (eds) (1997) *L'immigrazione silenziosa. Le comunità cinesi in Italia*. Turin: Edizioni della Fondazione Giovanni Agnelli.

Carchedi, F. and Ferri, M. (1998) 'The Chinese presence in Italy: dimensions and structural characteristics', in G. Benton, F.N. Pieke (eds) *The Chinese in Europe*. London: Macmillan, 261–80.

Ceccagno, A. (ed.) (1997) *Il caso delle comunità cinesi in Italia. Comunicazione interculturale ed istituzioni*. Roma: Armando.

Ceccagno, A. (1998) *Cinesi d'Italia*. Milan: Manifestolibri.

Ceccagno, A. (1999) 'Nei-Wai: interazioni con il tessuto socioeconomico e autoreferenzialità etnica nelle comunità cinesi in Italia', *Mondo Cinese*, 101: 75–93.

Ceccagno, A. (2003) *Migranti a Prato. Il distretto tessile multietnico*. Milan: Franco Angeli.

Ceccagno, A. and Omodeo, M. (1995) 'Essere cinese in Toscana', *Limes*, 1/1995: 213–20.

Ceccagno, A. and Rastrelli, R. (eds) (1999) *La presenza degli stranieri a Prato. Spunti di riflessione per l'Amministrazione. Attività del Centro 1998*. Prato: Centro di Ricerca e Servizi per l'Immigrazione del Comune di Prato.

Coleman, J.S. (1990) 'Social capital', in J.S. Coleman (ed.) *Foundations of Social Theory*. Cambridge MA and London: Belknap Press, 300–21.

Cologna, D. (1997a) 'Dal Zhejiang a Milano: profilo di una comunità in transizione', in A. Ceccagno (ed.) *Il caso delle comunità cinesi in Italia. Comunicazione interculturale e istituzioni*. Rome: Armando Editore, 23–35.

Cologna, D. (1997b) 'Un'economia etnica di successo', in P. Farina, D. Cologna, A.

Lanzani and L. Breveglieri (eds) *Cina a Milano. Famiglie, ambienti e lavori della popolazione cinese a Milano*. Milan: Abitare Segesta, 107–47.

Cologna, D. (1998a) 'Economic and social insertion strategies of Chinese immigrants in Italy', in E. Reyneri (ed.) *MIGRINF: Second Italian Report. Migrant Insertion in the Informal Economy, Deviant Behaviour and the Impact on Receiving Societies* Brussels: CE–DGXII–TSER.

Cologna, D. (1998b) 'The role of social and community norms in the insertion processes and social deviance of Chinese immigrants in Italy', in S. Palidda (ed.) 'Immigrant deviant behaviour in Italy and in particular in Milan', in E. Reyneri (ed.) *MIGRINF: Second Italian Report. Migrant Insertion in the Informal Economy, Deviant Behaviour and the Impact on Receiving Societies*. Brussels: CE–DGXII–TSER.

Cologna, D. (1999) *Dinamiche fondamentali dell'inserimento sociale, culturale ed economico degli immigrati cinesi in Provincia di Milano*. Milan: Synergia.

Cologna, D. (2000) 'I cinesi nella società milanese', in S. Palidda (ed.) *Socialità e inserimento degli immigrati nella società milanese*. Milan: Franco Angeli, 31–55.

Cologna, D. (2002a) *Bambini e famiglie cinesi a Milano. Materiali per la formazione degli insegnanti del materno infantile e della scuola dell'obbligo*. Milan: Franco Angeli.

Cologna, D. (2002b) *La Cina sotto casa. Convivenza e conflitti tra cinesi e italiani in due quartieri di Milano*. Milan: Franco Angeli.

Cologna, D. (2003) *Asia a Milano. Famiglie, ambienti e lavori delle popolazioni asiatiche di Milano*. Milan: Abitare Segesta.

Cologna, D. and Breveglieri, L. (2000) 'Immigrati imprenditori asiatici e africani a Milano'. Paper for the international conference 'Migrazioni. Scenari per il XXI secolo', Milan: 23–24 November.

Cologna, D. and Breveglieri, L. (2003) *I figli dell'immigrazione. Ricerca sull'integrazione*. Milan: Franco Angeli.

Cologna, D. and Mancini, L. (2000) 'Inserimento socioeconomico e percezione dei diritti di cittadinanza degli immigrati cinesi a Milano: una ricerca pilota', *Sociologia del Diritto*, 3: 53–94.

Colombi, M., Guercini, S. and Marsden, A. (2002) *L'imprenditoria cinese nel distretto industriale di Prato*. Florence: Leo S. Olschki.

Colombi, M., Marcetti, C., Omodeo, M. and Solimano, N. (1995) *Wenzhou–Firenze. Identità, imprese e modalità di insediamento dei cinesi in Toscana*, Florence: Angelo Pontecorboli.

Farina, P., Cologna, D., Lanzani, A. and Breveglieri, L. (1997) *Cina a Milano. Famiglie, ambienti e lavori della popolazione cinese a Milano*. Milan: Abitare Segesta.

Granovetter, M. (1985) 'Economic action and social structure: the problem of embeddedness', in *American Journal of Sociology*, 91 (3): 481–510.

Koo, A.Y.C. and Yeh, K.C. (1995) 'The impact of township, village, and private enterprises growth on state enterprises reform: three regional case studies', in US Congress Joint Economic Committee, *China's Economic Future. Challenges to US Policy*. Washington DC: US Congress (also in O. Schell and D. Shambaugh (eds) *The China Reader. The Reform Era*, New York: Vintage Books, 321–34).

Kloosterman, R. and Rath, J. (2001) 'Immigrant entrepreneurs in advanced economies: mixed embeddedness further explored', *Journal of Ethnic and Migration Studies*, 27 (2); 189–202.

Li, M. (1999) 'Xiangdui shiluo yu liansuo xiaoying: guanyu dandai Wenzhou diqu chuguo yiminchaode fenxi yu sikao' (Relative deprivation and chain migration:

study and analysis of Wenzhou outmigration), *Shehuixue Yanjiu* (Social Research), 5/1999: 83–93.

Marsden, A. (1994) *Cinesi e fiorentini a confronto*. Florence: Firenze Libri.

Massey, D.S., Arango, J., Hugo, G., Kouaouci, A., Pellegrino, A. and Taylor, J.E. (1998) *Worlds in Motion. Understanding International Migration at the End of the Millennium*. Oxford: Clarendon Press.

Miranda, A. (2002) 'Les Chinois dans la région de Naples. Ancrages et mouvement dans un monde local en mutation'. Paper presented at the seminar 'Economie de bazar dans les métropoles euroméditerranéennes', Laboratoire Méditerranéen de Sociologie–Maison Méditerranéenne des Sciences de l'Homme, Aix-en-Provence, 29–31 May.

Myers, W.H. III (1997) 'Of *Qinqing, Qinshu, Guanxi* and *Shetou*: the dynamic elements of Chinese irregular population movement', in P.J. Smith (ed.) *Human Smuggling. Chinese Migrant Trafficking and the Challenge to America's Immigration Tradition*. Washington DC: Center for Strategic and International Studies, 93–113.

Pina-Guerassimoff, C. (2003) 'Circulation de l'information migratoire et mobilité internationale des migrants chinois', *Migrations Société*, 15 (86): 9–22.

Pina-Guerassimoff, C. and Guerassimoff, É. (2003) 'La France, carrefour européen de la nouvelle migration chinoise', *Migrations Société*, 15 (89): 105–19.

Portes, A. (ed.) (1995) *The Economic Sociology of Immigration. Essays on Networks, Ethnicity and Entrepreneurship*. New York: Russell Sage Foundation.

Portes, A. (1998) 'Social capital: its origins and applications in modern sociology', *Annual Review of Sociology*, 24: 1–24.

Portes, A. and Manning, R.D. (1986) 'The immigrant enclave: theory and empirical examples', in O. Olzak and J. Nagel (eds) *Competitive Ethnic Relations*. Orlando FL: Academic Press.

Portes, A. and Sensenbrenner, J. (1993) 'Embeddedness and immigration: notes on the social determinants of economic action', *American Journal of Sociology*, 98 (6): 1320–50.

Thunø, M. (1996) 'Chinese emigration to Europe: combining European and Chinese sources', *Revue Européenne des Migrations Internationales*, 12: 275–96.

Waldinger, R. (1997) *Social Capital or Social Closure? Immigrant Networks in the Labour Market*. Lewis Center for Regional Policy Studies Working Paper 26 (www.sppsr.ucla.edu/lewis).

Waldinger, R., Aldrich, H. and Ward, R. (1990) *Ethnic Entrepreneurs. Immigrant Business in Industrial Societies*. Newbury Park CA: Sage.

Yang, M.M. (1994) *Gifts, Favors and Banquets. The Art of Social Relationships in China*. Ithaca NY: Cornell University Press.

Zhou, M. (1992) *Chinatown. The Socio-economic Potential of an Urban Enclave*. Philadelphia: Temple University Press.

15 Chinese entrepreneurship in Spain

The seeds of Chinatown

Joaquín Beltrán Antolín

Chinatown in Spain?

In the mid-1990s it seemed that Spain was a special case in the settlement pattern of Chinese immigrants owing to the size of their population and their economic activities (Beltrán, 1997). If we define Chinatown as a high concentration of Chinese population and business in a delimited geographical area (Light, 1972), then the Chinese community in Spain is now in the initial phase of concentration following a long process of dispersal settlement. This pattern of dispersal and concentration is in this case directly linked with the most important entrepreneurial activity of the Chinese, namely the catering trade.

In order to understand the process of the emergence of residential and business concentration of the Chinese in Spain during the 1990s it is necessary to analyse several interrelated factors: (1) the evolution of the Chinese population size, and their settlement patterns, directly linked with (2) their economic activities and changes of entrepreneurial strategies, that, at the same time, are in relation to (3) the Spanish economic structure and (4) immigration policy, and (5) the transnational strategies of the Chinese families resident in Spain, predominantly in the two cities of Madrid and Barcelona.

Before continuing with the argument of the relationship between economic activity and residential pattern, it might be helpful to make one semantic excursus. Chinatown is translated into Spanish by *Barrio chino*. That term has one special and specific meaning without any relation to Chinese residence within the spatial location so denominated. *Barrio chino* is the place in the cities where low-status prostitution and the retail drug traffic are concentrated, including nude spectacles, male and female travesty and carnal trades, next to small thieving activities. Usually it occupies some of the oldest parts of the city, in a degraded urban environment with very low living standards and hardly any public infrastructure. The Spanish are familiar with the meaning of *barrio chino* and they also know that no Chinese live there. At present it is usual to find the word 'Chinatown' (*barrio chino*) in many literary and journalistic works – so there is one word for two different realities.

However, is there such a thing as a Chinatown with Chinese immigrants in Spain? The first reference to a concentration of Chinese people is dated in the 1940s and referred to Hortaleza Street in Madrid, but it was without continuity (Beltrán, 1998). In the mid-1980s a rumour was spread in Madrid about a Chinese plan to create a Chinatown in the Batan district. It was said that Chinese were buying property to establish their residences and businesses, but apart from the rumour nothing was done. During the first half of the 1990s the presence of Chinese people in specific neighbourhoods of the two biggest Spanish cities, Madrid and Barcelona, became visible. In Madrid they were living in the Tetuan and Lavapies districts and in Barcelona in Eixample district. However, the total number of Chinese residents never reached more than one-third of the total population settled in any one of these districts. In the second half of the 1990s this trend continued. Moreover, some Chinese residential and business concentrations in metropolitan cities like Fuenlabrada in Madrid, and Santa Coloma de Gramenet and Badalona in Barcelona, have appeared (Beltrán and Sáiz, 2003a).

For a better understanding of the emergence of small Chinese residential concentrations in Spain we need to analyse the evolution of spatial patterns of settlement following Chinese economic activities. First, Chinese were almost exclusively in the restaurant business and dispersed all over Spain; second, the catering trade became saturated; third, Chinese entrepreneurs set up garment workshops as well as other commercial ventures (import/export companies, wholesale stores, gifts shops, etc.); fourth, the development of new Chinese economic sectors forged a new geographical concentration.

Chinese immigration and settlement

The Chinese have lived in Spain for more than a century, but their presence was not very conspicuous until the 1990s, particularly in the second half of the decade. In 2001 they constituted the fourth largest non-EU immigrant community in Spain after the Moroccan, Ecuadorian and Colombian populations. The first concentrations of Chinese were found in jail: they came from Cuba, where they had participated in the war of liberation during the second half of the nineteenth century. After capture, they were detained, judged and sentenced to prison in Spain. After their release, they settled in several southern cities and some of them set up the first Chinese businesses: lodging houses in Melilla, selling potable water in Huelva and pastry shops in Madrid. Some others served as cooks for the military. In the 1920s and 1930s Qingtian-Wenzhou peddlers were all around Spain, but most resided in the three cities with Chinese embassies and consular services, Madrid, Barcelona and Valencia. Furthermore, there were also several circuses from Shandong province whose European base was located in Madrid. At the time they were in touch with their kin living in other countries of Europe and they were highly mobile (Beltrán, 1996, 1997, 1998).

After the Spanish Civil War (1936–39) the Chinese almost disappeared from Spain, save for some who came as refugees fleeing from the Second World War in Europe. The post-Civil War period in Spain, with a political dictatorship and autarchic economic policies, was very hard. It meant food rationing, famine and economic restructuring. Spain was one of the poorest countries in Europe, and so it was not an attractive destination for Chinese migrants, who preferred the richer countries of northern Europe. The anti-communist Franco and Chiang Kai-shek established diplomatic relations with the help of the Catholic Church and subsequently the arrival of Tai-wanese students began. Some of these students eventually settled in Spain and took up professional activities. The first Chinese restaurants were opened in the 1950s, but it was not until the late 1960s and 1970s that the development of the tourist sector in the Spanish Mediterranean coast and islands took off and Chinese capital and workers came in from other Euro-pean countries. In 1961 there were 167 Chinese residents living in fifteen of the fifty-two Spanish provinces, with 78 per cent of them settled in only three provinces: Madrid (51 per cent), Barcelona (15 per cent) and Valencia (12 per cent). Ten years later (1971) Madrid and Barcelona had 57 per cent of the total Chinese residents, and the next two provinces were Las Palmas and Malaga that together have 19 per cent of the total. The Asian popu-lation in Spain in 1975 was 5.7 per cent of total foreign residents, and in 2001 Asian people represented 8.2 per cent. The Chinese population increased during the same period from 4.8 per cent of total Asian residents to 39.5 per cent. In 1986 they already surpassed 10 per cent (2,455 Chinese of 24,007 Asian residents) and in 1991 the 20 per cent (6,428 Chinese, 31,976 Asian). In 2003 there were 121,455 Asians in Spain of whom 56,086 or 46 per cent are Chinese.

The pattern of settlement turned the Chinese towards the Mediterranean coast and the island tourist destinations, where their catering business had an international clientele. Between 1971 and 1990 they set up restaurants and residence especially near the coast, and it was not till 1992 when they spread over all the Spanish provinces. During the 1990s the settlement pattern showed two trends: first, the Chinese settled in all the cities and townships of over 10,000 people and second, they concentrated again in Madrid and Barcelona, provinces that absorbed most of the new Chinese immigrants. The steady Chinese population growth in Spain has not been uniform. During 1995–96 the Chinese increased 18.1 per cent, but twenty-one provinces lost some Chinese residents. The same happened between 1998 and 1999, during which the Chinese showed 19.3 per cent national average growth, with fifteen provinces losing Chinese residents, and 2000–01 (25.9 per cent inter-annual growth, and seventeen provinces lost residents). This domestic movement and geographical redistribution of Chinese population has given way to the seeds of Chinatown, or to a relat-ively high Chinese residential and business concentration in Madrid and Barcelona.

The actual trend of the Chinese immigrants' spatial concentration in

Spain is a phenomenon related in part to developments within China: there 'peasant enclaves' developed in the cities due to different migration processes. One example is the Subei people in Shanghai analysed by Honig (1992). They were the objects of prejudice, who worked in special economic niches, and they were dependent on their own resources and social networks. More recent developments of this kind are the peasant enclaves in Beijing as the paradigmatic case of the Wenzhou village (Tomba, 1999; Li Zhang, 2001; Mallée, 2000; Ma and Xiang, 1998; Xiang, 1999). Chinese policies to control domestic population movements are very similar to the European international migration policies: to get a European visa for a Chinese sometimes is as difficult as to get an urban residence card for a rural resident in China. The result is that there are a lot of undocumented migrants in the Chinese big cities, and many undocumented Chinese migrants in European countries.

The rural migrants in the urban environment are discriminated against by the government and by the urban population. They tend to live together, and depend on their kin, friends and village networks to find jobs and residence. They used to engage in special economic niches such as construction, domestic service and leather and garment workshops. The Wenzhou–Qingtian people in Beijing, the same origin area of the larger part of the Chinese immigrants in Spain too, control several lines of businesses in the commercial and manufacturing sectors: they operate garment and shoe workshops and shops where they market their own produce. And, since they do not have a permit to live in the city, they have set up all kinds of services for themselves that include private education and medical care. Wenzhou people in Spain and in China show structural similarities such as their social position (foreigners or immigrants/outsiders without permits) and their entrepreneurial and economic strategies (family-based businesses, control of the production, distribution and selling of their manufactures). The ethnic enclave offers them ways to advance their social and economic position, and their reference world for prestige is almost limited exclusively to their home towns. They could be successful in Beijing or in Spain, but they display their wealth especially in their natal villages (see also Cologna, Chapter 12 in this volume).

The relationship between domestic and international migration from the Wenzhou area is strong. China has a continental size and to pass from one province to other is similar to going from one European country to another (Beltrán, 1996, 1997). They diversify their businesses, setting up new shops and workshops in different places, and in different countries, always looking for new profitable markets and spreading risks. They change, if necessary, to other lines of business. Wenzhou people in China invest a share of their profits to sponsor the international migration of family members. In Europe they work, for example, as sales agents of their workshops in China through import–export companies, storage activities and wholesale stores.

The small spatial concentrations of Chinese residents developed in Spain at the turn of the century. This relatively late development is the result of a complex process where the idiosyncrasies of the Spanish Chinese entrepreneurial strategies are of paramount importance and also have relationships with several developments within the Chinese world view on economic activity. For example, the family-based business is the key to the extraordinary development of Chinese economies like Hong Kong, Taiwan and, more recently, the rural industrialization boom in the People's Republic (Beltrán, 2000a; Greenhalgh, 1994; Niehoff, 1987). Most of the rural industries are family enterprises that have contributed in large part to the high growth rate of Chinese GNP during the last twenty years. Family-based businesses are also the origin of the most prosperous Chinese business conglomerates operating in South East Asia (Ong and Nonini, 1997; Redding, 1990). What is the economic ideology of this kind of development? In fact, it is something more than a migratory context or a Confucian capitalism paradigm (Redding, 1990). The Chinese petty-capitalism mode of production analysed by Gates (1996) could offer some clues. The Chinese have always adhered to a way of life emphasizing autonomy and self-sufficiency. The family has been the primordial focus of any social or economic endeavour, and the success and prosperity of their past, present and future members have been the horizon of all the family outlooks and initiatives. Self-reliance is the core value, sometimes even promoted by utopian Maoism to speed up the growth of the economy.

Some Chinese migrants become very wealthy and act as role models for others. They use networks and personal connections (*guanxi*) in a very pragmatic way to advance in their self-reliance and success, but always family-oriented (Greenhalgh, 1994; Chan and Chiang, 1994; Song, 1997). In unreceptive environments, usually in migratory contexts, they organize themselves for defensive matters and for the promotion of their business activities. The segregation and discriminative immigration policies they confront bring about the strengthening of their self-employment and ethnic networks for mutual profit, and, finally, to their residential concentration (Light, 1972).

The recent development of Chinese residential concentration in Spain goes along with two new lines of business: wholesale stores and garment workshops. These economic activities are related with previous Chinese developments and capital coming from other European countries and China. First, leather/plastic handbag workshops based in France and Italy sell their commodities to the Spanish Chinese wholesale stores, some of which are branches of the French and Italian ones. Second, the garment workshops developed more than fifteen years before in Italy and France (Carchedi and Ferri, 1998; EFCO, 1999; Ma Mung, 2000; Yu-Sion, 1998; Tomba, 1999) and they expanded their businesses to Spain with capital, information and know-how. Third, there is capital coming from China to set up wholesale stores as branches of Chinese domestic business.

In the case of the wholesale stores the transnational family-economic network connection is obvious, coming direct from France, Italy or China. The manufacture of the commodities for sale could be made in Spain, in other European countries or in China, or in all these places. These wholesale stores are also directly connected with the import–export business, and commercial storage activities. On the other hand, most of the garment workshops are small-scale and family operated. Often, family members manufacture the products with sewing machines in their home dwelling. They need less initial capital to begin this kind of business in comparison with the catering trade. Many workers go into this sector after unsuccessful ventures in others, lured by the new prospects of this industry, and by the anticipation of becoming self-reliant rapidly. Anyway, once they accumulate enough capital, they try to set up another kind of family-based business.

The Chinese position in the Spanish labour market

Between 1961 and 2001 the most important economic activity of the Chinese in Spain was the catering trade, for example Chinese restaurants. However, in the second half of the 1990s new economic niches developed that favoured the concentration of the population, unlike the dispersion strategy appropriate for the restaurants. The expansion of the Chinese restaurants till the 1990s was directly linked with the tourist industry, and then to the domestic increase of living standards and changes in the Spanish population's food habits. The opening of restaurants in tourist destinations began in the 1960s with the expansion of Chinese businesses based in other European countries that were looking for new places to invest their capital (Beltrán, 1997; Watson, 1975). Spain's joining the European Economic Community in 1986 was an important factor to attract Chinese migrants to a country with hitherto only a few Chinese restaurants. New opportunities were required as in several European countries their catering trade was reaching the saturation point of competence (Pieke, 1991; Pieke and Benton, 1998). As before, the arrival of new Chinese immigrants was oriented almost exclusively to restaurants, but during the second half of the 1990s this sub-sector reached its saturation point in Spain as well. Thus, it became necessary to explore new economic niches such as garment workshops, wholesale and retail stores, street vending and a large variety of service businesses catering to Chinese customers.

With regard to the saturation of the catering trade I will give two examples from the Autonomous Community of Catalonia. The first phase of the process was settling in smaller cities: according to the municipal census of 1991 there were Chinese residents in thirty-four Catalonian townships (26 per cent of them with less than 20,000 population). By 1996 they were residents in 108 townships (51 per cent of them with less than 20,000 population). The proportion of Chinese population residing in Barcelona city decreased from 61 per cent of the total Chinese population in Catalonia in 1991 to the 37 per cent in 1996. As in the rest of Spain, the geographical dispersion of the Chinese in Catalonia was a fact, but also very soon reached the limit. The second phase was the crisis of saturation due to fast growth: in Barcelona 500 restaurants were set up during three years (1990–93), 200 of which had to close within the next two years (Beltrán Antolin and Sáiz Lopez, 2001).

For some of the new trades, particularly those that give way to the concentration of Chinese people, it is of no consequence if businesses are established near one another. On the contrary, proximity is a factor promoting their development and profits. The concentration in a limited space, e.g. in one or several adjacent streets, or within one village, conforms to a traditional Chinese pattern for the development of specific types of commercial and industrial activities (Francis, 1996; Ma and Xiang, 1998; Ma Mung, 1992, 2000; Mallée, 2000; Honig, 1992). This new phenomenon manifests itself in Barcelona and Madrid and in some cities within their metropolitan areas.

The search for new economic activities by the Chinese immigrants began around 1990. At that moment, and during most of the decade, the new businesses also followed a pattern of spatial dispersion. The first garment workshops were located in many different places, trying not to attract attention. Almost all of them were unlicensed, informal economic activities linked with restaurant owners who invested their capital in new, and sometimes more profitable, ventures. The wholesale stores have a more recent setup and they appeared directly linked to their concentration. The retail shops specialized in gift and very cheap commodities[1] have followed the known dispersal pattern associated with restaurants, because in order to be successful they needed enough exclusive customers. These kinds of shops often have a short life span and close after a very short time. In this line of business, that requires a smaller amount of initial capital in comparison with restaurants, success is measured within a much shorter period of time (Beltrán, 2000a, b; Beltrán and Sáiz, 2003b).

The Chinese are one of the immigrant communities in Spain with the highest percentage of self-employed labour, with an average of 41.7 per cent of the total Chinese workers during 1988–98 (Table 15.1). This fact must take into account two related phenomena. First, many Chinese workers are not registered in the foreign worker statistics, and most of the unregistered are employees. Second, some of the self-employed are in fact employees, although they pay their own social security fees as self-employed workers. In any case, the Chinese have a strong compulsion to be entrepreneurs and turn out to be successful in their businesses (Beltrán, 2000b; Chan and Chiang, 1994; Harell, 1985; Niehoff, 1987; Oxfeld, 1991; Redding, 1990; Teixeira, 1998; Tomba, 1999). In Spain, this was evident from the very beginning: When they were peddlers, they were already small entrepreneurs, of which the most successful set up wholesale stores in the 1920s and 1930s, or other businesses with fixed address (Beltrán, 1996, 2003). At the end of the twentieth century, they built on their past entrepreneurial experience and

Table 15.1 Chinese workers by sex, employment status and sector of activity, 1988–98

Year	Workers	Sex				Employment status				Sector		
		Male	Female	%		Employed	%	Self-employed	%	Service	%
1988	1,260	1,060	200	18.9		660		600	47.6	1,171	92.9
1989	1,415	1,144	271	19.1		556		859	60.7	1,392	98.4
1990	1,651	1,324	327	19.8		616		1,035	62.7	1,621	98.2
1991	4,573	3,406	1,167	25.5		3,051		1,522	33.3	4,491	97.3
1992	5,712	4,263	1,449	25.3		3,734		1,978	34.8	5,598	98.0
1993	4,789	3,567	1,222	25.5		2,831		1,958	40.9	4,713	98.4
1994	5,672	4,081	1,591	28.0		3,253		2,419	42.6	5,556	97.9
1995	6,203	4,454	1,749	28.2		3,632		2,571	41.4	6,083	98.0
1996	8,205	5,682	2,523	30.7		5,068		3,137	38.7	7,967	97.1
1997	9,252	5,877	3,375	36.5		6,317		2,935	31.7	8,985	97.1
1998	11,933	8,525	3,408	28.5		7,747		4,186	35.1	11,480	96.2
1999	12,394	7,849	4,545	36.7		9,132		3,262	26.3	11,596	93.6

Source: Adapted from *Estadística de Permisos de Trabajo a Extranjeros*, Ministerio de Trabajo y Asuntos Sociales.

returned to economic activities already exploited several generations before. These consist of wholesale stores (garments/clothing, leather/plastic hand-bags, cheap gift commodities) together with street vending, or selling in retail shops, the articles bought in those stores or imported by themselves: old images with new people. Today, for most of the Chinese, street vending is a part-time, temporary or supplementary job for some members of the family, in particular women, youngsters and newly arrived immigrants. In this way, they earn some money (generally without having a selling licence) while waiting for better occupational opportunities, unlike the pre-war period when it was the only job for most of them.

The process of settlement and setting up a new business usually needs much time. First, many foreign workers enter Spain irregularly (Beltrán and Sáiz, 2003b). After several years they get work permits affirming their status as regular employees. Then, after some years, they can try to set up an indi-vidual (family) business, thereby changing to self-employment. The self-employment rate of Chinese workers is relatively high, but it decreases with the passage of time and the increasing numbers in the labour force. On 10 April 2002 there were 22,036 Chinese workers registered with the Spanish social security system, 6,141 of whom were self-employed (28 per cent of the total).

Chinese residents in Spain have a relatively balanced sex ratio (52.6 per cent male, 45.4 per cent female, 1.9 per cent gender not included), but their labour-force sex ratio is predominantly male. However, the proportion of Chinese women in the labour force has increased steadily from 18.9 per cent in 1988 to 36.7 per cent in 1999. Examining the distribution of men and women by employment status shows one significant characteristic of Chinese women in Spain (Table 15.2): the proportion of women who are self-employed workers is almost similar to that of men (Beltrán, 2000a; Colec-tivo IOE, 2001, 2002).

In fact, Chinese women in Spain have the highest self-employment rate of all the foreign women workers. How could we explain this trend? Sáiz (in press) indicates some clues that go along with the family-based migratory project of Chinese migrants in Spain. Chinese women are entrepreneurs, owners, have rights to proprietorship, and improve their position within the family when they become self-employed. Often, women are the first migrants of a household, and later they help their husbands and children to migrate by way of the family reunification mechanism. Women have the support of other family members (at a bilateral scope). Women are instru-mental in the family search for autonomy. Once she has a chance to migrate, she will do it with the approval of her family and will do her best for the future well-being and prosperity of the family as a whole. One case study introduced by Sáiz (in press) may illustrate this trend:

Mei Hua came to Spain in 1991, but she got her residence and work permit only in 1998. In 1999 she set up a shop in Santa Coloma

Table 15.2 Chinese and total foreign workers registered with the Spanish social security system, by sex and self-employment regime, 10 April 2002

Foreign workers	China	Total
Total workers		
Men		
No.	14,451	467,224
%	65.5	65.7
Women		
No.	7,615	244,192
%	34.5	34.3
Total self-employed workers		
Men		
No.	4,185	467,224
%	29.0	12.8
Women		
No.	1,956	24,032
%	25.7	9.8
Total	22,066	711,432

Source: Adapted from Tesorería de la Seguridad Social.

(Barcelona) where she sells different kinds of clothes, imported from Hong Kong, especially to Chinese's garment shops customers. In 2001 she started the procedures to reunite with her husband and two children. She used the service of one snakehead to travel, and the cost was paid by her husband, half before leaving China, and half at the moment of arrival in Spain. When she arrived in Spain, she was working in Madrid during several years, most of the time sewing clothes in garment shops. When her husband will arrive, she plans to work as a seamstress again in order to earn more money. Her husband works as a policeman in China and has a family of twenty-five members. Most of them live in the house that has been constructed in part with the money that she sent during the last few years. Besides family reunification, she has applied for a 'work demand' in order to help her brother to migrate. Mei Hua is instrumental in opening the way towards the migration of family and kin.

The Chinese are further one of the foreign communities with a high concentration of workers in the service sector (Table 15.3). Official statistics regarding the composition of the Chinese labour force by occupation indicate that the three largest categories in 1999 are 'catering worker' (50 per cent), 'business management' (23 per cent) and 'domestic employees and cleaning workers' (about 13 per cent) (Table 15.4). To understand the latter

Table 15.3 Chinese immigration contingent (quota), by sector of activity

Year	Total	Agriculture	Construction	Domestic service	Other services	Other activities
1997	1,579	38	13	982	546	–
1998	3,023	82	23	1,783	1,073	62
1999	3,788	120	87	2,086	1,310	185

Source: Adapted from Ministerio de Trabajo y Asuntos Sociales.

figure it is necessary to take into account the Spanish immigrant quota policy that has fixed and limited lines of economic activity. Many Chinese immigrants possess a 'domestic employee' permit, but *de facto* they are not working as such. It is only a strategy to get a work permit within a rigid immigrant policy. There are also many garment workers not registered as such in the statistics. Moreover, many people work without a permit at all.

The growing size of the Chinese immigrant population offers new opportunities to develop all kinds of ethnic services catering to Chinese customers. These ethnic businesses tend to concentrate in the neighbourhoods where most Chinese are living. The types of business Chinese entrepreneurs have set up vary considerably and include grocery stores, Chinese food supermarkets, hairdressing and beauty parlours, travel agencies, Chinese medicine clinics, *karaoke* bars, Chinese restaurants, video rental libraries, Chinese crop farms, state agents, construction teams, food processing, fast-photo shops, restaurant equipment, consumer electronics, sewing machines, accountancy and financial consultants.

Table 15.4 Chinese workers by occupation, 1995–99

Occupation	1995	1996	1997	1998	1999
Total	6,203	8,205	9,252	11,933	12,394
Business management	1,328	1,920	2,311	2,621	2,803
Technicians	132	147	168	186	214
Administrative employees	32	54	58	64	126
Catering workers	3,170	4,137	4,753	5,736	6,176
Sales workers	65	151	218	310	559
Qualified workers	18	17	20	59	140
Textile qualified workers	50	135	147	187	296
Domestic employees	1,313	1,501	1,419	2,486	1,628
Street vending	61	76	85	92	82
Non-qualified workers	13	40	44	149	289
Non-classifiable	22	27	29	43	81

Source: Adapted from Ministerio de Trabajo y Asuntos Sociales.

Spanish economic structure and immigration policy

Spain has a high unemployment rate and, just as other Mediterranean countries like Italy and Greece, one of the most developed informal economies in Europe. Workers pushed from the formal sector of the economy go to work in different kinds of informal economic activities. Sometimes the previous formal entrepreneurs became the new informal ones: they evade taxes and fail to offer the basic and regulated labour conditions to their employees. Entrepreneurs and workers get more direct profits, but without any kind of social benefits or welfare protection, that became a private matter. Strategies adapted to the flexible informal economy, such as subcontracting, putting out and working at home, are used (Ong and Nonini, 1997). A large proportion of the foreign labour force works in the informal sector. They have neither residence nor work permits, and the Spanish employers contract them on salaries far below the national average. Spanish politicians and journalists have labelled them 'illegal people' and 'persons without civil rights', making them more easily exploited, detained and expelled. The Chinese are a particular case because they, as entrepreneurs, almost exclusively contract a Chinese work force, formal or informally, something that apparently is more dangerous than if employer and employees were both Spanish (Colectivo IOE, 1999a, b; Smart and Smart, 1993; Song, 1997, Beltrán and Sáiz, 2003b).

The tourist industry is of great importance to Spain – and to Chinese immigrants. Spain is the one of the most important destinations in the world of international tourism. The service sector is highly developed. Chinese immigrants took advantage of this and became active exactly in this sector. Foreign capital investments in Spain are also welcomed by the State, and Chinese are known to invest in industrial and commercial activities in particular. Some of the new import–export business and wholesale stores have been funded with Chinese capital originating in China, but also from different European countries.

Spanish immigration policy is very restrictive and adheres to the Schengen accords. In reaction to this the Chinese have used a family chain-migration strategy to explore all the possibilities, resulting in the settlement of whole households. The demographic structure of the Chinese population shows a relative balance between women and men, including children of all ages, and older people. The percentage of Chinese people under sixteen years old is similar to the same age interval of the average Spanish population. However, the Chinese are also entering Spain through various non-official ways. They have been one of the immigrant collectives with the highest demand of regulation, both during the extraordinary regulation processes of immigrant workers and residents (in 1986, 1991–92, 1996, 2000, 2001), as in the annual contingents quotas established from 1993 on. These can be considered as mechanisms of cover regulation, giving work permits to people already living and working in Spain (Izquierdo, 1996; de Lucas *et al.*, 2001).

The rules and regulations are so inflexible that the Chinese sometimes have to give false information to the authorities in order to get an official work permit. For example, they say that they are domestic servants when in fact they are sales employees or work in manufacturing. Another important aspect is the specific financial strategies and economic culture that this immigrant group have. The informal networks and co-operatives that facilitate capital to set up new business are seen by the Spanish as criminal activities rather than alternative ways to accumulate capital outside the control of the formal financial institutions. The Chinese are always under suspicion, and police question them about the origin of their capital, out of ignorance of the specific Chinese arrangements linked with the mobilization, accumulation and movement of business capital (Beltrán and Sáiz, 2001, 2003b; Light, 1972).

The special characteristics of the Spanish economic structure and immigration policy have a lot in common with other southern European countries like Italy. Both have a high proportion of undocumented migrants because the official mechanisms to migrate are underdeveloped, and are very restrictive, while the preferred approach to regulate the migrant flows is a police-based one. What ensues is an increasingly socially excluded population working mainly in the informal sector of the economy. Periodically the government offers 'extraordinary' regulation procedures that eventually have become the 'ordinary' way to handle the immigration flows, for example, giving permits to people already living and working in Spain and Italy without any kind of recognition. The Spanish attempt to contract foreign workers in their country of origin has turned out to be a failure so far.

Madrid and Barcelona: Chinese entrepreneurship and residential concentration

Until the 1990s the total Chinese population in Spain was still modest, although its relative concentration in Madrid and Barcelona provinces has been a fact from the very beginning. From 1961 to 1982 these two provinces together had around half the total Chinese residents in Spain. In 1983 Barcelona lost its second provincial place but regained the position in 1992. Madrid has been the province with the largest Chinese population, but its proportion in the total population decreased from around 40 per cent during the period 1961–82 to 30 per cent during 1983–94 and 25 per cent between 1995 and 2000. During the years 2000 and 2001 Barcelona surpassed Madrid for the first time in terms of the number of Chinese residents (Table 15.5).

The growth of Chinese residents during 1961–90 in Madrid and Barcelona provinces was less than the Spanish total average increase, but during the 1990s this trend changed (with some exceptions) (Table 15.6). Together Madrid and Barcelona have had around 45 per cent to 49 per cent of Chinese residents in Spain from 1992 till 2000, but in 2001 it was an

Table 15.5 Chinese residents in Spain, Madrid and Barcelona provinces and inter-period variation ratio

Year	Spain (No.)	Inter-period variation ratio (%)	Madrid (No.)	Inter-period variation ratio (%)	Barcelona (No.)	Inter-period variation ratio (%)
1961	167	–	86	–	26	–
1971	439	162.87	176	104.65	77	196.15
1981	758	72.66	282	60.22	111	44.15
1991	6,482	58.48	–	–	–	–
1992	6,711	3.53	2,391		935	
1993	7,750	15.48	2,366	−1.04	1,271	35.93
1994	8,119	4.76	2,304	−2.62	1,393	9.59
1995	9,158	12.79	2,036	−11.37	1,981	42.21
1996	10,816	18.10	2,988	46.75	2,337	17.97
1997	15,754	45.65	4,067	36.11	3,033	29.78
1998	20,690	31.33	5,366	31.94	4,225	39.30
1999	24,693	19.34	6,553	22.11	5,619	33.00
2000	28,693	16.19	6,731	2.71	7,390	31.51
2001	36,143	25.96	10,354	53.82	10,527	42.44
2002	45,815	26.76	11,570	11.74	12,805	21.64
2003	56,086	22.42	w.d.	–	w.d.	–

Source: Adapted from Instituto Nacional de Estadística, Anuario de Migraciones, Anuario Estadístico de Extranjería.

Table 15.6 Absolute increase and inter-period variation ratio (%) of Chinese residents in Spain, Madrid and Barcelona provinces, 1990–99

Period	Spain		Madrid		Barcelona	
	Absolute increase	Inter-period increase ratio	Absolute increase	Inter-period increase ratio	Absolute increase	Inter-period increase ratio
1990–2001	32,053	783.69	9,342	923.12	10,266	3,933.33
1990–96	6,726	164.45	1,976	195.26	2,076	895.40
1996–2001	25,327	234.16	7,366	246.52	8,190	350.45

Source: Adapted from Instituto Nacional de Estadística, *Anuario de Migraciones*, Anuario Estadístico de Extranjería.

exceptional 57.7 per cent, a similar proportion to 1971. The most recent residents, as a group almost four times larger during 1996–2001 than during the previous period (1990–96), settled more in Madrid and Barcelona (9,342 in Madrid and 10,226 in Barcelona for 1990–2001) than in the other fifty provinces. In fact both provinces concentrated 61 per cent of the new residents in Spain during 1990–2001 (Beltrán, 2000b).

In the two biggest cities of Spain, the first, and still weak, economic and residential concentrations of Chinese developed after the catering trade and the accompanying geographical dispersion wore out. It would not be appropriate to consider them already as Chinatowns. The most noticeable aspect is the concentration of wholesale stores (garments and handbags) in several streets of Lavapies district in Madrid and Eixample district in Barcelona. In Lavapies there are around 150 Chinese businesses, and in Eixample sixty Chinese wholesale stores. In Lavapies district there are many Chinese, but they are also dispersed all over Madrid, with a special concentration in Tetuan district too. The difference between Lavapies and Tetuan is that in the former they settled later (during the 1990s) and set up wholesale stores, and other kinds of shops and businesses. This occurred in a relatively small space and short time span giving the impression of an especially high concentration of Chinese. A similar phenomenon has developed in Barcelona city: the concentration of Chinese wholesale stores in several nearby streets from 1997 on, leading to a great visibility in a place where not long ago Chinese were absent. The official statistics indicate that at the end of 2001 there were 10,527 Chinese in Barcelona province, and the Barcelona municipal census show that the number of Chinese increased from 1,309 in March 1999 to 5,272 in January 2003.[2]

The Barcelona municipal census from 1995 to 2003 shows that the highest concentration of Chinese has always been in the Eixample district, namely between 28 per cent and 34 per cent of the total Chinese living in Barcelona city (Table 15.7). However, the most recent settlement pattern does not point to a trend of increasing concentration. The six largest 'statis-

Table 15.7 Chinese in Barcelona by district

District	31 March 1995		31 March 1998		31 March 2000		1 January 2003	
	No.	%	No.	%	No.	%	No.	%
Total	705		950		1,929		5,272	
Ciutat Vella	39	5.5	52	5.5	181	9.4	588	11.2
Eixample	242	34.3	290	30.5	659	34.2	1,615	30.6
Sants-Montjuïc	76	18.8	114	12.0	213	11.0	604	11.5
Les Corts	52	7.4	53	5.6	87	4.5	173	3.3
Sarrià-Sant Gervasi	80	11.3	73	7.7	91	4.7	285	5.4
Gràcia	75	10.6	64	6.7	136	7.1	256	4.9
Horta-Guinardó	28	4.0	39	4.1	70	3.6	211	4.0
Nou Barris	14	2.0	51	5.4	108	5.6	342	6.5
Sant Andreu	39	5.5	76	8.0	128	6.6	358	6.8
Sant Martí	60	8.5	138	14.5	256	13.3	831	15.8
No data	–	–	–	–	–	–	9	0.2

Source: Adapted from Ajuntament de Barcelona.

tical zones' (administrative neighbourhoods) by size of Chinese residents in 1995 coalesced 53 per cent of the total residents, but in 2003 their sum was only 38 per cent of the total number of Chinese (Table 15.8). Year by year there are more Chinese in Barcelona but they become more dispersed, coupled with a geographical redistribution of their residence. The Chinese settlement in Eixample district is related to the fact that it is one of the most commercial areas of Barcelona, with more services. On the other hand, only 11.2 per cent of the Chinese are living in Ciutat Vella, the famous and traditional (in the sense explained at the beginning of the chapter) *barrio chino* of Barcelona, a district where almost half the total foreign residents in Barcelona live. Ciutat Vella district doubled its percentage of Chinese from 1995 to 2003 for several reasons. It is located at the centre of the city, some neighbourhoods have the cheapest and lowest-quality residential space, and, most important, it is adjacent to the actual Chinese commercial centre. The last factor has some relationship with a significant geographical redistribution: Sant Marti district has increased its percentage of total Chinese residents; at the same time it has decreased in Sants district. The new arrivals have fewer economic resources and settle in neighbourhoods where property prices are cheaper.

Barcelona and Madrid cities, in spite of the increasing number of Chinese population, and of some new business concentration (wholesale shops), still do not show a clear trend towards an actual residential concentration. This is because the Chinese still follow a dispersed pattern of settlement associated with other commercial activities. By contrast, in metropolitan cities they have started to concentrate in relation to the development of the production sub-sector of garment workshops.

The case of Santa Coloma is paradigmatic of this new trend. According to the municipal census in 1990 there were nine Chinese residents, increasing to fifty-one in 1996, 473 in 1999, 1,226 in 2001 and 1,888 in September 2002. About 37 per cent of them were living in VI District (named Fondo), located at the border with Badalona city, another place of Chinese residence and garment workshops (here there were 1,093 Chinese residents in 2001). Santa Coloma and Badalona are two Barcelona metropolitan conurbations that experienced a dramatic population growth in the 1960s and the 1970s due to domestic migration movements; now, however, their total population is decreasing. Santa Coloma, with the second highest density of the metropolitan cities of Barcelona (16,475 population per square kilometre) began to lose population in the year 1977, when it had a total of 143,232 residents. In 1991 the population was 133,650 and during the 1990s continued to decline to 116,974 in 2001. Simultaneously, foreigners are settling in increasing numbers in this city.

Santa Coloma property market prices are much lower than in Barcelona, and so are living standards. Santa Coloma and Badalona have passed through a textile industrial crisis. The traditional textile enterprises closed, giving way to informal economy activities. Chinese directly entered this economic

Table 15.8 Six largest statistical zones of Barcelona with Chinese (% of the total Chinese population)

31 March 1995	%	31 March 1998	%	31 March 2000	%	1 January 2003	%
Esquerra Eix.	18.0	Esquerra Eix.	13.4	Esquerra Eix.	13.9	Esquerra Eix.	12.2
Gràcia	9.6	Gràcia	6.5	Gràcia	6.9	Estació Nord	5.6
Sant Gervasi	7.9	Sagrada Fam.	5.9	Dreta Eix.	6.4	Sants	5.3
Les Corts	7.4	Sants	5.6	Sagrada Fam.	6.1	Sagrada Fam.	5.2
Sant Antoni	5.1	Les Corts	5.2	Estació Nord	5.5	Poble Nou	4.9
Sants	5.0	Sant Gervasi	4.9	Sants	5.4	Clot	4.9
Total Six	53.0		41.5		44.2		38.3

Source: Adapted from Ajuntament de Barcelona.

niche and competed with existing garment workshops run by Spanish and Moroccan entrepreneurs. The conflict has moved to the streets, many Chinese having been assaulted and robbed by gangs of Moroccan youngsters. Subsequently, in order to stop the assaults, the Chinese have organized themselves. A similar process of inter-ethnic conflict – between Chinese and Moroccans – has developed in the Lavapies district of Madrid (Beltrán and Sáiz, 2003a).

The Chinese in Santa Coloma comprised 1.6 per cent of the city total population in September 2002. In the VI District they form 4.6 per cent of the total population. If we include the Chinese not registered by the municipal census, maybe they represent around 9 per cent of the district population. This is the highest density of Chinese residence in any Spanish city or neighbourhood. At the end of 2002, and during the first half of 2003, there were signs that they were reaching a kind of saturation point. The population growth rate slowed a little, and in other cities farther away from Barcelona, but also with a textile industry tradition as Mataro, the number of Chinese and Chinese garment workshops increased dramatically. At the beginning of 2002 Mataró had 231 Chinese population according to the municipal census, and at the end of that year there were 521.

In the Fondo district of Santa Coloma there is a high, and increasing, concentration of Chinese business oriented to Chinese customers. Chinese workers need services that other Chinese can satisfy much better than non-Chinese. In Santa Coloma–Badalona there are around 300 Chinese garment shops, most of them family-based business. At the end of 2002 they had more than sixty services and commercial businesses, almost half for Chinese customers, and most of them located in Fondo district. They run restaurants, bars, photo studios, Internet cafés, telephone shops, beauty parlours, Chinese food supermarkets, Chinese herbal medicine shops, Chinese and Western medicine clinics, jewellers, video rental, and so on. They also have several Chinese community services, such as a Christian church, a Chinese weekend school, next to some associations, all of them established during the last few years.

The transnational links of the business owners and their families are indisputable. In a survey in Santa Coloma (Beltrán and Sáiz, 2003a, b), many Chinese running a business there had previously resided in the Netherlands, Portugal, France, Italy – indicating the existence of strong transnational networks. Some of them had lived in more than one European country before coming to Spain. Most of them still have kin abroad, while a few youngsters were born in France or Italy where their parents were working at the time.

There is also movement within Spain. Before arriving in Santa Coloma Chinese resided in several other Spanish cities, e.g. Alicante, Seville, Logroño, Gijon, Badajoz, Malaga, Valencia, Valladolid, Saragossa or Madrid. Some also came from other places in Catalonia, and from Barcelona. Several businesses have close linkages with similar businesses set up in other places in Spain (Madrid, Saragossa, Barcelona), or beyond (Paris, China). Some entrepreneurial families have interest in different economic sub-sectors, e.g. garment shops, catering services, commercial ventures, etc.

The main reason for Santa Coloma's attraction for Chinese is the relatively high demand for workers in the garment shops and the chances for the expansion and success inside the sector. Badalona has the second most important ready-made garment distribution centre in Europe (named Montigala), after Milan. Here, Spanish enterprises subcontract the sewing process to Chinese who offer a very competitive price for their work. The Chinese in this economic activity can relatively quickly attain the ideal of becoming independent. With a family enterprise the family owns the means of production (sewing machines) and have the work force (the family members). Garment shops are one of the enterprises that need less initial investment capital, which makes them attractive to Chinese entrepreneurs with little financial resources. This type of work is almost always considered temporary, i.e. as a way to accumulate as soon as possible sufficient capital to start a more prestigious and less hard-working venture. Thus, many new Chinese entrepreneurs in Spain were former garment shop workers.

Chinese garment shops are located all over Spain, including several districts of Madrid city (Usera, Vallecas). However, only Santa Coloma–Badalona has seen a real concentration of Chinese residence and business.

Conclusion

At present Chinatown, defined here as a high concentration of Chinese population and business in a delimited spatial context, is not a reality in Spain, although there are specific places that show the first signs of becoming so.

The Chinese registered population in Spain was 56,086 persons at the end 2003. During the 1990s Chinese settled specially in the most populous provinces and cities of Spain, Madrid and Barcelona, after decades of more dispersed settlement along the Mediterranean coast and islands. The high rate of Chinese workers in services (i.e. Chinese food restaurants) explain their settlement pattern in the 1970s and 1980s when Chinese businesses were oriented towards international tourism destinations.

After its expansion over the whole of Spain the Chinese restaurant sector became saturated and the Chinese diversified their investment pattern and developed activities in other economic sectors. Unlike the restaurant trade, these new economic niches did not require dispersed settlement to secure success; on the contrary, concentration could prove to be profitable. As a result, two significant concentrations of Chinese workers and businesses have emerged: around wholesale stores (garments, handbags) and around garment workshops.

Without doubt the most important concentration is the latter, especially near the largest ready-made garment distribution centre in metropolitan Barcelona. The subsequent increasing work force demand has been satisfied by a growing number of Chinese coming from different provinces and cities in Spain, and, farther afield, from Europe and China. The family-based garment shop is regarded as a fast transitional way to become autonomous, and a means to accumulate more capital to set up more prestigious and profitable enterprises.

To conclude, in order for a Chinatown to develop at least two conditions are necessary: first, a significant number of Chinese concentrated in a delimited area where they form the majority; second, access to employment in one economic sector that permits their concentration. In Spain the former is still not a real fact, and the second is a relatively later development associated with the garment shops. Anyway, considering the demand for Chinese labour in the Santa Coloma–Badalona textile industry it seems that it is nearing its saturation point, reinforced by the fact that the Spanish sector is in crisis. In the near future, perhaps that situation will put limits to the growing concentration of Chinese population and businesses. Thus, the emergence of Chinatowns in Spain at the turn of the century will maybe be discontinued before any real consolidation.

Another factor bearing on the historical development of Chinatowns is discriminatory and segregationist policies addressed against the Chinese population. Spain[3] has hitherto not developed special policies for the Chinese immigrants in the vein of the Exclusion Acts of the United States and Canada, or the discriminatory policies of colonial and postcolonial states of South East Asia. In these countries the most important Chinatowns in the world are located. When Chinese migrants have to face exclusionary policies and a xenophobic and racist environment, they tend to concentrate their residence for defensive, social welfare, promotion of business, and similar concerns. Therefore, if the Spanish government does not evolve discriminatory policies against international migrants, it will not give grounds for defensive behaviour. But if the actual governmental approach towards immigrants follows an exclusionary path, then it is possible that in a direct or indirect way the development of Chinatowns will be promoted.

Notes

1 *Todo a cien*, namely, all articles cost only 100 pesetas (€0.6), is the commonly used Spanish term for such businesses.
2 Usually the municipal census figure is lower than the figure based on resident permits, at least for the Chinese case.
3 Except when Spain was involved in the coolie trade in former Spanish colony of the Philippines.

References

Beltrán Antolín, J. (1996) 'Parentesco y organización social en los procesos de emigracion internacional chinos. Del sur de Zhejiang a Europa y España'. Doctoral thesis, Universidad Complutense de Madrid.

Beltrán Antolín, J. (1997) 'Immigrés chinois en Espagne ou citoyens européens?' *Revue Européenne des Migrations Internationales*, 13 (2): 63–79.

Beltrán Antolín, J. (1998) 'The Chinese in Spain', in G. Benton and F. Pieke (eds) *The Chinese in Europe*. London: Macmillan, 211–37.

Beltrán Antolín, J. (2000a) 'La empresa familiar: trabajo, redes sociales y familia en el colectivo chino', *Ofrim/Suplementos*, 6: 129–53.

Beltrán Antolín, J. (2000b) 'Expansión geográfica y diversificación económica. Pautas y estrategias del asentamiento chino en España'. Paper presented at the second Congreso sobre la Inmigración en España, Madrid, 5–7 October (www.imsersomigracion.upco.es/Documentos/Otros/congreso/datos/CDRom/FLU JOS/Otros%20documentos/JoaquinBeltranAntolin.PDF).

Beltrán Antolín, J. (2003) *Los ocho inmortales cruzan el mar. Chinos en Extremo Occidente*. Barcelona: Bellaterra.

Beltrán Antolín, J. and Sáiz López, A. (2001) *Els xinesos a Catalunya. Família, educació i integració*. Barcelona: Fundació Jaume Bofill.

Beltrán Antolín, J. and Sáiz López, A. (2002) 'Comunidades asiáticas en España', *Documentos CIDOB. Relaciones España–Asia* 3 (www.cidob.org/Castellano/Publicaciones/documentos%20cidob/pdf/beltran.pdf).

Beltrán Antolín, J. and Sáiz López, A. (2003a) *La comunidad china en Santa Coloma de Gramenet*. Santa Coloma de Gramenet, Barcelona: Diputación de Barcelona.

Beltrán Antolín, J. and Sáiz López, A. (2003b) 'Trabajadores y empresarios chinos en Cataluña'. Paper presented at the fourth Congres Catala de Sociologia, Reus, Tarragona, 5–6 April (www.iecat.net/acs/IV%20Congres%20sociologia/Informacio%20grups%20de%20treball/Grup%20de%20treball%2013/beltransaiz.doc).

Carchedi, F. and Ferri, M. (1988) 'The Chinese presence in Italy: dimensions and structural characteristics', in G. Benton and F.N. Pieke (eds) *The Chinese in Europe*. London: Macmillan.

Carchedi, F. and Marica, F. (1998) 'The Chinese presence in Italy: dimensions and structural characteristics', in G. Benton and F. Pieke (eds) *The Chinese in Europe*. London: Macmillan, 261–77.

Chan, K.B and Chiang, S.N.C. (1994) 'Cultural values and immigrant entrepreneurship', *Revue Européenne des Migrations Internationales*, 10 (2): 87–117.

Colectivo IOÉ (1999a) *Inmigrantes, trabajadores, ciudadanos. Una visión de las migraciones desde España*. Valencia: Patronat Sud-Nord, Universitat de València.

Colectivo IOÉ (1999b) *Inmigración y trabajo en España. Trabajadores inmigrantes en el sector de la hostelería*. Madrid: IMSERSO.

Colectivo IOÉ (2000) *Trabajadores extranjeros en la hostelería andaluza*. Seville: Consejería de Asuntos Sociales, Junta de Andalucía.

Colectivo IOÉ (2001) *Mujer, inmigración y trabajo*. Madrid: IMSERSO.

Colectivo IOÉ (2002) *Immigracio, escola y mercat de treball. Una radiografia actualitzada*. Estudios sociales 11. Barcelona: Fundacio La Caixa (www.estudis.Lacaixa.es).

de Lucas, J. *et al.* (2001) *Inmigrantes: una aproximación jurídica a sus derechos*. Alzira, Valencia: Germania.

European Federation of Chinese Organisations (1999) *The Chinese Community in Europe*. Amsterdam: EFCO.

Francis, C.B. (1996) 'Reproduction of *danwei* institutional features in the context of China's market economy: the case of Haidian district's high-tech sector', *The China Quarterly*, 47: 839–59.

Gates, H. (1996) *China's Motor. A Thousand Years of Petty Capitalism*. Ithaca NY: Cornell University Press.

Greenhalgh, S. (1994) 'De-orientalizing the Chinese family firm', *American Ethnologist*, 21: 746–75.

Harell, S. (1985) 'Why do the Chinese work so hard? Reflections on an entrepreneurial ethic', *Modern China*, 11: 203–26.

Honig, E. (1992) *Creating Chinese Ethnicity: Subei People in Shanghai, 1850–1980*. New Haven CT: Yale University Press.

Izquierdo, A. (1996) *La inmigración inesperada. La población extranjera en España 1991–1995*. Madrid: Trotta.

Li Zhang (2001) 'Migration and privatization of space and power in late socialist China', *American Ethnologist*, 20 (1): 179–203.

Light, I. (1972) *Ethnic Enterprises in America. Business Welfare among Chinese, Japanese, and Blacks*. Berkeley CA: University of California Press.

Ma, L.J.C. and Cartier, C. (eds) *Geographic Perspectives on the Chinese Diaspora: Space, Place, Mobility and Identity*. Lanham MD: Rowman & Littlefield, 1–49.

Ma, L.J.C. and Xiang, B. (1998) 'Native place, migration and the emergence of peasant enclaves in Beijing', *China Quarterly*, 155: 586–41.

Ma Mung, E. (1992) 'Dispositif économique et ressources spatiales: éléments d'une économie de diaspora', *Revue Européenne des Migrations Internationales*, 8 (3): 175–91.

Ma Mung, E. (2000) *La diaspora chinoise. Géographie d'une migration*. Paris: Ophrys.

Mallée, H. (2000) 'Migration, hukou and resistance in reform China', in E. Perry and M. Selden (eds) *Chinese Society. Change, Conflict and Resistence*. London: Routledge, 83–101.

Niehoff, J. (1987) 'The villager as industrialist: ideologies of household manufacturing in rural Taiwan', *Modern China*, 13: 278–309.

Ong, A. and Nonini, D. (eds) (1997) *Ungrounded Empires. The Cultural Politics of Modern Chinese Transnationalism*. London: Routledge.

Oxfeld, E. (1991) 'The sexual division of labor and the organization of family and firm in an overseas Chinese community', *American Ethnologist*, 18 (4): 700–18.

Pieke, F. (1991) 'Immigration et entreprenariat: les Chinois aux Pays-Bas', *Revue Européenne des Migrations Internationales*, 8 (3): 33–50.

Pieke, F. and Benton, G. (1998) 'The Chinese in the Netherlands', in G. Benton and F. Pieke (eds) *The Chinese in Europe*. London: Macmillan, 125–67.

Redding, S.G. (1990) *The Spirit of Chinese capitalism*. Berlin: de Gruyter.

Sáiz López, A. (in press) 'Mujeres empresarias chinas en un contexto migratorio: adaptación y continuidad', in Francisco Checa (ed.) *La mujer inmigrante en España*. Barcelona: Icaria.

Smart, J. and Smart, A. (1993) 'Obligation and control: employment of kin in capitalist labour management in China', *Critique of Anthropology*, 13 (1): 7–31.

Song, M. (1997) 'Children's labour in ethnic family business: the case of Chinese take-away business in Britain', *Ethnic and Racial Studies*, 20 (4): 690–716.

Teixeira, A. (1998) 'Entrepreneurs of the Chinese community in Portugal', in G. Benton and F. Pieke (eds) *The Chinese in Europe*. London: Macmillan, 238–60.

Thuno, M. (1997) 'Chinese Migration to Denmark. Catering and Ethnicity'. Ph.D. thesis, University of Copenhagen.

Thuno, M. (1998) 'Chinese in Denmark', in G. Benton and F. Pieke (eds) *The Chinese in Europe*. London: Macmillan, 168–96.

Tomba, L. (1999) 'Exporting the "Wenzhou model" to Beijing and Florence: labour and economic organization in two migrant communities', in F. Pieke and H. Mallée (eds) *Internal and International Migration. Chinese Perspectives*, Richmond: Curzon Press, 280–94.

Watson, J. (1975) *Emigration and the Chinese lineage. The Mans in Hong Kong and London*. Berkeley CA: University of California Press.

Weidenbaum, M. (1996) 'The Chinese family business enterprise', *California Management Review*, 38 (4): 141–56.

Xiang, B. (1999) 'Zhejiang village in Beijing: creating a visible non-state space through migration and marketized networks', in F. Pieke and H. Malle (eds) *Internal and International Migration. Chinese Perspectives*. Richmond: Curzon Press, 215–50.

Yu-Sion, L. (1998) 'The Chinese community in France: immigration, economic activity, cultural organization and representations', in G. Benton and F. Pieke (eds) *The Chinese in Europe*. London: Macmillan, 96–124.

16 The working of networking

Ethnic networks as social capital among Chinese migrant businesses in Germany

Maggi Leung

Migrants, or 'foreigners' (*Ausländer*) in official language, have been considered generally as a burden to the economy since the economic downturn in the mid-1970s in Germany. These 'foreigners' who mainly comprise the 'guest workers', ethnic Germans (*Aussiedler*) and asylum seekers are often represented in the press and media reports as undesirable and problematic, used as scapegoats for socio-economic problems, such as persistent unemployment, rising criminality and lack of economic growth, especially since the mid-1970s. The image of migrants has taken some new turns since the beginning of the twenty-first century, when the Social Democratic Party (SDP)-Green coalition brought forth the economic advantages of migrants and foreigners. These welcomed migrants range from the sorely needed information technology (IT) experts from Eastern Europe and South Asia to children or grandchildren of the 'guest workers' who are fortunately staying in Germany and making contributions to the shrinking retirement fund of the 'local' population.[1] One of the positive aspects highlighted in the media and literature is the rising rate of entrepreneurship among migrants. The number of self-employed of the largest sub-group, Turkish migrants, for example, rose from 1990 to 1999 by 42.4 per cent, accounting for 47,000 enterprises in 1999 (Wolber, 1999). Geographers Laux and Thieme (2002) further estimated that Turkish migrant enterprises generate an annual turnover of more than €28 billion and employ over 330,000 people. The increasing importance of migrant entrepreneurship in Germany indicates more than economic significance. It can also function socially and politically to raise the general respect for migrants and foreigners generally. As argued by the authors of a cover story 'Unsung heroes' in *Business Week* (February 2000) about successful migrant businesspeople in Europe, underlying the contribution of migrants to the European economy would be '[a] kind of hard-core economic argument [that] provide[s] one of the Continent's best weapons against the anti-immigrant rhetoric of politicians such as Haider' (Echikson *et al.*, 2000).[2]

Reflecting the composition of the migrant population, ethnic entrepreneurs originating from Italy, Turkey, Greece and the former Yugoslavia make up the majority of foreign businesses in Germany. Migrant entrepreneurship is most highly represented in business sectors such as the restaurant trade, retail shops (mainly grocery shops), manufacturing and construction (Wilpert, 1999). Understandably, most of the media and scholarship coverage of migrant entrepreneurship in the German discourse also concentrates on these migrant groups, especially Turkish businesses (for example, Hillmann, 1998; Kontos, 1997, 2000; Laux and Thieme, 2002; Özscan and Seifert, 2000; Pütz, 2000; Swiaczny, 1999; Wiebe, 1982; Yavuzcan, 2000).

In this chapter[3] I shall explore some cases of entrepreneurship among Chinese migrants[4] in Germany. In addition to bringing forth the diversity within the category 'migrant business' in terms of entrepreneurs' geographic origins, my examples here also highlight the variety of migrant entrepreneurship in terms of nature of business. More specifically, I shall present three case studies of Chinese migrant businesses, namely the 'typical' restaurant trade, and the 'atypical' computer sales and travel agencies. Particular emphasis will be placed on the ways in which entrepreneurs make use of ethnic networks that span different geographical levels (local, regional and transnational) strategically to strive and thrive. In the following, I shall first provide a review of the conceptualizations of ethnic networks and their applications in the research of migrant businesses. An overview of the situations of migrant entrepreneurship and Chinese migration in Germany will follow. These sections serve as a background to introduce the case studies that demonstrate the dynamic nature and diverse functions of ethnic networks among migrant entrepreneurs. Illustrations from my interviews and observations argue *against* crude generalizations of the structure and functions of ethnic networks, but *for* careful examinations of the sector- or firm-specific characteristics that account for the kind of networking observed.

Fieldwork for this study was conducted between 1999 and 2001 in various urban centres in Germany, including Hamburg, Frankfurt am Main, Düsseldorf, Bonn, Cologne and Mannheim (Figure 16.1). The findings are drawn from my interactions with owners and employees of Chinese restaurants, computer firms and travel agencies in these urban centres. Official statistical data on Chinese businesses are not readily available. The information presented here is largely based on interviews with Chinese entrepreneurs and careful reviews of community publications. In total, twenty-two restaurateurs (four female, eighteen male), twenty-two computer business owners (four female, eighteen male) and eleven travel agency operators (four female, seven male) were interviewed. In most cases, I first identified key informants such as community leaders, representatives of communal organizations and trade associations, as well as researchers in the fieldwork area. The sampling was then performed utilizing the snowball technique, i.e. by asking research partners to identify other individuals who might be interested in taking part in the research.[5] This follows the theoretical sampling method in which research partners are selected sequentially on the basis of earlier observations. Principally, sampling continued to the point of data saturation. Interviews were conducted in a semi-structured manner in Cantonese or

Figure 16.1 Germany: the main towns.

Mandarin Chinese with an average duration of ninety minutes to three hours. Information was gathered on the research partners' migration biographies, education and career histories. In particular, business decisions concerning financing, recruitment and personnel management as well as the entrepreneurs' embeddedness in ethnic networks were investigated. Interviews were mostly audio-recorded, transcribed and analysed. Findings from these interviews were also supplemented with observations in restaurants, computer firms, travel agencies and other community places.

Networks as social capital for migrant businesses

The importance of ethnic economic and social networks in migrant communities has been a research focus in migration studies and related disciplines. Earlier examples include the analyses by Light (1972), Light and Bonacich (1988), Wong (1988) and Waldinger *et al.* (1990). These earlier works examine the roles of ethnic economic networks in marginal labour-intensive sectors, such as garment manufacturing and ethnic retailing. These linkages are understood as a valuable mechanism to recruit labour, collect capital, as well as exchange expertise and information. In the current era of intensifying globalization, accelerated flexible accumulation and widespread transnational migration, a wave of newly found interest in ethnic networks has arisen. Networks in this era are also considered powerful devices that help capital and people circumvent laws and regulations set up by governments and other formal institutions. In addition, the emphasis of this new current of research extends beyond the labour-intensive and low-skilled sectors which have hitherto dominated in this area of research. The growing attention to the Chinese transnational business networks is one of the cases in point. I shall return to discuss this expanding body of literature in more details in the next section.

Ethnic networks are structures of social and cultural relationships or transactions based on various commonalities shared by a group of people. These commonalities may be some combinations of traits such as language, culture, religion and/or place-based identities (for example, home-town origin) (Mitchell, 2000). Ethnic networks structure, often in important ways, the social and economic organizations within migrant communities. Social and economic networks among co-ethnics should, however, not be assumed to be something predetermined on the basis of the cultural identities of a certain group. How these networks are created, organized and maintained, as well as their roles in linking members of ethnic communities, should be understood by a careful consideration of how their ethnic identities intersect with their other identities (e.g. gender, occupation, geographical background, religious and political affiliation) embedded in a specific context.

This chapter will focus more on the importance of ethnic networks in the economic realm. It is, however, necessary to underline that such linkages signify relationships that stretch beyond the economics. Kin and friendship networking has been recognized as a useful instrument in shaping and sustaining migration by reducing costs of migration and facilitating migrants' adaptation processes (Gurak and Caces, 1992). Assaad *et al.* (1997) list three broad roles social networks play in people's lives. First, they improve participants' access to resources, both tangible (e.g. physical and financial resources) as well as intangible (e.g. information). Second, they provide channels for moral support, political mobilization and popular empowerment. Finally, they are important mechanisms that facilitate and regulate formal socio-economic transactions. Thus, using network as an analytical framework to understand migrant economic activities liberates us from merely focusing on the economic side of the phenomenon as encouraged by traditional models (Mitchell, 2000). Rather a linkage between the economic and socio-cultural realms, which have often been depicted as separate and distinct, can be made.

This echoes the essence of the 'mixed embeddedness approach', first introduced by Kloosterman *et al.* (1999), in the analysis of migrant entrepreneurship. Their argument is simple, namely the logic that, for a comprehensive understanding of any case of (migrant) businesses, a broader framework that pays attention to the social, economic and institutional contexts is necessary. The authors contend that a migrant group's level of participation in entrepreneurship at a specific place and time results from the interplay between the socio-economic and ethno-social characteristics of the particular group and the opportunity structure in which it is embedded. This latter is a function of the state of the technology, costs of production, nature of the demand for the products or services as well as the institutional framework. Thus, the most important element in the mixed embeddedness approach is *context*. Kloosterman and Rath (2001) remind us that this is particularly important for the study of migrant businesses in Europe, because the scholarship has often simply reiterated the findings and theorization from the North American literature. The mixed embeddedness approach thus cautions against 'putting all our eggs in one basket' in the analysis of migrant businesses. With this approach in mind, one would depart automatically from cultural deterministic arguments which dominated this area of research before the 1980s. This previously popular framework views an ethnic business first and foremost as *ethnic* before it is considered as a *business*. It is argued that an ethnic group's cultural predisposition explains the types of businesses that attract its members and how they then perform relative to other cultural groups. Typically, subscribers to this school of thought reason that certain migrant groups are more hard-working, save more money, gravitate 'culturally' toward entrepreneurship and are more willing to devote their lives to their businesses (e.g. Light, 1972; Watson, 1977). 'The Chinese are good at business' would be a typical argument in this tradition (see Collins, 2001, for a counter-observation drawing upon his findings among ethnic Chinese entrepreneurs in Australia). Since the 1980s, studies of a more empirical nature have moved the general conceptualization beyond such purely cultural arguments. Few researchers would now argue that it is only because of migrants' culture and ethnicity that they enter certain businesses and employ particular strategies. Economic, sociological, political and geographical perspectives have been integrated into the body of literature contributing to a more solid understanding.

In the German literature on migrant self-employment and entrepreneurship, geographer Pütz (2000) has developed an explanatory framework which echoes the essence of the mixed embeddedness approach. He laid out different factors that together shape the emergence of migrant self-employment. Determining factors include market entry opportunities, labour market conditions, individual characteristics of the migrant and group resources (Figure 16.2).

Group resources are also referred to as 'ethnic resources', 'social capital' or 'network capital' in the literature to which ethnic networks belong. This concept is based on the observation that some migrant groups tend to engage in self-employment more often than others which seems to suggest that these migrant groups have access to specific resources that ease their path toward self-employment or entrepreneurship (Portes, 1995; Waldinger *et al.*, 1990). I echo, however, Pütz's emphasis that group resources should be considered as one of the interacting factors, and *not* be considered in isolation which would advocate a cultural deterministic impression.

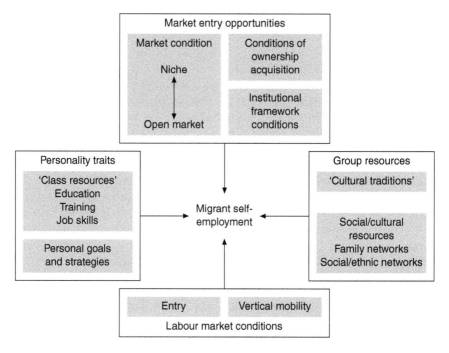

Figure 16.2 Explanatory framework for migrant self-employment (source: Modified from Pütz (2000)).

An 'ungrounded empire' of the 'bamboo networks'

The rapid growth and apparent economic success of ethnic Chinese business in the Asia-Pacific rim has aroused scholarly and journalistic interest which analyses the specificity of the ethnic Chinese business network – also known as the so-called 'bamboo network' (Weidenbaum and Hughes, 1996) or 'gift economy' (Yang, 1994) or '*guanxi* economy' (Smart, 1993, 1998) that stretches across the globe to form an 'invisible empire' (Seagrave, 1995) or 'ungrounded empire' (Ong and Nonini, 1997). According to these analyses, ethnic Chinese are linked by their shared languages, heritage and culture, which facilitate also their close economic connections that contribute to the expanding economic 'empire' in Asia and to a lesser extent around the globe. Weidenbaum and Hughes (1996: 8) revealed in their study how companies owned by ethnic Chinese families in Singapore, Malaysia, Thailand, Indonesia and the Philippines make up about 70 per cent of the private business sector in those countries, and that ethnic Chinese investments are also gaining influence in Vietnam and Australia.

Ever since the adoption of the 'open door' policy by the Chinese government in 1978, foreign ethnic Chinese investors have ventured into the huge land of cheap labour and sizeable markets, often taking advantage of their ethnic networks. Weidenbaum and Hughes (1996: 5) assert that these 'members of the bamboo network ... [provide] three essential ingredients lacking in a communist society – entrepreneurship, risk-taking capital investment, and business management capability'. Thus, the 'motherland' and the ethnic Chinese business people can be considered to share a symbiotic relationship in this new economic era. Indeed, the economic boom in southern China since the 1980s has largely been brought about by capital investment from overseas Chinese entrepreneurs.[6] Much beyond being a passive recipient of investment, the Chinese state has also taken active steps to incorporate their 'subjects overseas'. Zhejiang province, for example, has actively sought support from fellow Zhejiangese. Late in 1999 a delegation, headed by the Zhejiangese party secretary, visited Germany, praising the spirit of entrepreneurship and business success shown by his fellow Zhejiangnese and inviting them to return and invest in their home province (*Euro-Sino Blatt*, December 1999). In addition, seminars in the province have opened their doors to representatives from various migrant Chinese communities in North America, Japan, Australia and Europe to discuss investment opportunities, especially in the high-tech sector. According to a community newspaper report, 2,000 former overseas students had returned to Zhejiang to work; the returnees had established over 200 firms by the end of 2001 (*Chinesische Handelszeitung*, 15 December 2001). Singapore, for another example, having recognized the economic opportunities the millions of ethnic Chinese brothers (and very rarely sisters) worldwide represent, organized the first ever World Chinese Entrepreneurs Convention in August 1991. The event aimed to enable Chinese businessmen and technocrats to network and discuss issues of economic and social concern. While the role of the various states in nurturing the globalizing ethnic network economy is important, especially in the contemporary form of transnationalism (Portes *et al.*, 1999), this chapter focuses on the entrepreneurs as active agents in these projects.

Migrant businesses in Germany

Relative to other European countries, Germany has one of the lowest rates of self-employment. As unemployment grew in the 1980s and 1990s self-employment began to carry a more positive image as a potential remedy for the slow-growing economy. The rate of increase among migrants, moreover, has outrun that of German nationals. Table 16.1 charts the development of self-employment in Germany in the last decade in more detail. Migrants and foreigners, compared to their local counterparts, generally face more difficulties in the formal labour market as they have a significantly lower education level and are more often unemployed or underemployed in Germany.[7] The rising self-employment rate among migrants in recent years, as a result of the restructuring of the German economy, has pushed many of them to explore 'being one's own boss' as a way to secure a livelihood.

Table 16.1 Development of self-employment in Germany (×1,000)

Population	1991		1995		2000	
	Female	Male	Female	Male	Female	Male
Total	41,281	38,548	41,900	39,670	42,080	40,080
Total working	16,862	23,125	17,154	22,929	17,649	22,677
Total German nationals	38,746	35,443	38,703	35,757	39,463	36,285
Total working (German nationals)	15,885	21,183	15,900	20,528	16,355	20,425
Total non-German nationals	2,535	3,105	3,197	3,913	3,345	3,795
Total working (non-German nationals)	977	1,942	1,254	2,401	1,294	2,252
Self-employed (total)	780	2,257	880	2,456	1,012	2,631
Self-employed (German nationals)	739	2,124	827	2,270	944	2,442
Self-employed (non-German nationals)	41	133	53	186	68	189

Source: Statistisches Bundesamt micro census, with own calculations.

The changing demographics of Chinese migrant communities and Chinese businesses

According to official census data, 77,309 individuals with Chinese and Taiwanese nationalities were living in Germany by the end of 2002 (Statistisches Bundesamt, 2003). Scholars have estimated that there are about 100,000 ethnic Chinese living, including undocumented migrants and Chinese people from national origins other than China and Taiwan, such as Hong Kong, Singapore, Malaysia, Vietnam and Indonesia (Ma, 2003; Pang, 2002; Pieke, 2002). The Federation of Chinese Organizations in Europe estimates, however, that the number of ethnic Chinese living in the Federal Republic is closer to 150,000 persons (telephone interview on 31 October, 1999). These migrants have varied migration experiences. For instance, Chinese migrants from Indochina generally arrived in Germany as refugees in the 1970s while most of those from Taiwan have moved to Germany as students, professionals or business people. Albeit all being ethnic Chinese, the diverse background of these migrants account for the different types and amount of economic and social capital they are armed with, which in turns shape their potential to become entrepreneurs of a certain kind.

In the following three sections, I shall turn to each of the case studies to chart the nature and importance of ethnic networks used by entrepreneurs in different sectors. My analysis will emphasize the importance of geography in networking. Namely, the case studies consider the spatial span and functions of the networks used in different economic sectors. Let me start with the most typical branch of Chinese business in Germany: the restaurant business.

Case 1 The restaurant trade

When the term 'Chinese business' is heard, most people would associate it immediately with restaurants or grocery shops. Mirroring what is typical of many other overseas Chinese communities, the restaurant trade may be considered the most significant employment sector, both quantitatively and historically, of various businesses areas in which Chinese migrants are engaged in Germany.[8] Like Turkish kebab shops, Italian ice cream parlours and *pizzerias*, Chinese fast-food businesses, take-aways and restaurants have become expected features of the German urban landscape. Due to problems of definition and the complexities regarding registration, on the one hand, and turbulent market conditions on the other, an exact number of such Chinese businesses is not available. Although the chairman of the Federation of Chinese Organizations in Europe places the number at 13,000 to 14,000 such businesses in Germany, only about 3,000 were listed in *Schober Firmenadressen* (a German business directory), while approximately 4,300 restaurants are indexed in the Chinese business directory, *China-Branchenbuch*, in 2001. These establishments can be categorized into three main types:

restaurants, bistros and *Imbiß* (fast-food/take-away). Chinese restaurant businesses are found in all parts of Germany, but are clustered in the larger urban centres, such as Berlin, Hamburg, Frankfurt, Munich, Bonn, Cologne and Düsseldorf. (Figure 16.3).

Support from co-ethnics play an important role in the functioning of the restaurant business. Echoing much with the early studies, ethnic networks are utilized among restaurateurs in hiring or soliciting formal and informal labour, raising start-up capital, and gathering important information such as the modification of regulations or current business conditions. Here, I shall focus on two aspects, namely overseas hiring and raising start-up capital. Both areas of examination reflect the mixed embeddedness of the restaurant business, in particular, how economic, socio-cultural and institutional factors interact to shape the form and functions of networking.

Figure 16.3 Distribution of Chinese restaurants in Germany, 2001 (source: Schober Firmenadressen (2001)).

Before the advent of stringent and bureaucratic control imposed by the German authorities in 1997, hiring of Chinese cooks from overseas was commonly practised. Until the 1980s Chinese restaurateurs in Germany comprised mainly migrants from Hong Kong, Taiwan and South East Asian countries. Chain migration arranged via familial and friendship networks in these places provided the expanding industry with a continuous supply of labour. As the living standards of these 'dragon economies' grew, overseas hiring had also extended into China in the 1990s. This important hiring practice has been challenged by the changes in the German and Chinese regulatory frameworks. Aiming to tighten control over the inflow of Chinese restaurant workers, the German Central Employment Agency (Zentrale und Internationale Management- und Fachvermittlung für Hotel- und Gaststättenpersonal, ZIHOGA) reached an agreement with the Chinese Ministry of Foreign Trade and Economic Cooperation (MOFTEC) – a unique bilateral arrangement made in the catering sector, to formalize overseas hiring. This declared that, based upon the size of the restaurant, an owner is eligible to hire either one or two chefs possessing proven certification from a training college, at least three years' prior work experience as a chef, and German or English-language proficiency. The hiring of more than two cooks (a maximum of five) is possible only if a tax consultant attests that the gross annual turnover of the business per chef exceeds €75,000. In order to prevent these labour migrants from staying for a longer period in Germany, the work permit is first issued for one year with the possibility of a two-year renewal. A further extension is not allowed. Only after having left Germany and waiting three years may chefs once again apply for a work permit. Many of my research partners expressed their frustration with this recruitment regulation. Instead of the established practice, which restaurant owners preferred, whereby cooks were recruited through social networks in China (which was legal before 1997), hiring now must be conducted through one of the twenty-five Chinese state agencies, which in theory provide training and perform assessment. Many restaurant owners consider these agencies to be a *de facto* licence for Chinese officials to extort money.[9] By using their own transnational network in hiring, employers were able to engage a chef in twenty-five days. Now, the process takes at least three to five months and sometimes even longer. Some employers have also found that the 'speciality chefs' sent to Germany had apparently bought their certificates from corrupt officials and arrived without any prior cooking experience. The Chinese restaurateur communities in Germany have initiated a concerted effort to have this system removed. Numerous petitions have been filed with MOFTEC and the Chinese embassy. Their repeated complaints and attempts since 1997 have so far been without success. In September 2000, after a three-year trial period, the recruitment system was renewed until the year 2004. At the end of my fieldwork period in 2002, news was spread that the recruitment mechanism would be reversed – a change that is welcomed by the restaurateur community.[10]

The complicated and bureaucratic overseas hiring procedures have amplified the importance of family and friends as labour source. Soliciting such 'help' from family and friends extends beyond the local network. One of my research partners revealed his clever idea of inviting his parents for a three-month visit, since tourist visas have been waived for Hong Kong passport holders in the Schengen states. 'As long as they don't wear an apron, they would not be caught even if the police do check', said my research partner, half jokingly. By engaging family members and others from their social networks, which are sometimes transnational in scope, Chinese restaurateurs are able to reduce operational costs and increase flexibility. While this sort of 'helping out' is often considered as one of 'ethnic resources' armed among migrants, I would emphasize the importance in contextualizing this 'cultural act' of 'helping out' also in the broader institutional and economic framework as the above analysis has shown.

Capital pooling through money clubs (*biaohui*, *yuehui* or *hui*) among co-ethnics for business set-ups – another 'ethnic resource' often mentioned in the literature – has become less important in recent years, particularly since the high sums needed to establish a fancy restaurant are less frequently sought. A few of my research partners confirmed that such money clubs among co-ethnics exist, but not as many as could be found in the 1970s and 1980s. The lending of money to co-ethnics, however, should not be attributed to a special ethnic quality of benevolence. In addition to helping out co-ethnics, some of the money cycled through the community in the heyday of this credit system was, according to a few of my research partners, 'black money', earned from smuggling and drug dealing, which were then more common. Some restaurants (or other businesses) were established for money-laundering purposes; lending money to co-ethnics was yet another common way to 'get rid of the money'. Repeatedly, different restaurant owners told me that the business had been 'cleaned up' after more stringent controls were put into place in the 1990s. The drastic decrease in vice activities performed under the cover of the catering business, as well as the general economic downturn of the European economy, have combined to reduce 'free floating money' available to the community. Money pools have consequently become less important as a source of capital for newcomers to the business sector. The increased difficulty in raising capital, coupled with a highly competitive business environment, favours the choice of small *Imbiß* or bistros for those newcomers willing to chance the market while incurring minimal financial risk.

Networking among Chinese restaurateurs takes mostly informal shape. Communal institutions and occasions such as various Chinese cultural associations, Chinese-language weekend schools, social gatherings and annual festivals serve, in addition to their official functions, also as arenas where business information is exchanged, formal or informal recruitment is performed or even venture capital is raised. Chinese community and business newspapers are also important sources of information and advertising space for recruitment, as well as the purchase and sale of shop space and equipment.

Case 2   Computer wholesale and retail

In the last few decades, East Asia has witnessed a boom in information technology (IT) research, development and manufacturing. While Japan achieved its eminence in the field in the 1970s to early 1980s, Taiwan, South Korea, Hong Kong and Singapore gained momentum in this field in the late 1980s and throughout the 1990s. Beginning at the end of the 1990s, production in China has rapidly accelerated, becoming the centre of attention in East Asia, leading to expectations that it will 'displace Taiwan as the king of hardware' (Gay, 2001). In the last two decades, many ethnic Chinese computer firms,[11] especially Taiwanese producers, have become major players in the global market, by either producing equipment that bears their own trade name (such as Acer) or, more often, conducting original equipment manufacturing (OEM) – acting as behind-the-scenes hardware suppliers to big-name foreign computer companies.

Taiwanese own the majority of ethnic Chinese computer firms in Germany. Migrants from Hong Kong and Singapore also own firms, while those from China are increasingly making their presence felt. According to the Taiwanese Business Association (interview on 10 March, 2000), there were around 300 Taiwanese-owned computer firms in Germany, which accounted for about 80–90 per cent of all Taiwanese enterprises in the country in 2000. The largest concentration is located in the western part of Germany, the most important centre near Düsseldorf, followed by that near Hamburg. Most of these businesses were started in the 1990s. Although several were subsidized with a substantial amount of capital from (*de facto*) mother companies in East Asia, as well as being given considerable technical and personnel support, most followed the rather 'classic' way to entrepreneurship, i.e. financing their firms with personal and family savings, sometimes augmented by bank loans. My study here also focuses on these small to medium-scale businesses (ranging from none to thirty-seven employees in my interviewed sample) located in Hamburg and Düsseldorf area. Common among these entrepreneurs is their embeddedness in transnational networks with producers and distributors in Asia (earlier in Taiwan but now also in China) that facilitate timely and flexible shipment of products, an important mechanism in the fast-changing sector.

Contrary to accepted wisdom about ethnic enterprises or the case of the restaurant sector above, ethnic networks have not been important for employee recruitment among my research partners in the computer sales sector, for co-ethnic employees are not particularly desirable in providing marketing services mainly to German customers. All the Chinese firm owners and managers interviewed consider hiring German employees an essential component in the successful sale of their products. While there is no problem in engaging Chinese employees with proficiency in the German language, managers in this sector prefer German employees who 'know exactly what [the customers] mean ... when they speak to each other' and that '[the German customers] would rather see a German person who sells to them'.

What *is* interesting is the fact that networks are considered to be an important channel in recruiting non-Chinese personnel. In the literature, there has been a presumed connection between 'ethnic' and 'network', which this particular observation challenges. It is clear from the present case study that interactions between migrant entrepreneurs and their potential German employees likewise greatly rely upon networking and connections. Most of my research partners cited friends and acquaintances in the field as good sources for contact to potential employees, while formal channels, such as advertising in the print media or hiring through employment agencies and labour offices, play a significantly less important role.

Zhou (1996) highlights the importance of inter-firm linkages among Chinese computer firms in Los Angeles, asserting that their close co-operation helps them meet the special challenges of the rapidly changing computer industry. Saxenian (1999) also affirms the significant roles, similar to those of their mainstream American counterparts, that local social and professional networks play among Chinese and Indian immigrant entre-preneurs in Silicon Valley in supporting their business activities. Regarding the importance of inter-firm linkages, Tom Lee in his forties,[12] who had worked for the Thai branch of his computer company before being trans-ferred to Hamburg, offered me this take on his international experiences. He compared the role of the Taiwanese Business Association in connecting the Taiwanese in Europe with the situation in Asia (interview on 6 December, 1999):

> I think in Asia, in South East Asia, the Taiwanese Business Associations are of different nature than the ones in Europe. Because in Europe, mostly we sell, manufacturing is very rare. So we are mostly competi-tors. But in South East Asia, they manufacture, so they need a bigger network for sources and markets, they might work with each other more. It depends on the items. For computers, it is no longer so easy. Because for computers [produced] in Asia, they are now highly special-ized. Those who produce keyboards, they only do keyboards, or those who make monitors only monitors. They are all specialized. So they no longer use the Chinese network, but a global network.

Among retailers, however, the local network displays another value. Susanne Hong, whose business focuses on retail, technical support, after-sales main-tenance service and customized configurations, is well embedded in the local and regional ethnic networks as source of merchandise in small quantities (interview on 4 December, 1999):

We are small firms, you know? We can't keep a stock of everything. So we know who has what, and when I need only a small quantity, we'd rather get them from the other Chinese firms than making an order ourselves. They want to make a profit, of course. That is OK, as we get a good price here. That is how we work with each other. Or I would let them show their items here. You see, outside [the shop front], there are boxes of pamphlets. I let them distribute them for free. I could charge them money, but no, so this is free advertisement, and we can work with each other this way. You know? We are all small firms. No one wants to or can put so much in stock. Especially for this computer business, changing so fast. If we keep too much stock, we lose money.

On the other side, the Chinese network is also of great importance as a source of clientele among retailers. Among the few Chinese retailers working in the computer sales sector, Chinese clientele make up 20–90 per cent of their customers – as opposed to the case with wholesalers, for whom Chinese clientele represent less than 10 per cent of their sales in most cases in my interviewed sample. This economic potential explains the strong incentives among retailers in cultivating close connections with other co-ethnics for business purposes.

A careful comparison of the different kinds of computer sales businesses reveals that there are sector-specific opportunities and constraints that in turn demand varied kinds of ethnic network. Among computer wholesale companies, embeddedness in the transnational network is an important asset that guarantees more timely and flexible supply of the quickly evolving IT products from East Asia. On the other hand, retailers are keener to cultivate the local and regional networks as a source of client and potential business support.[13]

Case 3 Travel agencies

The intensification of transnationalism is marked by, among other developments, an increase in international travel. Migrants' visits between different homes, as well as to other places where families, friends and relatives around the globe live, provide one of the most cohesive forces that maintain transnational communities. Diaspora tourism has grown into a significant market niche in the tourism sector in recent years. Having said that, it is important to remember that ethnic travel agencies have no mandate to provide only 'diasporic' services. As international travel proliferates among all segments of the global community, ethnic travel agencies have also availed themselves of a market niche in which customers, migrants and non-migrants alike, travel for pleasure to destinations not necessarily related to their original migration biographies.

A complete list of travel agencies operated by Chinese migrants in Germany is not available. Although estimates from my research partners ranged from one to a few hundred travel agencies owned by ethnic Chinese in Germany, the addresses of only thirty-five such travel agencies were listed in the *China-Branchenbuch* published in 1998. In addition, a few other firms that put advertisements in the monthly Chinese business newspaper *Chinesische Handelszeitung* and other Chinese-language community publications were noted. The following analysis draws upon interviews with eleven travel agents in Hamburg, Frankfurt and Mainz. Similar to the computer retailers presented above, co-ethnics represent a valuable pool of clients for Chinese travel agencies. Li Kai, who has been very successful in his travel agency business in Hamburg, reflects on the importance of the local and regional ethnic networks as a basis of clientele when he explained to me his business rationale (interview on 3 December, 1999):

> Among German people, they might not once think of going away. But among Chinese, may they be from Hong Kong or China, they want to go home. So I believe that every single Chinese who is running on the street here could potentially become my customer.

The importance of the co-ethnic clientele explains the high proportion among the employees who are Chinese-speaking (70–100 per cent of the personnel among sampled agencies), through which they can offer services not provided by other travel agencies, namely doing business in their clients' language. Hiring is mainly conducted via friends and family networks as well as advertising in Chinese business newspapers. German employees are recruited mostly through posting in German-language newspapers and on university campuses.

Service providers, such as travel agencies, are generally in competition with each other. Occasionally, however, there are also business synergies between agencies; these probably account for the only 'co-operation' among such Chinese enterprises. Li Kai said (interview on 3 December, 1999):

> We just work independently. There are really no ways to co-operate. Sometimes, you know, we sell tickets in big quantity, and so our cost is lower. Because we are, if you only count Chinese agencies in northern Germany, the biggest customer for Lufthansa tickets going to Hong Kong and China, so we have a special deal with Lufthansa. So when other Chinese agencies need tickets, they might call us and buy from us. That is a common practice. For example, we sell to customers, adding DM100, and for them, I sell adding DM50 to the cost price. So they can also make a DM50 profit.

Inter-firm linkages across geographical borders are, however, a valuable resource, especially for smaller agencies that lack branch offices in the places where their tours visit. In order to satisfy the desire of tourists who come to visit Europe on 'seven days, seven countries' tours, transnational inter-firm networks come in handy. This also applies for the larger agencies, which have branches outside of Germany. Li's business has expanded much since he established it in 1992, with annual sales of DM 25 million in 1999. Now, he also has branch offices in major Chinese and European cities. For places where he has no branch offices, tours are subcontracted to partner travel agencies run by ethnic Chinese. As quoted earlier, he thinks 'Chinese agencies understand what the "strange" Chinese want' and that makes work that much easier. Good relationships with other Chinese travel agencies within Europe (and increasingly beyond Europe) are only one of Li's keys to success. Another is his previous 'special *guanxi*' in China, which enabled his travel agency to win sizeable contracts for providing services to Chinese tourists. In fact, a special relationship with officials in China is almost a requisite for anyone in the travel business, since tourism is still subject to heavy regulation by the state.[14]

In Germany, there are several agencies that have expanded rapidly in the last few years and established branches in various European countries and in China. In some cases, these 'branches' had previously been separate agencies. The 'expansion' was simply a business strategy, adopted by the handful of agencies, which held that, by grouping themselves together, they would have a better image, more suggestive of affluence and success. Certainly inter-firm linkages are important, whether bound by formal or informal contracts. Co-ethnic business networks have a structure that extends beyond the tourist agency sector. As mentioned earlier, package tour providers offer a sense of 'home away from home' to Chinese tourists by arranging overnight stays in Chinese-operated lodgings and meals at Chinese restaurants in the different cities through which they pass. Interviews with Chinese restaurateurs in various cities also confirm that Chinese tours have become an increasingly important clientele for them.

It is necessary, however, to remember that such business linkages are nothing intrinsically Chinese or migrant-specific. Rather they are common to the tourist industry throughout the world. Formal and informal business contracts are made between tourist agencies and specific restaurants or shops where tourists are encouraged to dine and make their purchases. What is interesting here is that, as Chinese migrant businesses increase in number and diversify in nature, ethnic business networks likewise change in structure and scale. As Rath (2002) has observed, migrants are increasingly involved in service provision related to the tourist industry in major cities all over the world. They work in restaurants, travel agencies, hotels, souvenir shops, telephone shops, Internet cafes and as organizers of cultural and entertainment performances. As Chinese migrants venture into a variety of tourism-related businesses, ranging from karaoke clubs to centres of traditional Chinese medicine with many more yet to come, it remains a fascinating research challenge to scrutinize ethnic networks as they dynamically unfold across transnational space.

Conclusion

By providing concrete examples of business practices in three Chinese migrant businesses in Germany, this chapter contributes to the discussion of ethnic business networks by challenging culturally deterministic arguments such as 'the Chinese stick together because they are Chinese'. I hope that my illustrations have demonstrated the need to pay careful attention to the branch specificities in order to understand the real working of networking among Chinese migrant business people. In particular, this chapter focuses on the geographic characteristics of these networks, that is, how networks stretching local, regional and transnational space serve to facilitate migrant businesses of varying nature and scale. My investigation of three different branches first serves to bring forth the diversity of Chinese migrant businesses in Germany, thus debunking the usually unquestioned assumption that Chinese businessmen (they are indeed mostly men) are restaurateurs or grocery store operators. Furthermore, the two 'atypical' sectors help to illustrate the dynamic and diverse nature of migrant entrepreneurship. This emphasis challenges popular practices that collapse different forms of entrepreneurship among migrants under the ethnic rubric, including the general discussion on the use of ethnic networks.

I have in this chapter mapped out the concrete and non-mystic ways how Chinese networks function in different sectors of migrant businesses among my research partners. Depending on the needs of the particular business concerned, entrepreneurs participate in different webs of networks. Among computer wholesalers, for example, transnational networks are an important mechanism for flexible and timely shipments, while computer retailers, like travel agents, consider local and regional networks also of high importance as a source of clientele. The attachment of high value to business linkages should not, however, be understood as intrinsically or exclusively Chinese. Networking is in fact a business strategy commonly used by many other entrepreneurs with or without a migration background.

The nature and intensity of ethnic networks is dynamic, taking new forms in response to the ever changing economic and institutional context. My example of the waning importance of transnational networks in overseas hiring, at least in the legal sense, among Chinese restaurateurs as a result of the tightening of labour immigration controls in Germany illustrates this point. For another example: Kwong, an experienced cook from Vietnam, who has worked in different Chinese restaurants in Hamburg, observed the community spirit had 'deteriorated' and meetings had become more decentralized (interview, 4 February, 2002). Reasons for these changes in the meaning of local networks include the increasingly complex geographical background of the restaurateurs, as well as the convenience brought by the increasingly globalized political economy, marked by more affordable connections between migrants and their homelands (in terms of physical visits, telecommunications or satellite television). None the less the intensification of globalization should not be interpreted as a definite cause of decreasing importance of local networking. Community activities and ethnic networks among Chinese, restaurateurs or otherwise, still serve as an important source of support, entertainment and a sense of belonging. These are in turn necessary ingredients of working long-term or establishing a new home overseas among these migrants. The nature, functions and geographical stretch of the ethnic Chinese networks are likely to display different faces in a fast speed as the Chinese economy opens up further more. It is important for observers to keep their analytical utensils sharpened to monitor developments of the dynamic Chinese ethnic networks across the globe.

Notes

1 According to migrant expert at the OECD, Garson, for every unemployed foreigner, eight more legal migrants work and provide jobs for other Europeans. Von Loeffelholz at the RWI Economic Institute in Essen also pays tribute to migrants in Europe for their economic contribution. He gave estimates that non-nationals in the EU earn at least US$461 billion a year and pay US$153 billion in taxes – far more than the estimated $92 billion immigrants receive in welfare (Weiss, 2000).

2 The Austrian politician Jörg Haider is infamous for his open statements implying support for the ideas of Nazism. His politics are widely viewed as neofascist.

3 This chapter draws upon my doctoral dissertation, now published as *Chinese Migration in Germany. Making Home in Transnational Space* (2004). The research was funded by the German Academic Exchange Service (DAAD). I am particularly grateful to my many research partners who generously shared their thoughts and life stories with me. Thanks are also due to the European Science Foundation for sponsoring me to attend the workshop on 'Asian Entrepreneurship in the European Community' (Nijmegen, Netherlands, May 2001), where I first came into contact with the editors of this volume.

4 Unless otherwise specified, the term 'Chinese' refers to ethnic Chinese throughout the chapter. 'Migrants' are defined as persons who personally, or whose ancestors have crossed borders and settled in Germany (thus excluding short-term visitors such as tourists), and that they share strong cultural and/or social ties with a country (or more than one) other than Germany. The usage is not defined by legal status.

5 Informants or interviewees are referred to as 'research partners' throughout this work. The choice of terminology underlines my belief that the researcher and those researched engage in a co-production of knowledge. I consider those who have let me share their stories as my partners in knowledge co-production, while nevertheless acknowledging my sole responsibility for the given interpretations.

6 Overseas Chinese and compatriot Chinese (an official term used in China to denote Chinese from Hong Kong, Macau and Taiwan) are estimated to account for close to 80 per cent of the total foreign direct investment (FDI) in China as of the mid-1990s (Tefft, 1994). Compatriot Chinese have contributed about two-thirds of the Chinese FDI; other overseas Chinese (mainly from South East Asia) made up the second largest contribution source (about 15 per cent).

7 According to the project Bildung PLUS, 77 per cent of those who are unemployed and without professional training were foreigners in Germany in 2000 (http://www.forum-bildung.de/bip_launch02/templates/imfokus_inhalt.php?artid=85).

8 According to a BBC report, up to 80 per cent of the Chinese work force are engaged in the restaurant sector in the United Kingdom, where 12,000 Chinese takeaways and 3,000 restaurants are in business (BBC News, 5 April, 2001).

9 While neither the MOFTEC nor ZIHOGA officially charges a fee for such recruitment service, chefs, who use the required services of a certified Chinese educational agency, have to pay a fee of €75 per month during their contract period in Germany. The fee covers social insurance contributions in China, the provision of training programmes and the administrative fees of the Chinese authorities. There are also indications that additional fees are charged unofficially. Peng (2001) reports that in one case, a chef was requested to pay RMB 30,000 (approximately €4,000) for an assignment. In addition, extra fees are levied for the issue of a passport, visa, language courses and skills assessment.

10 I have elsewhere provided a more detailed analysis of the Chinese restaurant trade in Germany (Leung, 2002). For a more thorough study of changes in the overseas hiring of Chinese speciality chefs in response to the reform of the labour recruitment policies particularly, see Leung (2003).

11 I consider firms based in Taiwan, Hong Kong, Singapore and China as 'ethnic Chinese firms' since the vast majority of the population in these countries are ethnic Chinese.

12 All research partners are given pseudonyms. It is possible, nonetheless, that their identities are discoverable by people sharing close relationships with them or who are familiar with the Chinese communities in Germany.

13 For a more detailed analysis of the Chinese computer trade sector in Germany, see Leung (2001).

14 One of my research partners, who does not work in the tourist sector, revealed that one agency had connections with certain officials in China. They could arrange undocumented immigration, bringing in 'tourists' who would eventually 'disappear' from the group and thus be able to start a new life in Europe.

References

Assaad, R., Zhou, Y. and Razzaz, O. (1997) *Why is Informality a Useful Analytical Category for Understanding Social Network and Institution?* MacArthur Consortium Working Paper Series. Minneapolis MN: MacArthur Interdisciplinary Program on Peace and International Cooperation.

BBC News (2001) 'Chinese fight foot-and-mouth claims', 5 April (http://news.bbc.co.uk/hi/english/uk/newsid_1260000/1260861.stm).

Chinesische Handelszeitung (2001) '2001 haiwai xueren huiguo chuangye fazhan yantaohui zai Hangzhou juxing', 15 December, 14.

Collins, J. (2001) 'Chinese entrepreneurs: the Chinese diaspora in Australia', *International Journal of Entrepreneurial Behaviour and Research*, 8 (1–2): 113–33.

Echikson, W., Schmidt, K., Dawley, K. and Bawden, A. (2000) 'Unsung heroes: Europe's immigrant entrepreneurs are creating thriving business – and thousands of jobs', *Business Week* (international edition), 28 February, 20–4.

Euro-Sino Blatt (1999) 'Zhengjiang shengweishuji zhangdejiang huijian lüde qiaoling', December, 8.

Gay, Daniel (2001) 'Slumping tigers driven dragon', in *Asiaweek.com*, 27 July–3 August (http://www.asiaweek.com/asiaweek/technology/article/0,8707,168236,00.html).

Gurak, D. and Caces, F. (1992) 'Migration networks and the shaping of migration systems', in M. Krtiz, L.L. Lim and H. Zlotnik (eds) *International Migration Systems. A Global Approach*. Oxford: Clarendon Press, 150–76.

Hillmann, F. (1998) 'Türkische Unternehmerinnen und Beschäftigte im berliner ethnischen Gewerbe'. Discussion paper, Berlin: Wissenschaftszentrum Berlin.

Kloosterman, R. and Rath, J. (2001) 'Immigrant entrepreneurs in advanced economies: mixed embeddedness further explored', *Journal of Ethnic and Migration Studies*, 27 (2): 189–201.

Kloosterman, R., Leun, J. van der and Rath, J. (1999) 'Mixed embeddedness: (in)formal economic activities and immigrant businesses in the Netherlands', *International Journal of Urban and Regional Research*, 23 (2): 252–65.

Kontos, M. (1997) 'Von der Gastarbeiterin zur Unternehmerin. Biographie-analytische Überlegungen zu einem sozialen Transformationsprozeß', *Deutsch lernen*, 4: 275–90.

Kontos, M. (2000) 'Self-employment of Migrant Women and Ethnic Structures'. Paper presented at the fifth Migcities-Network International conference 'Ethnic Neighborhoods in European Cities. Entrepreneurship, Employment and Social Order'. Cologne, April.

Laux, H-D. and Thieme, G. (2002) 'Motives and Strategies of Ethnic Entrepreneurship. The Case of the Turks in Germany'. Presented at the annual meeting of the Association of American Geographers, Los Angeles, 23 March.

Leung, M.W.H. (2001) 'Get it going: new ethnic Chinese business: a case of Taiwanese-owned computer firms in Hamburg', *Journal of Ethnic and Migration Studies*, 27 (2): 277–94.

Leung, M.W.H. (2002) 'From four-course Peking Duck to take-away Singapore Rice: an inquiry into the dynamics of the ethnic Chinese catering business in Germany', *International Journal of Entrepreneurial Behaviour and Research*, 8 (1–2): 134–47.

Leung, M.W.H. (2003) 'Beyond Chinese, beyond food: unpacking the Chinese catering business in Germany', *Entrepreneurship and Regional Development*, 15 (2): 103–18.

Leung, M.W.H. (2004) *Chinese Migration in Germany. Making Home in Transnational Space*. Frankfurt: IKO Verlag.

Light, I. (1972) *Ethnic Enterprise in America. Business and Welfare among Chinese, Japanese, and Blacks*. Berkeley CA: University of California Press.

Light, I. (1999) Globalization and migration networks, in J. Rath (ed.) *Immigrant Businesses. The Economic, Political and Social Environment*. Basingstoke: Macmillan, 162–81.

Light, I. and Bonacich, E. (1988) *Immigrant Entrepreneurs. Koreans in Los Angeles, 1965–1982*. Berkeley CA: University of California Press.

Ma, L. (2003) 'Space, place and transnationalism: the Chinese diaspora as a geographic system', in L. Ma and C. Cartier (eds) *Geographic Perspectives on the Chinese Diaspora. Space, Place, Mobility and Identity*. Lanham MD: Rowman & Littlefield, 1–49.

Mitchell, K. (2000) 'Networks of Ethnicity', in E. Sheppard and T.J. Barnes (eds) *A Companion to Economic Geography*. Oxford and Malden MA: Blackwell, 392–407.

Olds, K. and Yeung, H.W.C. (1999) '(Re)shaping "Chinese" business networks in a globalising era', *Environment and Planning D. Society and Space*, 7 (5): 535–55.

Ong, A. and Nonini, D. (1997) *Ungrounded Empires. The Cultural Politics of Modern Chinese Transnationalism*. New York and London: Routledge.

Özscan, V. and Seifert, W. (2000) 'Selbständigkeit von Immigranten in Deutsch-land. Ausmerzung oder Weg der Integration?' *Soziale Welt*, 51: 289–302.

Pang, C.L. (2002) 'Business opportunity or food pornography? Chinese restaurant ventures in Antwerp', *International Journal of Entrepreneurial Behaviour and Research*, 8 (1/2): 148–61.

Peng, X. (2001) 'Zhongguo dangju de maiguo xieding shi guonei zhuanzhi yu fubai zai haiwai de yanshen', *Chinesische Allgemeine Zeitung*, 20 August.

Pieke, F. (2002) *Recent Trends in Chinese Migration to Europe. Fujianese Migration in Perspective.* Geneva: International Organization for Migration.

Portes, A. (ed.) (1995) *The Economic Sociology of Immigration. Essays on Networks, Ethnicity and Entrepreneurship.* New York: Russell Sage Foundation.

Portes, A., Guarnizo, L.E. and Landolt, P. (eds) (1999) *Transnational Communities*, special issue of *Ethnic and Racial Studies*, 22 (2).

Pütz, R. (2000) 'Von der Nische zum Markt? Türkische Einzelhändler im Rhein-Main Gebiet', in A. Escher (ed.) *Ausländer in Deutschland. Probleme einer transkultuerllen Gesellschaft aus geographischer Sicht.* Mainz: Mainzer Kontaktstudium Geographie, 6: 27–39.

Rath, J. (2002) 'Immigrants and the Tourist Industry'. Paper presented at the Association of American Geographers conference, session 'Migrant businesses', Los Angeles, April.

Redding, S.G. (1990) *The Spirit of Chinese Capitalism.* New York: de Gruyter.

Saxenian, A. (1999) *Silicon Valley's New Immigrant Entrepreneurs.* San Francisco: Public Policy Institute of California.

Seagrave, S. (1995) *Lords of the Rim. The Invisible Empire of the Overseas Chinese.* New York: Putnam.

Smart, A. (1993) 'Gifts, bribes, and *guanxi*: a reconsideration of Bourdieu's social capital', *Social Anthropology*, 8 (3): 388–408.

Smart, A. (1998) Review essay: *guanxi*, gifts and learning from China, *Anthropos*, 93 (4–6): 559–64.

Statistisches Bundesamt (2003) *Statistisches Jahrbuch für die Bundesrepublik Deutschland.* Bonn: Statistisches Bundesamt.

Swiaczny, F. (1999) 'Ökonomische Aktivitäten von Ausländern im Kontext von Migration und Integration am Beispiel der Stadt Mannheim'. Paper presented at the session 'Migration und Bevölkerungsentwicklung in Städten' of the fifty-second Deutscher Geographentag, Hamburg, October.

Tefft, S. (1994) 'The rootless Chinese: the repatriates transform economy, yet endure persistent resentment', *Christian Science Monitor*, 30 March.

Waldinger, R., Aldrich, H. and Ward, R. (1990) 'Opportunities, group character-istics and strategies', in R. Waldinger, H. Aldrich, R. Ward and Associates (eds) *Ethnic Entrepreneurs. Immigrant Business in Industrial Societies.* Newbury Park CA: Sage, 13–48.

Watson, J. (1977) 'The Chinese: Hong Kong villagers in the British catering trade', in J. Waston (ed.) *Between Two Cultures. Migrants and Minorities in Britain*, Manchester: Manchester University Press, 198–211.

Weidenbaum, M. and Hughes, S. (1996) *The Bamboo Network. How Expatriate Chinese Entrepreneurs are creating a New Economic Superpower in Asia.* New York: Martin Kessler.

Weiss, M. (2000) 'Welcome to Europe!' *CSIS Prospectus*, 1 (2), http://www.csis.org/pubs/prospectus/00summerWeiss.html.

Wiebe, D. (1982) 'Sozialgeographische Aspekte ausländischer Gewerbetätigkeiten in Kiel', *Zeitschrift für Wirtschaftsgeographie*, 26 (3): 69–78.

Wilpert, C. (1999) 'A Review of Research on Immigrant Business in Germany'. Paper presented at the conference 'Working on the Fringes. Immigrant Business, Economic Integration and Informal Practices', Amsterdam, 7–9 October.

Wolber, C. (1999) 'Ausländische Selbständige sind wichtiger Wirtschaftsfaktor', *Die Welt*, 8 January.

Wong, B. (1988) *Patronage, Brokerage, Entrepreneurship and the Chinese Community of New York*. New York: AMS Press.

Yang, M.Mh. (1994) *Gifts, Favors and Banquets. The Art of Social Relationships in China*. Ithaca NY: Cornell University Press.

Yavuzcan, I. (2000) 'Turkish Entrepreneurship in Cologne, Germany'. Paper presented at the fifth Migcities-Network International conference 'Ethnic Neighborhoods in European Cities. Entrepreneurship, Employment and Social Order', Cologne, April.

Yeung, H.W.C. (1997) 'Business networks and transnational corporations: a study of Hong Kong firms in the ASEAN region', *Economic Geography*, 71 (1): 1–25.

Yeung, H.W.C. (1999) 'The internationalization of ethnic Chinese business firms from South East Asia: strategies, processes and competitive advantage', *International Journal of Urban and Regional Research*, 23 (1): 103–27.

Zhou, Y. (1996) 'Inter-firm linkages, ethnic networks, and territorial agglomeration: Chinese computer firms in Los Angeles', *Journal of the Regional Science Association International*, 75 (3): 265–91.

Index

Subject

For Product Safety Concerns and Information please contact our EU
representative GPSR@taylorandfrancis.com
Taylor & Francis Verlag GmbH, Kaufingerstraße 24, 80331 München, Germany